FREE Test Taking Tips DVD Offer

To help us better serve you, we have developed a Test Taking Tips DVD that we would like to give you for FREE. **This DVD covers world-class test taking tips that you can use to be even more successful when you are taking your test.**

All that we ask is that you email us your feedback about your study guide. Please let us know what you thought about it – whether that is good, bad or indifferent.

To get your **FREE Test Taking Tips DVD**, email freedvd@studyguideteam.com with "FREE DVD" in the subject line and the following information in the body of the email:

    a. The title of your study guide.

    b. Your product rating on a scale of 1-5, with 5 being the highest rating.

    c. Your feedback about the study guide. What did you think of it?

    d. Your full name and shipping address to send your free DVD.

If you have any questions or concerns, please don't hesitate to contact us at freedvd@studyguideteam.com.

Thanks again!

TASC Test Prep

TASC Exam Book and Practice Test Questions
[3rd Edition]

TPB Publishing

Interested in buying more than 10 copies of our product? Contact us about bulk discounts:
bulkorders@studyguideteam.com

ISBN 13: 9781628459210
ISBN 10: 1628459212

Table of Contents

Quick Overview

As you draw closer to taking your exam, effective preparation becomes more and more important. Thankfully, you have this study guide to help you get ready. Use this guide to help keep your studying on track and refer to it often.

This study guide contains several key sections that will help you be successful on your exam. The guide contains tips for what you should do the night before and the day of the test. Also included are test-taking tips. Knowing the right information is not always enough. Many well-prepared test takers struggle with exams. These tips will help equip you to accurately read, assess, and answer test questions.

A large part of the guide is devoted to showing you what content to expect on the exam and to helping you better understand that content. In this guide are practice test questions so that you can see how well you have grasped the content. Then, answer explanations are provided so that you can understand why you missed certain questions.

Don't try to cram the night before you take your exam. This is not a wise strategy for a few reasons. First, your retention of the information will be low. Your time would be better used by reviewing information you already know rather than trying to learn a lot of new information. Second, you will likely become stressed as you try to gain a large amount of knowledge in a short amount of time. Third, you will be depriving yourself of sleep. So be sure to go to bed at a reasonable time the night before. Being well-rested helps you focus and remain calm.

Be sure to eat a substantial breakfast the morning of the exam. If you are taking the exam in the afternoon, be sure to have a good lunch as well. Being hungry is distracting and can make it difficult to focus. You have hopefully spent lots of time preparing for the exam. Don't let an empty stomach get in the way of success!

When travelling to the testing center, leave earlier than needed. That way, you have a buffer in case you experience any delays. This will help you remain calm and will keep you from missing your appointment time at the testing center.

Be sure to pace yourself during the exam. Don't try to rush through the exam. There is no need to risk performing poorly on the exam just so you can leave the testing center early. Allow yourself to use all of the allotted time if needed.

Remain positive while taking the exam even if you feel like you are performing poorly. Thinking about the content you should have mastered will not help you perform better on the exam.

Once the exam is complete, take some time to relax. Even if you feel that you need to take the exam again, you will be well served by some down time before you begin studying again. It's often easier to convince yourself to study if you know that it will come with a reward!

Test-Taking Strategies

1. Predicting the Answer

When you feel confident in your preparation for a multiple-choice test, try predicting the answer before reading the answer choices. This is especially useful on questions that test objective factual knowledge. By predicting the answer before reading the available choices, you eliminate the possibility that you will be distracted or led astray by an incorrect answer choice. You will feel more confident in your selection if you read the question, predict the answer, and then find your prediction among the answer choices. After using this strategy, be sure to still read all of the answer choices carefully and completely. If you feel unprepared, you should not attempt to predict the answers. This would be a waste of time and an opportunity for your mind to wander in the wrong direction.

2. Reading the Whole Question

Too often, test takers scan a multiple-choice question, recognize a few familiar words, and immediately jump to the answer choices. Test authors are aware of this common impatience, and they will sometimes prey upon it. For instance, a test author might subtly turn the question into a negative, or he or she might redirect the focus of the question right at the end. The only way to avoid falling into these traps is to read the entirety of the question carefully before reading the answer choices.

3. Looking for Wrong Answers

Long and complicated multiple-choice questions can be intimidating. One way to simplify a difficult multiple-choice question is to eliminate all of the answer choices that are clearly wrong. In most sets of answers, there will be at least one selection that can be dismissed right away. If the test is administered on paper, the test taker could draw a line through it to indicate that it may be ignored; otherwise, the test taker will have to perform this operation mentally or on scratch paper. In either case, once the obviously incorrect answers have been eliminated, the remaining choices may be considered. Sometimes identifying the clearly wrong answers will give the test taker some information about the correct answer. For instance, if one of the remaining answer choices is a direct opposite of one of the eliminated answer choices, it may well be the correct answer. The opposite of obviously wrong is obviously right! Of course, this is not always the case. Some answers are obviously incorrect simply because they are irrelevant to the question being asked. Still, identifying and eliminating some incorrect answer choices is a good way to simplify a multiple-choice question.

4. Don't Overanalyze

Anxious test takers often overanalyze questions. When you are nervous, your brain will often run wild, causing you to make associations and discover clues that don't actually exist. If you feel that this may be a problem for you, do whatever you can to slow down during the test. Try taking a deep breath or counting to ten. As you read and consider the question, restrict yourself to the particular words used by the author. Avoid thought tangents about what the author *really* meant, or what he or she was *trying* to say. The only things that matter on a multiple-choice test are the words that are actually in the question. You must avoid reading too much into a multiple-choice question, or supposing that the writer meant something other than what he or she wrote.

5. No Need for Panic

It is wise to learn as many strategies as possible before taking a multiple-choice test, but it is likely that you will come across a few questions for which you simply don't know the answer. In this situation, avoid panicking. Because most multiple-choice tests include dozens of questions, the relative value of a single wrong answer is small. As much as possible, you should compartmentalize each question on a multiple-choice test. In other words, you should not allow your feelings about one question to affect your success on the others. When you find a question that you either don't understand or don't know how to answer, just take a deep breath and do your best. Read the entire question slowly and carefully. Try rephrasing the question a couple of different ways. Then, read all of the answer choices carefully. After eliminating obviously wrong answers, make a selection and move on to the next question.

6. Confusing Answer Choices

When working on a difficult multiple-choice question, there may be a tendency to focus on the answer choices that are the easiest to understand. Many people, whether consciously or not, gravitate to the answer choices that require the least concentration, knowledge, and memory. This is a mistake. When you come across an answer choice that is confusing, you should give it extra attention. A question might be confusing because you do not know the subject matter to which it refers. If this is the case, don't eliminate the answer before you have affirmatively settled on another. When you come across an answer choice of this type, set it aside as you look at the remaining choices. If you can confidently assert that one of the other choices is correct, you can leave the confusing answer aside. Otherwise, you will need to take a moment to try to better understand the confusing answer choice. Rephrasing is one way to tease out the sense of a confusing answer choice.

7. Your First Instinct

Many people struggle with multiple-choice tests because they overthink the questions. If you have studied sufficiently for the test, you should be prepared to trust your first instinct once you have carefully and completely read the question and all of the answer choices. There is a great deal of research suggesting that the mind can come to the correct conclusion very quickly once it has obtained all of the relevant information. At times, it may seem to you as if your intuition is working faster even than your reasoning mind. This may in fact be true. The knowledge you obtain while studying may be retrieved from your subconscious before you have a chance to work out the associations that support it. Verify your instinct by working out the reasons that it should be trusted.

8. Key Words

Many test takers struggle with multiple-choice questions because they have poor reading comprehension skills. Quickly reading and understanding a multiple-choice question requires a mixture of skill and experience. To help with this, try jotting down a few key words and phrases on a piece of scrap paper. Doing this concentrates the process of reading and forces the mind to weigh the relative importance of the question's parts. In selecting words and phrases to write down, the test taker thinks about the question more deeply and carefully. This is especially true for multiple-choice questions that are preceded by a long prompt.

9. Subtle Negatives

One of the oldest tricks in the multiple-choice test writer's book is to subtly reverse the meaning of a question with a word like *not* or *except*. If you are not paying attention to each word in the question, you can easily be led astray by this trick. For instance, a common question format is, "Which of the following is...?" Obviously, if the question instead is, "Which of the following is not...?," then the answer will be quite different. Even worse, the test makers are aware of the potential for this mistake and will include one answer choice that would be correct if the question were not negated or reversed. A test taker who misses the reversal will find what he or she believes to be a correct answer and will be so confident that he or she will fail to reread the question and discover the original error. The only way to avoid this is to practice a wide variety of multiple-choice questions and to pay close attention to each and every word.

10. Reading Every Answer Choice

It may seem obvious, but you should always read every one of the answer choices! Too many test takers fall into the habit of scanning the question and assuming that they understand the question because they recognize a few key words. From there, they pick the first answer choice that answers the question they believe they have read. Test takers who read all of the answer choices might discover that one of the latter answer choices is actually *more* correct. Moreover, reading all of the answer choices can remind you of facts related to the question that can help you arrive at the correct answer. Sometimes, a misstatement or incorrect detail in one of the latter answer choices will trigger your memory of the subject and will enable you to find the right answer. Failing to read all of the answer choices is like not reading all of the items on a restaurant menu: you might miss out on the perfect choice.

11. Spot the Hedges

One of the keys to success on multiple-choice tests is paying close attention to every word. This is never truer than with words like almost, most, some, and sometimes. These words are called "hedges" because they indicate that a statement is not totally true or not true in every place and time. An absolute statement will contain no hedges, but in many subjects, the answers are not always straightforward or absolute. There are always exceptions to the rules in these subjects. For this reason, you should favor those multiple-choice questions that contain hedging language. The presence of qualifying words indicates that the author is taking special care with his or her words, which is certainly important when composing the right answer. After all, there are many ways to be wrong, but there is only one way to be right! For this reason, it is wise to avoid answers that are absolute when taking a multiple-choice test. An absolute answer is one that says things are either all one way or all another. They often include words like *every*, *always*, *best*, and *never*. If you are taking a multiple-choice test in a subject that doesn't lend itself to absolute answers, be on your guard if you see any of these words.

12. Long Answers

In many subject areas, the answers are not simple. As already mentioned, the right answer often requires hedges. Another common feature of the answers to a complex or subjective question are qualifying clauses, which are groups of words that subtly modify the meaning of the sentence. If the question or answer choice describes a rule to which there are exceptions or the subject matter is complicated, ambiguous, or confusing, the correct answer will require many words in order to be expressed clearly and accurately. In essence, you should not be deterred by answer choices that seem excessively long. Oftentimes, the author of the text will not be able to write the correct answer without

offering some qualifications and modifications. Your job is to read the answer choices thoroughly and completely and to select the one that most accurately and precisely answers the question.

13. Restating to Understand

Sometimes, a question on a multiple-choice test is difficult not because of what it asks but because of how it is written. If this is the case, restate the question or answer choice in different words. This process serves a couple of important purposes. First, it forces you to concentrate on the core of the question. In order to rephrase the question accurately, you have to understand it well. Rephrasing the question will concentrate your mind on the key words and ideas. Second, it will present the information to your mind in a fresh way. This process may trigger your memory and render some useful scrap of information picked up while studying.

14. True Statements

Sometimes an answer choice will be true in itself, but it does not answer the question. This is one of the main reasons why it is essential to read the question carefully and completely before proceeding to the answer choices. Too often, test takers skip ahead to the answer choices and look for true statements. Having found one of these, they are content to select it without reference to the question above. Obviously, this provides an easy way for test makers to play tricks. The savvy test taker will always read the entire question before turning to the answer choices. Then, having settled on a correct answer choice, he or she will refer to the original question and ensure that the selected answer is relevant. The mistake of choosing a correct-but-irrelevant answer choice is especially common on questions related to specific pieces of objective knowledge. A prepared test taker will have a wealth of factual knowledge at his or her disposal, and should not be careless in its application.

15. No Patterns

One of the more dangerous ideas that circulates about multiple-choice tests is that the correct answers tend to fall into patterns. These erroneous ideas range from a belief that B and C are the most common right answers, to the idea that an unprepared test-taker should answer "A-B-A-C-A-D-A-B-A." It cannot be emphasized enough that pattern-seeking of this type is exactly the WRONG way to approach a multiple-choice test. To begin with, it is highly unlikely that the test maker will plot the correct answers according to some predetermined pattern. The questions are scrambled and delivered in a random order. Furthermore, even if the test maker was following a pattern in the assignation of correct answers, there is no reason why the test taker would know which pattern he or she was using. Any attempt to discern a pattern in the answer choices is a waste of time and a distraction from the real work of taking the test. A test taker would be much better served by extra preparation before the test than by reliance on a pattern in the answers.

FREE DVD OFFER

Don't forget that doing well on your exam includes both understanding the test content and understanding how to use what you know to do well on the test. We offer a completely FREE Test Taking Tips DVD that covers world class test taking tips that you can use to be even more successful when you are taking your test.

All that we ask is that you email us your feedback about your study guide. To get your **FREE Test Taking Tips DVD**, email freedvd@studyguideteam.com with "FREE DVD" in the subject line and the following information in the body of the email:

- The title of your study guide.
- Your product rating on a scale of 1-5, with 5 being the highest rating.
- Your feedback about the study guide. What did you think of it?
- Your full name and shipping address to send your free DVD.

Introduction to the TASC Exam

Function of the Test

The Test Assessing Secondary Completion (TASC) Testing Program is a partnership between Data Recognition Corporation and the state that chooses to use the TASC test for those who wish to gain their high school equivalency diploma. The TASC test is offered in California, Colorado, Illinois, Indiana, Michigan, Mississippi, Nevada, New Jersey, New York, North Carolina, Ohio, Oklahoma, South Carolina, Texas, and West Virginia, and is used to replace the current General Educational Development (GED) exam. Those who take the TASC test in these states must be at least sixteen-years-old, must not be enrolled in high school, must not have graduated from high school, and must meet all of the requirements for taking the test for their state. Taking the TASC test is beneficial for those who have the knowledge and skills of a high-school equivalent grad and who want to enter a desired job or training program, or any post-secondary course that offers credit hours.

Test Administration

The TASC website has a page specifically designed to locate testing centers in your area. Once in your state portal, you can schedule exams electronically, access test scores, and find a testing center near you. If you take the test and fail any of the subtests, you are allowed to retake two free subtests, but any other retakes beyond two will cost $10.40 each. Any additional fees will be applied on a state-by-state basis, so check with your state before retesting. For those with disabilities, the Data Recognition Corporation offers special testing accommodations, including tools, procedures, and materials, as well as alternative formats.

Test Format

The TASC test is available in an online format for those who know how to use a keyboard and how to navigate with a computer mouse. For those who do not feel comfortable taking the test online, a paper version is offered. The online and paper version of the test are the same, just different formats. Since every testing center is different, make sure you confirm whether you will be taking the online version or the paper version with your local testing center before the day of the test.

The following is a table that shows the time structure of the TASC test:

Subtest	Time
Math Part 1	55 minutes
15 minute break	
Math Part 2	50 minutes
Writing	110 minutes
Reading	85 minutes
Science	75 minutes
Social Studies	75 minutes

Note that the math subtest is split into two parts. In part I, you will be able to use a calculator, but no calculator is permitted while answering part II of the math subtest. Most of the questions in each subtest will be in multiple-choice format; however, there are some variations that are displayed below:

Math Subtest:
- 37 multiple-choice
- 12 gridded-response
- 2 technology-enhanced

Writing Subtest:
- 42 multiple choice
- 2 technology-enhanced
- 1 writing prompt
- Up to 3 shared stimuli

Reading Subtest:
- 42 multiple-choice
- 2 technology-enhanced
- Up to 8 passages

Science Subtest:
- 42 multiple-choice
- 2 technology-enhanced
- Up to 5 shared stimuli

Social Studies Subtest:
- 42 multiple-choice
- 2 technology-enhanced
- Up to 5 shared stimuli

Scoring

The passing score for each subtest is 500. The passing score for the writing prompt is a score of at least 2, ranging up to 8. In order to pass the TASC as a whole, students must pass each of the subtests with a score of at least 500. Computer-based scores are released within 24 hours, and paper-based scores are released up to ten days after the test is taken.

Recent/Future Developments

The TASC test is a fairly new option for test-takers in certain states beginning in January 2014 who are in need of obtaining a high-school equivalency diploma. There are no new developments concerning the TASC test, although more and more states are offering the test each year alongside (or in place of) the GED. Check with your state for more information.

Reading Literacy

Reading Informational Texts

Informational texts are a category of texts within the genre of nonfiction. Their intent is to inform, and while they do convey a point of view and may include literary devices, they do not utilize other literary elements, such as characters or plot. An informational text also reflects a **thesis**—an implicit or explicit statement of the text's intent and/or a **main idea**—the overarching focus and/or purpose of the text, generally implied. Some examples of informational texts are informative articles, instructional/how-to texts, factual reports, reference texts, and self-help texts.

Nonfiction Works

Nonfiction works are best characterized by their subject matter, which must be factual and real, describing true life experiences. There are several common types of literary non-fiction.

Biography
A **biography** is a work written about a real person (historical or currently living). It involves factual accounts of the person's life, often in a re-telling of those events based on available, researched factual information. The re-telling and dialogue, especially if related within quotes, must be accurate and reflect reliable sources. A biography reflects the time and place in which the person lived, with the goal of creating an understanding of the person and his/her human experience. Examples of well-known biographies include *The Life of Samuel Johnson* by James Boswell and *Steve Jobs* by Walter Isaacson.

Autobiography
An **autobiography** is a factual account of a person's life written by that person. It may contain some or all of the same elements as a biography, but the author is the subject matter. An autobiography will be told in first person narrative. Examples of well-known autobiographies in literature include *Night* by Elie Wiesel and *Margaret Thatcher: The Autobiography* by Margaret Thatcher.

Memoir
A **memoir** is a historical account of a person's life and experiences written by one who has personal, intimate knowledge of the information. The line between memoir, autobiography, and biography is often muddled, but generally speaking, a memoir covers a specific timeline of events as opposed to the other forms of nonfiction. A memoir is less all-encompassing. It is also less formal in tone and tends to focus on the emotional aspect of the presented timeline of events. Some examples of memoirs in literature include *Angela's Ashes* by Frank McCourt and *All Creatures Great and Small* by James Herriot.

Journalism
Some forms of **journalism** can fall into the category of literary non-fiction—e.g., travel writing, nature writing, sports writing, the interview, and sometimes, the essay. Some examples include Elizabeth Kolbert's "The Lost World, in the Annals of Extinction series for *The New Yorker* and Gary Smith's "Ali and His Entourage" for *Sports Illustrated*.

Interpreting Textual Evidence in Informational Text

Literal and Figurative Meanings

It is important when evaluating informational texts to consider the use of both literal and figurative meanings. The words and phrases an author chooses to include in a text must be evaluated. How does the word choice affect the meaning and tone? By recognizing the use of literal and figurative language, a reader can more readily ascertain the message or purpose of a text. Literal word choice is the easiest to analyze as it represents the usual and intended way a word or phrase is used. It is also more common in informational texts because it is used to state facts and definitions. While figurative language is typically associated with fiction and poetry, it can be found in informational texts as well. The reader must determine not only what is meant by the figurative language in context, but also how the author intended it to shape the overall text.

Inference in Informational Text

Inference refers to the reader's ability to understand the unwritten text, i.e., "read between the lines" in terms of an author's intent or message. The strategy asks that a reader not take everything he or she reads at face value but instead, add his or her own interpretation of what the author seems to be trying to convey. A reader's ability to make inferences relies on his or her ability to think clearly and logically about the text. It does not ask that the reader make wild speculation or guess about the material but demands that he or she be able to come to a sound conclusion about the material.

An author's use of less literal words and phrases requires readers to make more inference when they read. Since inference involves **deduction**—deriving conclusions from ideas assumed to be true—there's more room for interpretation. Still, critical readers who employ inference, if careful in their thinking, can still arrive at the logical, sound conclusions the author intends.

Textual Evidence in Informational Text

Once a reader has determined an author's thesis or main idea, he or she will need to understand how textual evidence supports interpretation of that thesis or main idea. Test takers will be asked direct questions regarding an author's main idea and may be asked to identify evidence that would support those ideas. This will require test takers to comprehend literal and figurative meanings within the text passage, be able to draw inferences from provided information, and be able to separate important evidence from minor supporting detail. It's often helpful to skim test questions and answer options prior to critically reading informational text; however, test takers should avoid the temptation to solely look for the correct answers. Just trying to find the "right answer" may cause test takers to miss important supporting textual evidence. Making mental note of test questions is only helpful as a guide when reading.

After identifying an author's thesis or main idea, a test taker should look at the **supporting details** that the author provides to back up his or her assertions, identifying those additional pieces of information that help expand the thesis. From there, test takers should examine the additional information and related details for credibility, the author's use of outside sources, and be able to point to direct evidence that supports the author's claims. It's also imperative that test takers be able to identify what is strong support and what is merely additional information that is nice to know but not necessary. Being able to make this differentiation will help test takers effectively answer questions regarding an author's use of supporting evidence within informational text.

Understanding Organizational Patterns and Structures

Organizational Structure within Informational Text

Informational text is specifically designed to relate factual information, and although it is open to a reader's interpretation and application of the facts, the structure of the presentation is carefully designed to lead the reader to a particular conclusion or central idea. When reading informational text, it is important that readers are able to understand its organizational structure as the structure often directly relates to an author's intent to inform and/or persuade the reader.

The first step in identifying the text's structure is to determine the thesis or main idea. The thesis statement and organization of a work are closely intertwined. A **thesis statement** indicates the writer's purpose and may include the scope and direction of the text. It may be presented at the beginning of a text or at the end, and it may be explicit or implicit.

Once a reader has a grasp of the thesis or main idea of the text, he or she can better determine its organizational structure. Test takers are advised to read informational text passages more than once in order to comprehend the material fully. It is also helpful to examine any text features present in the text including the table of contents, index, glossary, headings, footnotes, and visuals. The analysis of these features and the information presented within them, can offer additional clues about the central idea and structure of a text. The following questions should be asked when considering structure:

- How does the author assemble the parts to make an effective whole argument?
- Is the passage linear in nature and if so, what is the timeline or thread of logic?
- What is the presented order of events, facts, or arguments? Are these effective in contributing to the author's thesis?
- How can the passage be divided into sections? How are they related to each other and to the main idea or thesis?
- What key terms are used to indicate the organization?

Next, test takers should skim the passage, noting the first line or two of each body paragraph—the **topic sentences**—and the conclusion. Key **transitional terms**, such as *on the other hand*, *also*, *because*, *however*, *therefore*, *most importantly*, and *first*, within the text can also signal organizational structure. Based on these clues, readers should then be able to identify what type of organizational structure is being used. The following organizational structures are most common:

- **Problem/solution**—organized by an analysis/overview of a problem, followed by potential solution(s)

- **Cause/effect**—organized by the effects resulting from a cause or the cause(s) of a particular effect

- **Spatial order**—organized by points that suggest location or direction—e.g., top to bottom, right to left, outside to inside

- **Chronological/sequence order**—organized by points presented to indicate a passage of time or through purposeful steps/stages

- **Comparison/Contrast**—organized by points that indicate similarities and/or differences between two things or concepts

- **Order of importance**—organized by priority of points, often most significant to least significant or vice versa

Workplace and Community Documents

Workplace and **community documents** help employers to communicate within the business world and foster positive community relations outside of it. Workplace communications typically craft a specific message to a targeted audience while community documents send a broader or more generic message to a wider range of recipients.

<u>Workplace Documents</u>
Even though workplace-related documents are generated in a multitude of paper and electronic formats—memorandums, bulletin boards, presentations, web conferencing, instant messaging, and e-mails—in general all effective business communications share relevant information concisely, accurately, and purposefully. Supervisors rely on workplace documents to communicate expectations to subordinates (downward communication), and subordinates rely on workplace documents to submit progress reports, ask questions, and address concerns with their supervisors (upward communication).

- **Memorandums:** Designed to communicate information to a wide audience, memorandums inform staff of company-wide policy changes. Similar to an e-mail, a memorandum has a header near the top which identifies the intended audience, the author of the memo, the subject of the memo, and the date it was issued. Unlike an e-mail, though, memorandums are longer, can be submitted in paper or electronic form, and contain an introduction that identifies the topic or problem, a body that expands on the topic, and a conclusion that suggests a course of action or solution.

- **Bulletin Boards:** Regardless of whether they are in paper or electronic form, bulletin boards provide a less formal setting for supervisors and staff to communicate. Bulletin boards are a perfect medium to post federal and state regulations, employee incentive initiatives, volunteer opportunities, and company news. While paper bulletin boards are limited to a specific office and personnel, electronic bulletin boards have the capability of broadcasting information nationally and even globally.

- **Presentations:** Presentations can be created with a variety of software—*PowerPoints*, *Google Slides*, and *Prezi*—and are given extemporaneously. Typically, presentations have an introductory slide, informational slides, and a concluding slide that gives the presenter's audience the opportunity to ask questions or create a dialogue. Presentations relay information in a media-rich format: graphics, tables, and hyperlinked documents and videos are easily imbedded within the slides.

- **Web Conferencing:** Web conferencing allows for employees to collaborate on projects and tasks. Employees are able to talk or videoconference from different locations, making it possible for remote workers from around the globe to participate simultaneously in one meeting. Web conferencing can be done via telephone with visuals (*PowerPoint* or *Microsoft Word* documents) or via video camera programs like *WebEx®*, *PGI GlobalMeet®*, and *Skype®*.

- **Instant Messaging:** Many staff outside of centralized locations who cannot communicate verbally with coworkers or supervisors rely, instead, on instant messaging programs. Instant messaging programs can deliver messages one-on-one or in circles and groups, and some software even provides screen-sharing capabilities. Since instant messaging is faster than e-mail, for many employees it has become the preferred method of communicating over long distances.

- **E-Mails:** In today's fast-paced business world, e-mails are heavily relied upon because they provide a platform that is perfect for quickly communicating brief, concise messages to targeted audiences. E-mails not only detail who the sender and receiver are but also provide a date, time, and subject line. Unlike a memorandum, sender and receiver can communicate back and forth, and, more importantly, they may do so over long distances. Many businesses today rely on *Google*, *Yahoo*, and *Outlook* mail severs.

Community Documents

Where business documents target a specific audience and often contain a higher level of proprietary, confidential, or sensitive information, community documents invite larger groups to discuss business matters in a less restrictive environment. Because community documents often help bridge the gap between businesses and community, an effective newsletter, discussion board, blog, website, and app have the power to influence public perception of private companies.

- **Newsletters:** Newsletters are used to provide the public information about the business. It can be used to generate excitement, inform or persuade staff or consumers, or give tips on how the public can contact or work with businesses. Newsletters can be mailed or sent electronically. Generally, newsletters are sent weekly, monthly, or quarterly. They can be interactive, providing the community a glance at what is going on with a business that they interact with, and, moreover, newsletters can communicate a company's mission, values, and priorities.

- **Discussion Boards:** Discussion boards offer a place to go to share information on a specific topic. Discussion boards are organized on menus, submenus, and discussion threads. People visit discussion boards to find out more about a topic. Discussion board members are granted greater access to the site and have greater power to publish and comment, but, still, visitors are welcome, and they often get difficult or obscure questions resolved.

- **Blogs:** A blog, which is usually centered on a specific topic or theme, is a website where individuals post and update information constantly. Blogs tend to feature the newest posts first while archiving older ones. Articles, editorials, images, videos, surveys, and social media (just to name a few) can all be imbedded within blogs. Menus, sidebars, recent posts, and search boxes help visitors wade through a dizzying array of media formats and topics. Though individuals can hire a web designer to create a blog for them, most people use existing platforms like *WordPress®*, *Blogger®*, and *Tumbler®*.

- **Websites:** As the world becomes more technologically savvy, a website can be used to house community documents, giving consumers instant access to tools they will need to interact with the business. It can house forms, contact information, discussion boards, surveys, and blogs. It is a one-stop-shop that can assist in the interaction between the business and the consumer.

Websites are advantageous because they can be accessed via computer, tablet, or mobile phone.

- **Apps:** Similar to a website, an application (app) gives businesses yet another method to reach individuals or segments of a community. Apps provide instant access to forms and other documentation. Designed to be downloaded on tablets or mobile phones, apps are streamlined and intuitive, allowing consumers on the go to access information at their convenience. The primary operating systems for apps include *iOS®* and *Android®*.

Reading Literature

Classifying literature involves an understanding of the concept of genre. A **genre** is a category of literature that possesses similarities in style and in characteristics. Based on form and structure, there are four basic genres.

Fictional Prose

Fictional prose consists of fictional works written in standard form with a natural flow of speech and without poetic structure. Fictional prose primarily utilizes grammatically complete sentences and a paragraph structure to convey its message. Fiction written in prose can be further broken down into **fiction genres**—types of fiction. Some of the more common genres of fiction are as follows:

- **Classical fiction:** a work of fiction considered timeless in its message or theme, remaining noteworthy and meaningful over decades or centuries—e.g., Charlotte Brontë's *Jane Eyre*, Mark Twain's *Adventures of Huckleberry Finn*

- **Fables:** short fiction that generally features animals, fantastic creatures, or other forces within nature that assume human-like characters and has a moral lesson for the reader—e.g., *Aesop's Fables*

- **Fairy tales:** children's stories with magical characters in imaginary, enchanted lands, usually depicting a struggle between good and evil, a sub-genre of folklore—e.g., Hans Christian Anderson's *The Little Mermaid*, *Cinderella* by the Brothers Grimm

- **Fantasy:** fiction with magic or supernatural elements that cannot occur in the real world, sometimes involving medieval elements in language, usually includes some form of sorcery or witchcraft and sometimes set on a different world—e.g., J.R.R. Tolkien's *The Hobbit*, J.K. Rowling's *Harry Potter and the Sorcerer's Stone*, George R.R. Martin's *A Game of Thrones*

- **Folklore:** types of fiction passed down from oral tradition, stories indigenous to a particular region or culture, with a local flavor in tone, designed to help humans cope with their condition in life and validate cultural traditions, beliefs, and customs—e.g., William Laughead's *Paul Bunyan and The Blue Ox*, the Buddhist story of "The Banyan Deer"

- **Mythology:** closely related to folklore but more widespread, features mystical, otherworldly characters and addresses the basic question of why and how humans exist, relies heavily on allegory and features gods or heroes captured in some sort of struggle—e.g., Greek myths, Genesis I and II in the Bible, Arthurian legends

- **Science fiction:** fiction that uses the principle of extrapolation—loosely defined as a form of prediction—to imagine future realities and problems of the human experience—e.g., Robert

Heinlein's *Stranger in a Strange Land*, Ayn Rand's *Anthem*, Isaac Asimov's *I, Robot*, Philip K. Dick's *Do Androids Dream of Electric Sheep?*

- **Short stories**: short works of prose fiction with fully-developed themes and characters, focused on mood, generally developed with a single plot, with a short period of time for settings—e.g., Edgar Allan Poe's "Fall of the House of Usher," Shirley Jackson's "The Lottery," Isaac Bashevis Singer's "Gimpel the Fool"

Drama

Drama is fiction that is written to be performed in a variety of media, intended to be performed for an audience, and structured for that purpose. It might be composed using poetry or prose, often straddling the elements of both in what actors are expected to present. Action and dialogue are the tools used in drama to tell the story. Like prose fiction, drama has several genres. The following are the most common ones:

- **Comedy**: a humorous play designed to amuse and entertain, often with an emphasis on the common person's experience, generally resolved in a positive way—e.g., Richard Sheridan's *School for Scandal*, Shakespeare's *Taming of the Shrew*, Neil Simon's *The Odd Couple*

- **History**: a play based on recorded history where the fate of a nation or kingdom is at the core of the conflict—e.g., Christopher Marlowe's *Edward II*, Shakespeare's *King Richard III*, Arthur Miller's *The Crucible*

- **Tragedy**: a serious play that often involves the downfall of the protagonist. In modern tragedies, the protagonist is not necessarily in a position of power or authority—e.g., Jean Racine's *Phèdre*, Arthur Miller's *Death of a Salesman*, John Steinbeck's *Of Mice and Men*

- **Melodrama**: a play that emphasizes heightened emotion and sensationalism, generally with stereotypical characters in exaggerated or realistic situations and with moral polarization—e.g., Jean-Jacques Rousseau's *Pygmalion*

- **Tragi-comedy**: a play that has elements of both tragedy—a character experiencing a tragic loss—and comedy—the resolution is often positive with no clear distinctive mood for either— e.g., Shakespeare's *The Merchant of Venice*, Anton Chekhov's *The Cherry Orchard*

Poetry

Poetry is fiction in verse that has a unique focus on the rhythm of language and focuses on intensity of feeling. It is not an entire story, though it may tell one; it is compact in form and in function. Poetry can be considered as a poet's brief word picture for a reader. Poetic structure is primarily composed of lines and stanzas. Together, poetic structure and devices are the methods that poets use to lead readers to feeling an effect and, ultimately, to the interpretive message.

Different poetic structures and devices are used to create the various major forms of poetry. Some of the most common forms are discussed in the following chart.

Type	Poetic Structure	Example
Ballad	A poem or song passed down orally which tells a story and in English tradition usually uses an ABAB or ABCB rhyme scheme	William Butler Yeats' "The Ballad of Father O'Hart"
Epic	A long poem from ancient oral tradition which narrates the story of a legendary or heroic protagonist	Homer's *The Odyssey* Virgil's *The Aeneid*
Haiku	A Japanese poem of three unrhymed lines with five, seven, and five syllables (in English) with nature as a common subject matter	Matsuo Bashō "An old silent pond . . . A frog jumps into the pond, splash! Silence again."
Limerick	A five-line poem written in an AABBA rhyme scheme, with a witty focus	From Edward Lear's *Book of Nonsense*: "There was a Young Person of Smyrna Whose grandmother threatened to burn her . . ."
Ode	A formal lyric poem that addresses and praises a person, place, thing, or idea	Edna St. Vincent Millay's "Ode to Silence"
Sonnet	A fourteen-line poem written in iambic pentameter	Shakespeare's Sonnets 18 and 130

Literary Nonfiction

Literary nonfiction is prose writing that is based on current or past real events or real people and includes straightforward accounts as well as those that offer opinions on facts or factual events. The TASC exam distinguishes between literary nonfiction—a form of writing that incorporates literary styles and techniques to create factually-based narratives—and informational texts.

Key Ideas and Details

Central Ideas and Themes

Topic, Main Idea, Supporting Details, and Themes

The **topic** of a text is the overall subject, and the **main idea** more specifically builds on that subject. Consider a paragraph that begins with the following: "The United States government is made of up three branches: executive, judicial, and legislative." If this sentence is divided into its essential components, there is the topic (United States Government) and the main idea (the three branches of government).

A main idea must be supported with details, which usually appear in the form of quotations, paraphrasing, or analysis. Authors should connect details and analysis to the main point. Readers should always be cautious when accepting the validity of an argument and look for logical fallacies, such as slippery slope, straw man, and begging the question. It's okay for a reader to disagree with an author, because arguments may seem sound, but further analysis often reveals they are flawed.

It is important to remember that when most authors write, they want to make a point or send a message. This point, or the message of a text, is known as the **theme**. Authors may state themes

explicitly, like in *Aesop's Fables*. More often, especially in modern literature, readers must infer the theme based on textual details. Usually, after carefully reading and analyzing an entire text, the theme emerges. Typically, the longer the piece, the more numerous its themes, though often one theme dominates the rest, as evidenced by the author's purposeful revisiting of it throughout the passage.

Cultural Differences in Themes

Regardless of culture, place, or time, certain themes are universal to the human condition. Because all humans experience certain feelings and engage in similar experiences—birth, death, marriage, friendship, finding meaning, etc.—certain themes span cultures. However, different cultures have different norms and general beliefs concerning these themes. For example, the theme of maturing and crossing from childhood to adulthood is a global theme; however, the literature from one culture might imply that this happens in someone's twenties, while another culture's literature might imply that it happens in the early teenage years.

It's important for the reader to be aware of these differences. Readers must avoid being **ethnocentric**, which means believing the aspects of one's own culture to be superior to those of other cultures.

Summarizing Information Accurately

Analyzing Topics and Summary Sentences

Good writers get to the point quickly. This is accomplished by developing a strong and effective topic sentence that details the author's purpose and answers questions such as *What does the author intend to explain or impress?* or *What does the author want the reader to believe?* The **topic sentence** is normally found at the beginning of a supporting paragraph and usually gives purpose to a single paragraph. When reading, critical readers should find the topic sentence in each paragraph. If all information points back to one sentence, it's the topic sentence.

Summary sentences offer a recap of previously discussed information before transitioning to the next point or proceeding to the closing thoughts. Summary sentences can be found at the end of supporting paragraphs and in the conclusion of a text.

Drawing Logical Inferences and Conclusions

Identifying Logical Conclusions

Determining conclusions requires being an active reader, as a reader must make a prediction and analyze facts to identify a conclusion. A reader should identify key words in a passage to determine the logical conclusion from the information presented. Consider the passage below:

> Lindsay, covered in flour, moved around the kitchen frantically. Her mom yelled from another room, "Lindsay, we're going to be late!"

Readers can conclude that Lindsay's next steps are to finish baking, clean herself up, and head off somewhere with her baked goods. It's important to note that the conclusion cannot be verified factually. Many conclusions are not spelled out specifically in the text; thus, they have to be inferred and deduced by the reader.

Evaluating a Passage

Readers draw **conclusions** about what an author has presented. This helps them better understand what the writer has intended to communicate and whether or not they agree with what the author has offered. There are a few ways to determine a logical conclusion, but careful reading is the most important. It's helpful to read a passage a few times, noting details that seem important to the piece. Sometimes, readers arrive at a conclusion that is different than what the writer intended or they may come up with more than one conclusion.

Textual evidence within the details helps readers draw a conclusion about a passage. **Textual evidence** refers to information—facts and examples—that support the main point. Textual evidence will likely come from outside sources and can be in the form of quoted or paraphrased material. In order to draw a conclusion from evidence, it's important to examine the credibility and validity of that evidence as well as how (and if) it relates to the main idea.

If an author presents a differing opinion or a **counterargument**, in order to refute it, the reader should consider how and why this information is being presented. It is meant to strengthen the original argument and shouldn't be confused with the author's intended conclusion, but it should also be considered in the reader's final evaluation.

Sometimes, authors explicitly state the conclusion that they want readers to understand. Alternatively, a conclusion may not be directly stated. In that case, readers must rely on the implications to form a logical conclusion:

> On the way to the bus stop, Michael realized his homework wasn't in his backpack. He ran back to the house to get it and made it back to the bus just in time.

In this example, although it's never explicitly stated, it can be inferred that Michael is a student on his way to school in the morning. When forming a conclusion from implied information, it's important to read the text carefully to find several pieces of evidence to support the conclusion.

Summarizing is an effective way to draw a conclusion from a passage. A **summary** is a shortened version of the original text, written by the reader in his or her own words. Focusing on the main points of the original text and including only the relevant details can help readers reach a conclusion. It's important to retain the original meaning of the passage.

Like summarizing, **paraphrasing** can also help a reader fully understand different parts of a text. Paraphrasing calls for the reader to take a small part of the passage and list or describe its main points. However, paraphrasing is more than rewording the original passage; it should be written in the reader's own words, while still retaining the meaning of the original source. This will indicate an understanding of the original source, yet still help the reader expand on his or her interpretation.

Craft and Structure

Analyzing an Author's Rhetorical Choices

Authors utilize a wide range of techniques to tell a story or communicate information. Readers should be familiar with the most common of these techniques. Techniques of writing are also known as **rhetorical devices**.

In nonfiction writing, authors employ argumentative techniques to present their opinions to readers in the most convincing way. Persuasive writing usually includes at least one type of **appeal**: an appeal to logic (**logos**), emotion (**pathos**), or credibility and trustworthiness (**ethos**). When writers appeal to logic, they are asking readers to agree with them based on research, evidence, and an established line of

reasoning. An author's argument might also appeal to readers' emotions, perhaps by including personal stories and **anecdotes** (a short narrative of a specific event). A final type of appeal—appeal to authority—asks the reader to agree with the author's argument on the basis of their expertise or credentials. Three different approaches to arguing the same opinion are exemplified below:

Logic (Logos)

> Our school should abolish its current ban on cell phone use on campus. This rule was adopted last year as an attempt to reduce class disruptions and help students focus more on their lessons. However, since the rule was enacted, there has been no change in the number of disciplinary problems in class. Therefore, the rule is ineffective and should be done away with.

The above is an example of an appeal to logic. The author uses evidence to disprove the logic of the school's rule (the rule was supposed to reduce discipline problems, but the number of problems has not been reduced; therefore, the rule is not working) and to call for its repeal.

Emotion (Pathos)

An author's argument might also appeal to readers' emotions, perhaps by including personal stories and anecdotes.

The next example presents an appeal to emotion. By sharing the personal anecdote of one student and speaking about emotional topics like family relationships, the author invokes the reader's empathy in asking them to reconsider the school rule.

> Our school should abolish its current ban on cell phone use on campus. If they aren't able to use their phones during the school day, many students feel isolated from their loved ones. For example, last semester, one student's grandmother had a heart attack in the morning. However, because he couldn't use his cell phone, the student didn't know about his grandmother's accident until the end of the day—when she had already passed away, and it was too late to say goodbye. By preventing students from contacting their friends and family, our school is placing undue stress and anxiety on students.

Credibility (Ethos)

Finally, an appeal to authority includes a statement from a relevant expert. In this case, the author uses a doctor in the field of education to support the argument. All three examples begin from the same opinion—the school's phone ban needs to change—but rely on different argumentative styles to persuade the reader.

> Our school should abolish its current ban on cell phone use on campus. According to Dr. Bartholomew Everett, a leading educational expert, "Research studies show that cell phone usage has no real impact on student attentiveness. Rather, phones provide a valuable technological resource for learning. Schools need to learn how to integrate this new technology into their curriculum." Rather than banning phones altogether, our school should follow the advice of experts and allow students to use phones as part of their learning.

Rhetorical Questions

Another commonly used argumentative technique is asking **rhetorical questions**, which are questions that do not actually require an answer but that push the reader to consider the topic further.

> I wholly disagree with the proposal to ban restaurants from serving foods with high sugar and sodium contents. Do we really want to live in a world where the government can control what we eat? I prefer to make my own food choices.

Here, the author's rhetorical question prompts readers to put themselves in a hypothetical situation and imagine how they would feel about it.

Figurative Language

Similes and **metaphors** are types of figurative language that are used as rhetorical devices. Both are comparisons between two things, but their formats differ slightly. A simile says that two things are similar and makes a comparison using "like" or "as"—*A is like B*, or *A is as [some characteristic] as B*— whereas a metaphor states that two things are exactly the same—*A is B*. In both cases, similes and metaphors invite the reader to think more deeply about the characteristics of the two subjects and consider where they overlap. Sometimes the poet develops a complex metaphor throughout the entire poem; this is known as an extended metaphor. An example of metaphor can be found in the sentence: "His pillow was a fluffy cloud". An example of simile can be found in the first line of Robert Burns' famous poem:

> My love is like a red, red rose

This is comparison using "like," and the two things being compared are love and a rose. Some characteristics of a rose are that it is fragrant, beautiful, blossoming, colorful, vibrant—by comparing his love to a red, red rose, Burns asks the reader to apply these qualities of a rose to his love. In this way, he implies that his love is also fresh, blossoming, and brilliant.

In addition to rhetorical devices that play on the *meanings* of words, there are also rhetorical devices that use the sounds of words. These devices are most often found in poetry, but may also be found in other types of literature and in nonfiction writing like texts for speeches.

Alliteration and **assonance** are both varieties of sound repetition. Other types of sound repetition include: **anaphora**—repetition that occurs at the beginning of the sentences; **epiphora**—repetition occurring at the end of phrases; antimetabole—repetition of words in a succession; and antiphrasis—a form of denial of an assertion in a text.

Alliteration refers to the repetition of the first sound of each word. Recall Robert Burns' opening line:

> My love is like a red, red rose

This line includes two instances of alliteration: "love" and "like" (repeated *L* sound), as well as "red" and "rose" (repeated *R* sound). Next, assonance refers to the repetition of vowel sounds, and can occur anywhere within a word (not just the opening sound). Here is the opening of a poem by John Keats:

> When I have fears that I may cease to be

> Before my pen has glean'd my teeming brain

Assonance can be found in the words "fears," "cease," "be," "glean'd," and "teeming," all of which stress the long *E* sound. Both alliteration and assonance create a harmony that unifies the writer's language.

Another sound device is **onomatopoeia**—words whose spelling mimics the sound they describe. Words such as "crash," "bang," and "sizzle" are all examples of onomatopoeia. Use of onomatopoetic language adds auditory imagery to the text.

Readers are probably most familiar with the technique of using a **pun**. A pun is a play on words, taking advantage of two words that have the same or similar pronunciation. Puns can be found throughout Shakespeare's plays, for instance:

> Now is the winter of our discontent
>
> Made glorious summer by this son of York

These lines from *Richard III* contain a play on words. Richard III refers to his brother—the newly crowned King Edward IV—as the "son of York," referencing their family heritage from the house of York. However, while drawing a comparison between the political climate and the weather (times of political trouble were the "winter," but now the new king brings "glorious summer"), Richard's use of the word "son" also implies another word with the same pronunciation, "sun"—so Edward IV is also like the sun, bringing light, warmth, and hope to England. Puns are a clever way for writers to suggest two meanings at once.

Some examples of figurative language are included in the following table:

	Definition	Example
Simile	Compares two things using "like" or "as"	Her hair was like gold.
Metaphor	Compares two things as if they are the same	He was a giant teddy bear.
Idiom	Using words with predictable meanings to create a phrase with a different meaning	The world is your oyster.
Alliteration	Repeating the same beginning sound or letter in a phrase for emphasis	The busy baby babbled.
Personification	Attributing human characteristics to an object or an animal	The house glowered menacingly with a dark smile.
Foreshadowing	Giving an indication that something is going to happen later in the story	I wasn't aware at the time, but I would come to regret those words.
Symbolism	Using symbols to represent ideas and provide a different meaning	The ring represented the bond between us.
Onomatopoeia	Using words that imitate sound	The tire went off with a bang and a crunch.
Imagery	Appealing to the senses by using descriptive language	The sky was painted with red and pink and streaked with orange.
Hyperbole	Using exaggeration not meant to be taken literally	The girl weighed less than a feather.

Analyzing Text Structure

Analyzing and Evaluating Text Structure

Depending on what the author is attempting to accomplish, certain formats or text structures work better than others. For example, a sequence structure might work for narration but not when identifying similarities and differences between dissimilar concepts. Similarly, a comparison-contrast structure is not useful for narration. It's the author's job to put the right information in the correct format.

Readers should be familiar with the five main literary structures:

1. **Sequence** structure (sometimes referred to as the order structure) is when the order of events proceeds in a predictable manner. In many cases, this means the text goes through the plot elements: exposition, rising action, climax, falling action, and resolution. Readers are introduced to characters, setting, and conflict in the exposition. In the rising action, there's an increase in tension and suspense. The climax is the height of tension and the point of no return. Tension decreases during the falling action. In the resolution, any conflicts presented in the exposition are solved, and the story concludes. An informative text that is structured sequentially will often go in order from one step to the next.

2. In the **problem-solution** structure, authors identify a potential problem and suggest a solution. This form of writing is usually divided into two paragraphs and can be found in informational texts.

For example, cell phone, cable, and satellite providers use this structure in manuals to help customers troubleshoot or identify problems with services or products.

3. When authors want to discuss similarities and differences between separate concepts, they arrange thoughts in a **comparison-contrast** paragraph structure. **Venn diagrams** are an effective graphic organizer for comparison-contrast structures because they feature two overlapping circles that can be used to organize and group similarities and differences. A comparison-contrast essay organizes one paragraph based on similarities and another based on differences. A comparison-contrast essay can also be arranged with the similarities and differences of individual traits addressed within individual paragraphs. Words such as *however*, *but*, and *nevertheless* help signal a contrast in ideas.

4. The **descriptive** writing structure is designed to appeal to one's senses. Much like an artist who constructs a painting, good descriptive writing builds an image in the reader's mind by appealing to the five senses: sight, hearing, taste, touch, and smell. However, overly descriptive writing can become tedious; whereas sparse descriptions can make settings and characters seem flat. Good authors strike a balance by applying descriptions only to passages, characters, and settings that are integral to the plot.

5. Passages that use the **cause and effect** structure are simply asking *why* by demonstrating some type of connection between ideas. Words such as *if*, *since*, *because*, *then*, or *consequently* indicate relationship. By switching the order of a complex sentence, the writer can rearrange the emphasis on different clauses. Saying *If Sheryl is late, we'll miss the dance* is different from saying, *We'll miss the dance if Sheryl is late*. One emphasizes Sheryl's tardiness while the other emphasizes missing the dance. Paragraphs can also be arranged in a cause and effect format. Since the format—before and after—is sequential, it is useful when authors wish to discuss the impact of choices. Researchers often apply this paragraph structure to the scientific method.

Authorial Purpose and Perspective

No matter the genre or format, all authors are writing to persuade, inform, entertain, or express feelings. Often, these purposes are blended, with one dominating the rest. It's useful to learn to recognize the author's intent.

Persuasive writing is used to persuade or convince readers of something. It often contains two elements: the argument and the counterargument. The **argument** takes a stance on an issue, while the **counterargument** pokes holes in the opposition's stance. Authors rely on logic, emotion, and writer credibility to persuade readers to agree with them. If readers are opposed to the stance before reading, they are unlikely to adopt that stance. However, those who are undecided or committed to the same stance are more likely to agree with the author.

Informative writing tries to teach or inform. Workplace manuals, instructor lessons, statistical reports and cookbooks are examples of informative texts. Informative writing is usually based on facts and is often without emotion and persuasion. Informative texts generally contain statistics, charts, and graphs. Although most informative texts lack a persuasive agenda, readers must examine the text carefully to determine whether one exists within a given passage.

Stories or **narratives** are designed to entertain. When people go to the movies, they often want to escape for a few hours, not necessarily to think critically. **Entertaining** writing is designed to delight and engage the reader. However, sometimes this type of writing can be woven into more serious materials, such as persuasive or informative writing, to hook the reader before transitioning into a more scholarly discussion.

Emotional writing works to evoke the reader's feelings, such as anger, euphoria, or sadness. The connection between reader and author is an attempt to cause the reader to share the author's intended emotion or tone. Sometimes, in order to make a text more poignant, the author simply wants readers to feel the emotions that the author has felt. Other times, the author attempts to persuade or manipulate the reader into adopting their stance. While it's okay to sympathize with the author, readers should be aware of the individual's underlying intent.

Interpreting Authorial Decisions Rhetorically

There are a few ways for readers to engage actively with the text, such as making inferences and predictions. An **inference** refers to a point that is implied (as opposed to directly-stated) by the evidence presented:

> Bradley packed up all of the items from his desk in a box and said goodbye to his coworkers for the last time.

From this sentence, although it is not directly stated, readers can infer that Bradley is leaving his job. It's necessary to use inference in order to draw conclusions about the meaning of a passage. When making an inference about a passage, it's important to rely only on the information that is provided in the text itself. This helps readers ensure that their conclusions are valid.

Readers will also find themselves making predictions when reading a passage or paragraph. **Predictions** are guesses about what's going to happen next. This is a natural tendency, especially when reading a good story or watching a suspenseful movie. It's fun to try to figure out how it will end. Authors intentionally use suspenseful language and situations to keep readers interested:

> A cat darted across the street just as the car came careening around the curve.

One unfortunate prediction might be that the car will hit the cat. Of course, predictions aren't always accurate, so it's important to read carefully to the end of the text to determine the accuracy of one's predictions.

Readers should pay attention to the **sequence**, or the order in which details are laid out in the text, as this can be important to understanding its meaning as a whole. Writers will often use transitional words to help the reader understand the order of events and to stay on track. Words like *next, then, after*, and *finally* show that the order of events is important to the author. In some cases, the author omits these transitional words, and the sequence is implied. Authors may even purposely present the information out of order to make an impact or have an effect on the reader. An example might be when a narrative writer uses **flashback** to reveal information.

Drawing conclusions is also important when actively reading a passage. **Hedge phrases** such as *will, might, probably*, and *appear to be* are used by writers who want to cover their bases and show there are exceptions to their statements. **Absolute phrasing**, such as *always* and *never*, should be carefully considered, as the use of these words and their intended meanings are often incorrect.

Differentiating Between Various Perspectives and Sources of Information

Identifying the Appropriate Source for Locating Information

With a wealth of information at people's fingertips in this digital age, it's important to know not only the type of information one is looking for, but also in what medium he or she is most likely to find it. Information needs to be specific and reliable. For example, if someone is repairing a car, an encyclopedia would be mostly useless. While an encyclopedia might include information about cars, an owner's manual will contain the specific information needed for repairs. Information must also be reliable or credible so that it can be trusted. A well-known newspaper may have reliable information, but a peer-reviewed journal article will have likely gone through a more rigorous check for validity. Determining **bias** can be helpful in determining credibility. If the information source (person, organization, or company) has something to gain from the reader forming a certain view on a topic, it's likely the information is skewed. For example, if trying to find the unemployment rate, the Bureau of Labor Statistics is a more credible source than a politician's speech.

Primary sources are best defined as records or items that serve as evidence of periods of history. To be considered primary, the source documents or objects must have been created during the time period in which they reference. Examples include diaries, newspaper articles, speeches, government documents, photographs, and historical artifacts. In today's digital age, primary sources, which were once in print, are often embedded in secondary sources. **Secondary sources**—such as websites, history books, databases, or reviews—contain analysis or commentary on primary sources. Secondary sources borrow information from primary sources through the process of quoting, summarizing, or paraphrasing.

Today's students often complete research online through **electronic sources**. Electronic sources offer advantages over print, and can be accessed on virtually any computer, while libraries or other research centers are limited to fixed locations and specific catalogs. Electronic sources are also efficient and yield massive amounts of data in seconds. The user can tailor a search based on key words, publication years, and article length. Lastly, many **databases** provide the user with instant citations, saving the user the trouble of manually assembling sources for a bibliography.

Although electronic sources yield powerful results, researchers must use caution. While there are many reputable and reliable sources on the internet, just as many are unreliable or biased sources. It's up to the researcher to examine and verify the reliability of sources. *Wikipedia*, for example, may or may not be accurate, depending on the contributor. Many databases, such as *EBSCO* or *SIRS*, offer peer-reviewed articles, meaning the publications have been reviewed for the quality and accuracy of their content.

Identifying Literary Elements

There is no one, final definition of what literary elements are. They can be considered features or characteristics of fiction, but they are really more of a way that readers can unpack a text for the purpose of analysis and understanding the meaning. The elements contribute to a reader's literary interpretation of a passage as to how they function to convey the central message of a work. The most common literary elements used for analysis are the presented below.

Point of View

The **point of view** is the position the narrator takes when telling the story in prose. If a narrator is incorporated in a drama, the point of view may vary; in poetry, point of view refers to the position the speaker in a poem takes.

First Person

The first person point of view is when the writer uses the word "I" in the text. Poetry often uses first person, e.g., William Wordsworth's "I Wandered Lonely as a Cloud." Two examples of prose written in first person are Suzanne Collins' *The Hunger Games* and Anthony Burgess's *A Clockwork Orange*.

Second Person

The second person point of view is when the writer uses the pronoun "you." It is not widely used in prose fiction, but as a technique, it has been used by writers such as William Faulkner in *Absalom, Absalom!* and Albert Camus in *The Fall*. It is more common in poetry—e.g., Pablo Neruda's "If You Forget Me."

Third Person

Third person point of view is when the writer utilizes pronouns such as him, her, or them. It may be the most utilized point of view in prose as it provides flexibility to an author and is the one with which readers are most familiar. There are two main types of third person used in fiction. **Third person omniscient** uses a narrator that is all-knowing, relating the story by conveying and interpreting thoughts/feelings of all characters. In **third person limited**, the narrator relates the story through the perspective of one character's thoughts/feelings, usually the main character.

Plot

The **plot** is what happens in the story. Plots may be singular, containing one problem, or they may be very complex, with many sub-plots. All plots have exposition, a conflict, a climax, and a resolution. The **conflict** drives the plot and is something that the reader expects to be resolved. The plot carries those events along until there is a resolution to the conflict.

Tone

The **tone** of a story reflects the author's attitude and opinion about the subject matter of the story or text. Tone can be expressed through word choice, imagery, figurative language, syntax, and other details. The emotion or mood the reader experiences relates back to the tone of the story. Some examples of possible tones are humorous, somber, sentimental, and ironic.

Setting

The **setting** is the time, place, or set of surroundings in which the story occurs. It includes time or time span, place(s), climates, geography—man-made or natural—or cultural environments. Emily Dickinson's poem "Because I could not stop for Death" has a simple setting—the narrator's symbolic ride with Death through town towards the local graveyard. Conversely, Leo Tolstoy's *War and Peace* encompasses numerous settings within settings in the areas affected by the Napoleonic Wars, spanning 1805 to 1812.

Characters

Characters are the story's figures that assume primary, secondary, or minor roles. **Central** or **major characters** are those integral to the story—the plot cannot be resolved without them. A central character can be a **protagonist** or hero. There may be more than one protagonist, and he/she doesn't always have to possess good characteristics. A character can also be an **antagonist**—the force against a protagonist.

Character development is when the author takes the time to create dynamic characters that add uniqueness and depth to the story. *Dynamic* characters are characters that change over the course of the plot time. **Stock characters** are those that appear across genres and embrace stereotypes—e.g., the

cowboy of the Wild West or the blonde bombshell in a detective novel. A **flat character** is one that does not present a lot of complexity or depth, while a **rounded character** does. Sometimes, the **narrator** of a story or the **speaker** in a poem can be a character—e.g., Nick Carraway in F. Scott Fitzgerald's *The Great Gatsby* or the speaker in Robert Browning's "My Last Duchess." The narrator might also function as a character in prose, though not be part of the story—e.g., Charles Dickens' narrator of *A Christmas Carol*.

Understanding Poetic Devices and Structure

Poetic Devices

Rhyme is the poet's use of corresponding word sounds in order to create an effect. Most rhyme occurs at the ends of a poem's lines, which is how readers arrive at the **rhyme scheme**. Each line that has a corresponding rhyming sound is assigned a letter—A, B, C, and so on. When using a rhyme scheme, poets will often follow lettered patterns. Robert Frost's *"The Road Not Taken"* uses the ABAAB rhyme scheme:

Two roads diverged in a yellow wood,	A
And sorry I could not travel both	B
And be one traveler, long I stood	A
And looked down one as far as I could	A
To where it bent in the undergrowth;	B

Another important poetic device is **rhythm**—metered patterns within poetry verses. When a poet develops rhythm through **meter**, he or she is using a combination of stressed and unstressed syllables to create a sound effect for the reader.

Rhythm is created by the use of **poetic feet**—individual rhythmic units made up of the combination of stressed and unstressed syllables. A line of poetry is made up of one or more poetic feet. There are five standard types in English poetry, as depicted in the chart below.

Foot Type	Rhythm	Pattern
Iamb	buh Buh	Unstressed/stressed
Trochee	Buh buh	Stressed/unstressed
Spondee	Buh Buh	Stressed/stressed
Anapest	buh buh Buh	Unstressed/unstressed/stressed
Dactyl	Buh buh buh	Stressed/unstressed/unstressed

Structure

Poetry is most easily recognized by its structure, which varies greatly. For example, a structure may be strict in the number of lines it uses. It may use rhyming patterns or may not rhyme at all. There are three main types of poetic structures:

- **Verse**—poetry with a consistent meter and rhyme scheme
- **Blank verse**—poetry with consistent meter but an inconsistent rhyme scheme
- **Free verse**—poetry with inconsistent meter or rhyme

Verse poetry is most often developed in the form of **stanzas**—groups of word lines. Stanzas can also be considered **verses**. The structure is usually formulaic and adheres to the protocols for the form. For example, the English **sonnet** form uses a structure of fourteen lines and a variety of different rhyming patterns. The English **ode** typically uses three ten-line stanzas and has a particular rhyming pattern.

Poets choose poetic structure based on the effect they want to create. Some structures—such as the ballad and haiku—developed out of cultural influences and common artistic practice in history, but in more modern poetry, authors choose their structure to best fit their intended effect.

Identifying Major Literary Works and Authors

The TASC test will have a wide range of poetic, dramatic, fictional, and nonfiction works. In most cases, the test taker will be presented with a literary passage and be required to answer one or more questions about it. This may involve having to identify the literary work presented from a list of options. The following chart offers some examples of major works in addition to those listed elsewhere in this guide, but the list not exhaustive.

American
Fictional Prose Before 1920
- Nathaniel Hawthorne | *The Scarlett Letter* (1850)
- Harriet Beecher Stowe | *Uncle Tom's Cabin* (1852)
- Louisa May Alcott | *Little Women* (1868)
- Henry James | *Daisy Miller* (1878)
- Jack London | *The Call of the Wild* (1903)
- William Faulkner | *The Sound and the Fury* (1929)

Fictional Prose After 1920 and Before 1960
- Zora Neale Hurston | *Their Eyes Were Watching God* (1937)
- John Steinbeck | *Grapes of Wrath* (1939)
- Ernest Hemingway | *For Whom the Bell Tolls* (1940)
- J.D. Salinger | *Catcher in the Rye* (1951)
- Herman Melville | *Moby Dick* (1956)

Fictional Prose After 1960
- Harper Lee | *To Kill a Mockingbird* (1960)
- N. Scott Momaday | *The Way to Rainy Mountain* (1969)
- Marilynne Robinson | *Housekeeping* (1980)
- Alice Walker | *The Color Purple* (1982)
- Toni Morrison | *Beloved* (1987)

Drama
- Thornton Wilder I *Our Town* (1938)
- Tennessee Williams | *A Streetcar Named Desire* (1947)
- Eugene O'Neill |*Long Day's Journey into Night* (1956)
- Lorraine Hansberry | *A Raisin in the Sun* (1959)
- Edward Albee | *Who's Afraid of Virginia Woolf?* (1962)
- Amiri Baraka | *Dutchman* (1964)
- Sam Shephard | *Buried Child* (1978)

Poetry

- Anne Bradstreet | "In Reference to her Children, 23 June 1659"
- Phillis Wheatley | "On Being Brought from Africa to America" (1768)
- Edgar Allen Poe | "The Raven" (1845)
- Walt Whitman | "Song of Myself" (1855)
- Emily Dickinson | "Because I could not stop for Death" (1863)
- Robert Frost | "Wild Grapes" (1920)
- Langston Hughes | "Harlem" (1951)
- Richard Wilbur | "Love Calls Us to the Things of This World" (1956)
- Sylvia Plath | "Mirror" (1961)
- Eamon Grennan | "Cat Scat" (1988)
- Sandra Cisneros | "Loose Woman" (1994)
- Anne Carson | "The Glass Essay" (1994)
- Tupac Shakur | "The Rose that Grew from Concrete" (1999)

Literary Non-fiction

- Frederick Douglass | *My Bondage and My Freedom* (1855)
- Anne Frank | *The Diary of Anne Frank* (1947)
- Truman Capote | *In Cold Blood* (1966)
- Maya Angelou | *I Know Why the Caged Bird Sings* (1969)
- Sherman Alexie | *The Absolutely True Diary of a Part-Time Indian* (2007)
- Cynthia Levinson | *We've Got a Job* (2012)
- Malala Yousafzai and Christina Lamb | *I am Malala* (2013)
- Philip Hoose | *The Boys who Challenged Hitler* (2015

Identifying Literary Contexts

Understanding that works of literature emerged either because of a particular context—or perhaps despite a context—is key to analyzing them effectively.

Historical Context

The **historical context** of a piece of literature can refer to the time period, setting, or conditions of living at the time it was written as well as the context of the work. For example, Hawthorne's *The Scarlet Letter* was published in 1850, though the setting of the story is 1642-1649. Historically, then, when Hawthorne wrote his novel, the United States found itself at odds as the beginnings of a potential Civil War were in view. Thus, the historical context is potentially significant as it pertains to the ideas of traditions and values, which Hawthorne addresses in his story of Hester Prynne in the era of Puritanism.

Cultural Context

The **cultural context** of a piece of literature refers to cultural factors, such as the beliefs, religions, and customs that surround and are in a work of literature. The Puritan's beliefs, religion, and customs in Hawthorne's novel would be significant as they are at the core of the plot—the reason Hester wears the A and why Arthur kills himself. The customs of people in the Antebellum Period, though not quite as restrictive, were still somewhat similar. This would impact how the audience of the time received the novel.

Literary Context

Literary context refers to the consideration of the genre, potentially at the time the work was written. In 1850, Realism and Romanticism were the driving forces in literature in the U.S., with depictions of life as it was at the time in which the work was written or the time it was written *about* as well as some works celebrating the beauty of nature. Thus, an audience in Hawthorne's time would have been well satisfied with the elements of both offered in the text. They would have been looking for details about everyday things and people (Realism), but they also would appreciate his approach to description of nature and the focus on the individual (American Romanticism). The contexts would be significant as they would pertain to evaluating the work against those criteria.

Here are some questions to use when considering context:

- When was the text written?
- What was society like at the time the text was written, or what was it like, given the work's identified time period?
- Who or what influenced the writer?
- What political or social influences might there have been?
- What influences may there have been in the genre that may have affected the writer?

Additionally, test takers should familiarize themselves with literary periods such as Old and Middle English, American Colonial, American Renaissance, American Naturalistic, and British and American Modernist and Post-Modernist movements. Most students of literature will have had extensive exposure to these literary periods in history, and while it is not necessary to recognize every major literary work on sight and associate that work to its corresponding movement or cultural context, the test taker should be familiar enough with the historical and cultural significance of each test passage in order to be able to address test questions correctly.

The following brief description of some literary contexts and their associated literary examples follows. It is not an all-inclusive list. The test taker should read each description, then follow up with independent study to clarify each movement, its context, its most familiar authors, and their works.

Metaphysical Poetry

Metaphysical poetry is the descriptor applied to 17th century poets whose poetry emphasized the lyrical quality of their work. These works contain highly creative poetic conceits or metaphoric comparisons between two highly dissimilar things or ideas. Metaphysical poetry is characterized by highly prosaic language and complicated, often layered, metaphor.

Poems such as John Donne's "The Flea," Andrew Marvell's "To His Coy Mistress," George Herbert's "The Collar," Henry Vaughan's "The World," and Richard Crashaw's "A Song" are associated with this type of poetry.

British Romanticism

British Romanticism was a cultural and literary movement within Europe that developed at the end of the 18th century and extended into the 19th century. It occurred partly in response to aristocratic, political, and social norms and partly in response to the Industrial Revolution of the day. Characterized by intense emotion, major literary works of British Romanticism embrace the idea of aestheticism and the beauty of nature. Literary works exalted folk customs and historical art and encouraged spontaneity of artistic endeavor. The movement embraced the heroic ideal and the concept that heroes would raise the quality of society.

Authors who are classified as British Romantics include Samuel Taylor Coleridge, John Keats, George Byron, Mary Shelley, Percy Bysshe Shelley, and William Blake. Well-known works include Samuel Taylor Coleridge's "Kubla Khan," John Keats' "Ode on a Grecian Urn," George Byron's "Childe Harold's Pilgrimage," Mary Shelley's *Frankenstein*, Percy Bysshe Shelley's "Ode to the West Wind," and William Blake's "The Tyger."

American Romanticism

American Romanticism occurred within the American literary scene beginning early in the 19th century. While many aspects were similar to British Romanticism, it is further characterized as having gothic aspects and the idea that individualism was to be encouraged. It also embraced the concept of the *noble savage*—the idea that indigenous culture uncorrupted by civilization is better than advanced society.

Well-known authors and works include Nathanial Hawthorne's *The House of the Seven Gables*, Edgar Allan Poe's "The Raven" and "The Cask of Amontillado," Emily Dickinson's "I Felt a Funeral in My Brain" and James Fenimore Cooper's *The Last of the Mohicans*.

Transcendentalism

Transcendentalism was a movement that applied to a way of thinking that developed within the United States, specifically New England, around 1836. While this way of thinking originally employed philosophical aspects, transcendentalism spread to all forms of art, literature, and even to the ways people chose to live. It was born out of a reaction to traditional rationalism and purported concepts such as a higher divinity, feminism, humanitarianism, and communal living. Transcendentalism valued intuition, self-reliance, and the idea that human nature was inherently good.

Well-known authors include Ralph Waldo Emerson, Henry David Thoreau, Louisa May Alcott, and Ellen Sturgis Hooper. Works include Ralph Waldo Emerson's "Self-Reliance" and "Uriel," Henry David Thoreau's *Walden* and *Civil Disobedience*, Louisa May Alcott's *Little Women*, and Ellen Sturgis Hooper's "I Slept, and Dreamed that Life was Beauty."

The Harlem Renaissance

The **Harlem Renaissance** is the descriptor given to the cultural, artistic, and social boom that developed in Harlem, New York, at the beginning of the 20th century, spanning the 1920s and 1930s. Originally termed *The New Negro Movement*, it emphasized African-American urban cultural expression and migration across the United States. It had strong roots in African-American Christianity, discourse, and intellectualism. The Harlem Renaissance heavily influenced the development of music and fashion as well. Its singular characteristic was to embrace Pan-American culturalisms; however, strong themes of the slavery experience and African-American folk traditions also emerged. A hallmark of the Harlem Renaissance was that it laid the foundation for the future Civil Rights Movement in the United States.

Well-known authors and works include Zora Neale Hurston's *Their Eyes Were Watching God*, Richard Wright's *Native Son*, Langston Hughes' "I, Too," and James Weldon Johnson's "God's Trombones: Seven Negro Sermons in Verse" and *The Book of American Negro Poetry*.

Integration of Knowledge and Ideas

Understanding Authors' Claims

The goal of most persuasive and informative texts is to make a claim and support it with evidence. A **claim** is a statement made as though it is fact. Many claims are opinions; for example, "stealing is wrong." While this is generally true, it is arguable, meaning it is capable of being challenged. An initial reaction to "stealing is wrong" might be to agree; however, there may be circumstances in which it is warranted. If it is necessary for the survival of an individual or their loved ones (i.e., if they are starving and cannot afford to eat), then this assertion becomes morally ambiguous. While it may still be illegal, whether it is "wrong" is unclear.

When an assertion is made within a text, it is typically reinforced with supporting details as is exemplified in the following passage:

> The extinction of the dinosaurs has been a hot debate amongst scientists since the discovery of fossils in the eighteenth century. Numerous theories were developed in explanation, including extreme climate change, an epidemic of disease, or changes in the atmosphere. It wasn't until the late 1970s that a young geochemist, named Walter Alvarez, noticed significant changes in the soil layers of limestone he was studying in Italy. The layers contained fossilized remains of millions of small organisms within the layer that corresponded with the same period in which the dinosaurs lived. He noticed that the soil layer directly above this layer was suddenly devoid of any trace of these organisms. The soil layer directly above *this* layer was filled with an entirely new species of organisms. It seemed the first species had disappeared at the exact same time as the dinosaurs!

> With the help of his father, Walter Alvarez analyzed the soil layer between the extinct species and the new species and realized this layer was filled with an abnormal amount of *iridium* – a substance that is abundant in meteorites but almost never found on Earth. Unlike other elements in the fossil record, which take a long time to deposit, the iridium had been laid down very abruptly. The layer also contained high levels of soot, enough to account for all of the earth's forests burning to the ground at the same time. This led scientists to create the best-supported theory that the tiny organisms, as well as the dinosaurs and countless other species, had been destroyed by a giant asteroid that had slammed into Earth, raining tons of iridium down on the planet from a giant cosmic cloud.

Supporting Claims

Before embarking on answering these questions, readers should summarize each. This will help in locating the supporting evidence. These summaries can be written down or completed mentally; full sentences are not necessary.

Paragraph 1: Layer of limestone shows that a species of organisms disappeared at same time as the dinosaurs

Paragraph 2: Layer had high amounts of iridium and soot – scientists believe dinosaurs destroyed by asteroid.

Simply by summarizing the text, it has been plainly outlined where there will be answers to relevant questions. Although there are often claims already embedded within an educational text, a claim will most likely be given, but the evidence to support it will need to be located. Take this example question:

> Q: What evidence within the text best supports the theory that the dinosaurs became extinct because of an asteroid?

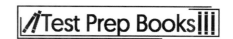

The claim here is that the <u>dinosaurs went extinct because of an asteroid</u>. Because the text is already outlined in the summaries, it is easy to see that the evidence supporting this theory is in the second paragraph:

> With the help of his father, they analyzed the soil layer between the extinct species and the new species and realized <u>this layer was filled with an abnormal amount of *iridium*</u> — a substance that is <u>abundant is meteorites</u> but almost never found on Earth. Unlike other elements in the fossil record, which takes a long time to deposit, the iridium had been laid down very abruptly. <u>The layer also contained high levels of soot</u>, enough to account for all of the earth's forests burning to the ground at the same time. <u>This led scientists to create the best-supported theory</u> that the tiny organisms, as well as the dinosaurs and countless other species, had been <u>destroyed by a giant asteroid</u> that had slammed into Earth, <u>raining tons of iridium down on the planet</u> from a giant cosmic cloud.

Now that the evidence within the text that best supports the theory has been located, the answer choices can be evaluated:

    a. Changes in climate and atmosphere caused an asteroid to crash into Earth
    b. Walter and Luis Alvarez studied limestone with fossilized organisms
    c. A soil layer lacking organisms that existed at the same time as the dinosaurs showed low levels of iridium
    d. A soil layer lacking organisms that existed at the same time as the dinosaurs showed high levels of iridium

Answer choice (a) is clearly false as there is nothing within the text that claims that climate changes caused an asteroid to crash into Earth. This kind of answer choice displays an incorrect use of detail. Although the passage may have contained the words "change," "climate," and "atmosphere," these terms were manipulated to form an erroneous answer.

Answer choice (b) is incorrect because while the scientists did study limestone with fossilized organisms, and in doing so they discovered evidence that led to the formation of the theory, this is not the actual evidence itself. This is an example of an out-of-scope answer choice: a true statement that may or may not have been in the passage, but that isn't the whole answer or isn't the point.

Answer choice (c) is incorrect because it is the opposite of the correct answer. Assuming the second paragraph was summarized correctly, it is already known that the soil layer contained *high* levels of iridium, not low levels. Even if the paragraph was not summarized that way, the final sentence states that "tons of iridium rained down on the planet." So, answer choice (c) is false.

Answer choice (d) is correct because it matches the evidence found in the second paragraph.

Differentiating Between Facts and Opinions

Fact and Opinion, Biases, and Stereotypes

It is important to distinguish between facts and opinions when reading a piece of writing. When an author presents **facts**, such as statistics or data, readers should be able to check those facts to verify that they are accurate. When authors share their own thoughts and feelings about a subject, they are expressing their **opinions**.

Authors often use words like *think, feel, believe,* or *in my opinion* when expressing an opinion, but these words won't always appear in an opinion piece, especially if it is formally written. An author's opinion may be backed up by facts, which gives it more credibility, but that opinion should not be taken as fact. A critical reader should be suspect of an author's opinion, especially if it is only supported by other opinions.

Fact	Opinion
There are nine innings in a game of baseball.	Baseball games run too long.
James Garfield was assassinated on July 2, 1881.	James Garfield was a good president.
McDonald's® has stores in 118 countries.	McDonald's® has the best hamburgers.

Critical readers examine the facts used to support an author's argument. They check the facts against other sources to be sure those facts are correct. They also check the validity of the sources used to be sure those sources are credible, academic, and/or peer-reviewed. When an author uses another person's opinion to support his or her argument, even if it is an expert's opinion, it is still only an opinion and should not be taken as fact. A strong argument uses valid, measurable facts to support ideas. Even then, the reader may disagree with the argument.

An authoritative argument may use the facts to sway the reader. In the example of global warming, many experts differ in their opinions of which alternative fuels can be used to aid in offsetting it. Because of this, a writer may choose to only use the information and experts' opinions that supports his or her viewpoint. For example, if the argument is that wind energy is the best solution, the author will use facts that support this idea. That same author may leave out relevant facts on solar energy. The way the author uses facts can influence the reader, so it's important to consider the facts being used, how those facts are being presented, and what information might be left out.

Authors can also demonstrate **bias** if they ignore an opposing viewpoint or present their side in an unbalanced way. A strong argument considers the opposition and finds a way to refute it. Critical readers should look for an unfair or one-sided presentation of the argument and be skeptical, as a bias may be present. Even if this bias is unintentional, if it exists in the writing, the reader should be wary of the validity of the argument.

Readers should also look for the use of stereotypes that refer to specific groups. **Stereotypes** are often negative connotations about a person or place and should always be avoided. When a critical reader finds stereotypes in a piece of writing, he or she should immediately be critical of the argument and consider the validity of anything the author presents. Stereotypes reveal a flaw in the writer's thinking and may suggest a lack of knowledge or understanding about the subject.

Using Evidence to Make Connections Between Different Texts

When analyzing two or more texts, there are several different aspects that need to be considered, particularly the styles (or the artful way in which the authors use diction to deliver a theme), points of view, and types of argument. In order to do so, one should compare and contrast the following elements between the texts:

- Style: narrative, persuasive, descriptive, informative, etc.
- Tone: sarcastic, angry, somber, humorous, etc.
- Sentence structure: simple (1 clause) compound (2 clauses), complex-compound (3 clauses)
- Punctuation choice: question marks, exclamation points, periods, dashes, etc.

- Point of view: first person, second person, third person
- Paragraph structure: long, short, both, differences between the two
- Organizational structure: compare/contrast, problem/solution, chronological, etc.

The following two passages concern the theme of death and are presented to demonstrate how to evaluate the above elements:

Passage I

Death occurs in several stages. The first stage is the pre-active stage, which occurs a few days to weeks before death, in which the desire to eat and drink decreases, and the person may feel restless, irritable, and anxious. The second stage is the active stage, where the skin begins to cool, breathing becomes difficult as the lungs become congested (known as the "death rattle"), and the person loses control of their bodily fluids.

Once death occurs, there are also two stages. The first is clinical death, when the heart stops pumping blood and breathing ceases. This stage lasts approximately 4-6 minutes, and during this time, it is possible for a victim to be resuscitated via CPR or a defibrillator. After 6 minutes however, the oxygen stores within the brain begin to deplete, and the victim enters biological death. This is the point of no return, as the cells of the brain and vital organs begin to die, a process that is irreversible.

Passage II

It was her sister Josephine who told her, in broken sentences; veiled hints that revealed in half concealing. Her husband's friend Richards was there, too, near her. It was he who had been in the newspaper office when intelligence of the railroad disaster was received, with Brently Mallard's name leading the list of "killed." He had only taken the time to assure himself of its truth by a second telegram, and had hastened to forestall any less careful, less tender friend in bearing the sad message.

She did not hear the story as many women have heard the same, with a paralyzed inability to accept its significance. She wept at once, with sudden, wild abandonment, in her sister's arms. When the storm of grief had spent itself she went away to her room alone. She would have no one follow her.

There stood, facing the open window, a comfortable, roomy armchair. Into this she sank, pressed down by a physical exhaustion that haunted her body and seemed to reach into her soul.

Excerpt from "The Story of an Hour" by Kate Chopin

Now, using the outline above, the similarities and differences between the two passages are considered:

1. **Style:** Passage I is an expository style, presenting purely factual evidence on death, completely devoid of emotion. Passage II is a narrative style, where the theme of death is presented to us by the reaction of the loved ones involved. This narrative style is full of emotional language and imagery.

2. **Tone:** Passage I has no emotionally-charged words of any kind, and seems to view death simply as a process that happens, neither welcoming nor fearing it. The tone in this passage, therefore, is neutral. Passage II does not have a neutral tone—it uses words like "disaster," "killed," "sad," "wept," "wild abandonment," and "physical exhaustion," implying an anxiety toward the theme of death.

3. **Sentence Structure:** Passage I contains many complex-compound sentences, which are used to accommodate lots of information. The structure of these sentences contributes to the overall informative nature of the selection. Passage II has several compound sentences and complex sentences on their own. It's also marked by the use of many commas in a single sentence, separating modifying words. Perhaps this variety is meant to match the sporadic emotion of the character's discovery of her husband's death.

4. **Punctuation Choice:** Passage I uses only commas and periods, which adds to the overall neutral tone of the selection. Passage II mostly uses commas and periods, and then one semicolon. Again, the excess of commas and semicolon in the first sentence may be said to mirror the character's anxiety.

5. **Point of View:** Passage I uses third-person point of view, as it avoids any first- or second-person pronouns. Passage II also uses third-person point of view, as the story is being told by a narrator about characters separate from the narrator.

6. **Paragraph Structure:** The first passage is told in an objective way, and each paragraph is focused on the topic brought up in the first sentence. The second passage has no specific topic per paragraph. It is organized in a sequential way, so the paragraphs flow into the next in a chronological order.

7. **Organizational Structure:** The structure of Passage I is told in a very objective, organized way. The first paragraph tells of the stages before death, and the second paragraph tells of the stages after death. The second passage is told in chronological order, as a sequence of events, like in a fictional story.

When analyzing the different structures, it may be helpful to make a table and use single words to compare and contrast the texts:

Elements	Passage I	Passage II
Style	Expository	Narrative
Tone	Neutral	Emotional
Sentence Structure	Long	Long/Sporadic
Punctuation Choice	.	. and ,
Point of View	Third	Third
Paragraph Structure	Focused	Sequential
Organizational	Objective/Logical	Chronological

Structure		

The main differences between the two selections are style, tone, and structure. Possibly the most noticeable difference is the style and tone, as one tone is more neutral, and the other tone is more emotional. This is due to the word choice used and how each passage treats the topic of death. These are only a handful of the endless possible interpretations the reader could make.

Analyzing How Authors Construct Arguments

Constructing Arguments Through Evidence

Using only one form of supporting evidence is not nearly as effective as using a variety to support a claim. Presenting only a list of statistics can be boring to the reader but providing a true story that's both interesting and humanizing helps. In addition, one example isn't always enough to prove the writer's larger point, so combining it with other examples in the writing is extremely effective. Thus, when reading a passage, readers should not just look for a single form of supporting evidence.

For example, although most people can't argue with the statement, "Seat belts save lives", its impact on the reader is much greater when supported by additional content. The writer can support this idea by:

- Providing statistics on the rate of highway fatalities alongside statistics of estimated seat belt usage.

- Explaining the science behind car accidents and what happens to a passenger who doesn't use a seat belt.

- Offering anecdotal evidence or true stories from reliable sources on how seat belts prevent fatal injuries in car crashes.

Another key aspect of supporting evidence is a **reliable source**. Does the writer include the source of the information? If so, is the source well-known and trustworthy? Is there a potential for bias? For example, a seat belt study done by a seat belt manufacturer may have its own agenda to promote.

Logical Sequence

Even if the writer includes plenty of information to support his or her point, the writing is only effective when the information is in a logical order. **Logical sequencing** is really just common sense, but it's also an important writing technique. First, the writer should introduce the main idea, whether for a paragraph, a section, or the entire text. Then he or she should present evidence to support the main idea by using transitional language. This shows the reader how the information relates to the main idea and to the sentences around it. The writer should then take time to interpret the information, making sure necessary connections are obvious to the reader. Finally, the writer can summarize the information in the closing section.

NOTE: Although most writing follows this pattern, it isn't a set rule. Sometimes writers change the order for effect. For example, the writer can begin with a surprising piece of supporting information to grab the reader's attention, and then transition to the main idea. Thus, if a passage doesn't follow the logical order, readers should not immediately assume it's wrong. However, most writing that has a nontraditional beginning usually settles into a logical sequence.

Vocabulary Acquisition and Use

Meaning of Words and Phrases

Another useful vocabulary skill is being able to understand meaning in context. A word's *context* refers to other words and information surrounding it, which can have a big impact on how readers interpret that word's meaning. Of course, many words have more than one definition. For example, consider the meaning of the word "engaged." The first definition that comes to mind might be "promised to be married," but consider the following sentences:

a. The two armies engaged in a conflict that lasted all night.

b. The three-hour lecture flew by because students were so engaged in the material.

c. The busy executive engaged a new assistant to help with his workload.

Were any of those sentences related to marriage? In fact, "engaged" has a variety of other meanings. In these sentences, respectively, it can mean: "battled," "interested or involved," and "appointed or employed." Readers may wonder how to decide which definition to apply. The appropriate meaning is prioritized based on context. For example, sentence *C* mentions "executive," "assistant," and "workload," so readers can assume that "engaged" has something to do with work—in which case, "appointed or employed" is the best definition for this context. Context clues can also be found in sentence *A*. Words like "armies" and "conflicts" show that this sentence is about a military situation, so in this context, "engaged" is closest in meaning to "battled." By using context clues—the surrounding words in the sentence—readers can easily select the most appropriate definition.

Context clues can also help readers when they don't know *any* meanings for a certain word. Test writers will deliberately ask about unfamiliar vocabulary to measure your ability to use context to make an educated guess about a word's meaning.

Which of the following is the closest in meaning to the word "loquacious" in the following sentence?

The *loquacious* professor was notorious for always taking too long to finish his lectures.
a. knowledgeable
b. enthusiastic
c. approachable
d. talkative

Even if the word "loquacious" seems completely new, it's possible to utilize context to make a good guess about the word's meaning. Grammatically, it's apparent that "loquacious" is an adjective that modifies the noun "professor"—so "loquacious" must be some kind of quality or characteristic. A clue in this sentence is "taking too long to finish his lectures." Readers should then consider qualities that might cause a professor's lectures to run long. Perhaps he's "disorganized," "slow," or "talkative"—all words that might still make sense in this sentence. Choice *D*, therefore, is a logical choice for this sentence— the professor talks too much, so his lectures run late. In fact, "loquacious" means "talkative or wordy."

One way to use context clues is to think of potential replacement words before considering the answer choices. You can also turn to the answer choices first and try to replace each of them in the sentence to see if the sentence is logical and retains the same meaning.

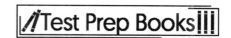

Another way to use context clues is to consider clues in the word itself. Most students are familiar with prefixes, suffixes, and root words—the building blocks of many English words. A little knowledge goes a long way when it comes to these components of English vocabulary, and these words can point readers in the right direction when they need help finding an appropriate definition.

Identifying Roots

By analyzing and understanding Latin, Greek, and Anglo-Saxon word roots and structure, authors better convey the thoughts they want to express to the readers of their words and help them to determine their meanings within the flow and without their missing a beat. For instance, **context**—how words are used in sentences—is from the Latin for *contextus*, which means "together" + "to weave," and gives readers a graphic for the minds' eyes to see the coming together of their usage. Like every other topic discussed herein, context is needed for understanding. This element actually has a second, crucial meaning. Context is not only the *how*, but the revealed moment of the *why* a writing has been composed; it is the "Aha" moment.

The way *how* words are used in sentences is important because it also gives meaning and cohesion from sentence to sentence, paragraph to paragraph, and page after page. In other words, it gives the document continuity.

Another upside of the how side is that readers have opportunities to understand new words with which they are unfamiliar. Of course, people can always look words up if a dictionary or thesaurus, if available, but meaning might be gleaned on the spot in a piece that is well-written. **Synonyms** (words or phrases that mean about the same) and **antonyms** (words or phrases that mean the opposite of the specific word) in context give clues to meanings, and sometimes reiteration of a word might add clarification. Repetition, wisely used, can also serve as a part of how a piece flows.

The revealed moment of the *why* is important because context, up to that moment, has determined the shape of the text. This is, essentially, to bring out what it is all about.

Prefixes

A **prefix** is a word, letter, or number that is placed before another. It adjusts or qualifies the original word's meaning.

Four prefixes represent 97 percent of English words with prefixes. They are:

- *dis-* means "not" or "opposite of"; *dis*abled
- in-, im-, il-, ir- mean "not"; illiterate
- *re-* means "again"; *re*turn
- *un-* means "not"; *un*predictable

Other commons prefixes include:

- *anti-* means "against"; antibacterial
- *fore-* means "before"; forefront
- *mis-* means "wrongly"; misunderstand
- *non-* means "not"; nonsense
- *over-* means "over"; overabundance
- *pre-* means "before"; preheat
- *super-* means "above"; superman

Suffixes

The official definition of a **suffix** is "a morpheme added at the end of a word to form a derivative." In English, that means a suffix is a letter or group of letters added at the end of a word to form another word. The word created with the addition is either a different tense of the same word (*help + ed = helped*) or a new word (*help + ful = helpful*).

They are:

- *-ed* is used to make present tense verbs into past tense verbs; wash*ed*
- *-ing* is used to make a present tense verb into a present participle verb; wash*ing*
- *-ly* is used to make characteristic of; love*ly*
- *-s* or *-es* are used to make more than one; chair*s* or box*es*

Other common suffixes include:

- *-able* means can be done; deplor*eable*
- *-al* means having characteristics of; comic*al*
- *-est* means comparative; great*est*
- *-ful* means full of; wonder*ful*
- *-ism* means belief in; commun*eism*
- *-less* means without; faith*less*
- *-ment* means action or process; accomplish*ment*
- *-ness* means state of; happ*yiness*
- *-ize* means to render, to make; terror*ize*, steril*ize*
- *-ise* means ditto, only this is primarily the British variant of *-ize*; surpr*ise*, advert*ise*
- *-ced* means go; spelling variations include -cede (concede, recede); -ceed (only three: proceed, exceed, succeed); -sede (the only one: supersede)

(Note: In some of the examples above, the *e* has been deleted.)

Practice Questions

Questions 1-6 are based on the following passage:

Dana Gioia argues in his article that poetry is dying, now little more than a limited art form confined to academic and college settings. Of course, poetry remains healthy in the academic setting, but the idea of poetry being limited to this academic subculture is a stretch. New technology and social networking alone have contributed to poets and other writers' work being shared across the world. YouTube has emerged to be a major asset to poets, allowing live performances to be streamed to billions of users. Even now, poetry continues to grow and voice topics that are relevant to the culture of our time. Poetry is not in the spotlight as it may have been in earlier times, but it's still a relevant art form that continues to expand in scope and appeal.

Furthermore, Gioia's argument does not account for live performances of poetry. Not everyone has taken a poetry class or enrolled in university—but most everyone is online. The Internet is a perfect launching point to get all creative work out there. An example of this was the performance of Buddy Wakefield's *Hurling Crowbirds at Mockingbars*. Wakefield is a well-known poet who has published several collections of contemporary poetry. One of my favorite works by Wakefield is *Crowbirds*, specifically his performance at New York University in 2009. Although his reading was a campus event, views of his performance online number in the thousands. His poetry attracted people outside of the university setting.

Naturally, the poem's popularity can be attributed both to Wakefield's performance and the quality of his writing. *Crowbirds* touches on themes of core human concepts such as faith, personal loss, and growth. These are not ideas that only poets or students of literature understand, but all human beings: "You acted like I was hurling crowbirds at mockingbars / and abandoned me for not making sense. / Evidently, I don't experience things as rationally as you do" (Wakefield 15-17). Wakefield weaves together a complex description of the perplexed and hurt emotions of the speaker undergoing a separation from a romantic interest. The line "You acted like I was hurling crowbirds at mockingbars" conjures up an image of someone confused, seemingly out of their mind . . . or in the case of the speaker, passionately trying to grasp at a relationship that is fading. The speaker is looking back and finding the words that described how he wasn't making sense. This poem is particularly human and gripping in its message, but the entire effect of the poem is enhanced through the physical performance.

At its core, poetry is about addressing issues/ideas in the world. Part of this is also addressing the perspectives that are exiguously considered. Although the platform may look different, poetry continues to have a steady audience due to the emotional connection the poet shares with the audience.

1. Which one of the following best explains how the passage is organized?
   a. The author begins with a long definition of the main topic, and then proceeds to prove how that definition has changed over the course of modernity.
   b. The author presents a puzzling phenomenon and uses the rest of the passage to showcase personal experiences in order to explain it.
   c. The author contrasts two different viewpoints, then builds a case showing preference for one over the other.
   d. The passage is an analysis of another theory that the author has no stake in.

2. The author of the passage would likely agree most with which of the following?
   a. Buddy Wakefield is a genius and is considered at the forefront of modern poetry.
   b. Poetry is not irrelevant; it is an art form that adapts to the changing time while containing its core elements.
   c. Spoken word is the zenith of poetic forms and the premier style of poetry in this decade.
   d. Poetry is on the verge of vanishing from our cultural consciousness.

3. Which one of the following words, if substituted for the word *exiguously* in the last paragraph, would LEAST change the meaning of the sentence?
   a. Indolently
   b. Inaudibly
   c. Interminably
   d. Infrequently

4. Which of the following is most closely analogous to the author's opinion of Buddy Wakefield's performance in relation to modern poetry?
   a. Someone's refusal to accept that the Higgs Boson will validate the Standard Model.
   b. An individual's belief that soccer will lose popularity within the next fifty years.
   c. A professor's opinion that poetry contains the language of the heart, while fiction contains the language of the mind.
   d. A student's insistence that psychoanalysis is a subset of modern psychology.

5. What is the primary purpose of the passage?
   a. To educate readers on the development of poetry and describe the historical implications of poetry in media.
   b. To disprove Dana Gioia's stance that poetry is becoming irrelevant and is only appreciated in academia.
   c. To inform readers of the brilliance of Buddy Wakefield and to introduce them to other poets that have influenced contemporary poetry.
   d. To prove that Gioia's article does have some truth to it and to shed light on its relevance to modern poetry.

6. What is the author's main reason for including the quote in the passage?
   a. The quote opens up opportunity to disprove Gioia's views.
   b. To demonstrate that people are still writing poetry even if the medium has changed in current times.
   c. To prove that poets still have an audience to write for even if the audience looks different than it did centuries ago.
   d. The quote illustrates the complex themes poets continue to address, which still draws listeners and appreciation.

Questions 7-14 are based on the following passage:

In the quest to understand existence, modern philosophers must question if humans can fully comprehend the world. Classical western approaches to philosophy tend to hold that one can understand something, be it an event or object, by standing outside of the phenomena and observing it. It is then by unbiased observation that one can grasp the details of the world. This seems to hold true for many things. Scientists conduct experiments and record their findings, and thus many natural phenomena become comprehendible. However, several of these observations were possible because humans used tools in order to make these discoveries.

This may seem like an extraneous matter. After all, people invented things like microscopes and telescopes in order to enhance their capacity to view cells or the movement of stars. While humans are still capable of seeing things, the question remains if human beings have the capacity to fully observe and see the world in order to understand it. It would not be an impossible stretch to argue that what humans see through a microscope is not the exact thing itself, but a human interpretation of it.

This would seem to be the case in the "Business of the Holes" experiment conducted by Richard Feynman. To study the way electrons behave, Feynman set up a barrier with two holes and a plate. The plate was there to indicate how many times the electrons would pass through the hole(s). Rather than casually observe the electrons acting under normal circumstances, Feynman discovered that electrons behave in two totally different ways depending on whether or not they are observed. The electrons that were observed had passed through either one of the holes or were caught on the plate as particles. However, electrons that weren't observed acted as waves instead of particles and passed through both holes. This indicated that electrons have a dual nature. Electrons seen by the human eye act like particles, while unseen electrons act like waves of energy.

This dual nature of the electrons presents a conundrum. While humans now have a better understanding of electrons, the fact remains that people cannot entirely perceive how electrons behave without the use of instruments. We can only observe one of the mentioned behaviors, which only provides a partial understanding of the entire function of electrons. Therefore, we're forced to ask ourselves whether the world we observe is objective or if it is subjectively perceived by humans. Or, an alternative question: can man understand the world only through machines that will allow them to observe natural phenomena?

Both questions humble man's capacity to grasp the world. However, those ideas don't consider that many phenomena have been proven by human beings without the use of machines, such as the discovery of gravity. Like all philosophical questions, whether man's reason and observation alone can understand the universe can be approached from many angles.

7. The word *extraneous* in paragraph two can be best interpreted as referring to which one of the following?
    a. Indispensable
    b. Bewildering
    c. Superfluous
    d. Exuberant

8. What is the author's motivation for writing the passage?
    a. To bring to light an alternative view on human perception by examining the role of technology in human understanding.
    b. To educate the reader on the latest astroparticle physics discovery and offer terms that may be unfamiliar to the reader.
    c. To argue that humans are totally blind to the realities of the world by presenting an experiment that proves that electrons are not what they seem on the surface.
    d. To reflect on opposing views of human understanding.

9. Which of the following most closely resembles the way in which paragraph four is structured?
    a. It offers one solution, questions the solution, and then ends with an alternative solution.
    b. It presents an inquiry, explains the details of that inquiry, and then offers a solution.
    c. It presents a problem, explains the details of that problem, and then ends with more inquiry.
    d. It gives a definition, offers an explanation, and then ends with an inquiry.

10. For the classical approach to understanding to hold true, which of the following must be required?
    a. A telescope
    b. A recording device
    c. Multiple witnesses present
    d. The person observing must be unbiased

11. Which best describes how the electrons in the experiment behaved like waves?
    a. The electrons moved up and down like actual waves.
    b. The electrons passed through both holes and then onto the plate.
    c. The electrons converted to photons upon touching the plate.
    d. Electrons were seen passing through one hole or the other.

12. The author mentions "gravity" in the last paragraph in order to do what?
    a. To show that different natural phenomena test man's ability to grasp the world.
    b. To prove that since man has not measured it with the use of tools or machines, humans cannot know the true nature of gravity.
    c. To demonstrate an example of natural phenomena humans discovered and understood without the use of tools or machines.
    d. To show an alternative solution to the nature of electrons that humans have not thought of yet.

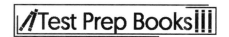

13. Which situation best parallels the revelation of the dual nature of electrons discovered in Feynman's experiment?

    a. A man is born color-blind and grows up observing everything in lighter or darker shades. With the invention of special goggles he puts on, he discovers that there are other colors in addition to different shades.

    b. The coelacanth was thought to be extinct, but a live specimen was just recently discovered. There are now two living species of coelacanth known to man, and both are believed to be endangered.

    c. In the Middle Ages, blacksmiths added carbon to iron, thus inventing steel. The consequences of this important discovery would have its biggest effects during the industrial revolution.

    d. In order to better examine and treat broken bones, the x-ray machine was invented and put to use in hospitals and medical centers.

14. Which statement about technology would the author likely disagree with?

    a. Technology can help expand the field of human vision.

    b. Technology renders human observation irrelevant.

    c. Developing tools used in observation and research indicates growing understanding of our world itself.

    d. Studying certain phenomena necessitates the use of tools and machines.

Questions 15-19 are based on the following passage:

The Middle Ages were a time of great superstition and theological debate. Many beliefs were developed and practiced, while some died out or were listed as heresy. Boethianism is a Medieval theological philosophy that attributes sin to gratification and righteousness with virtue and God's providence. Boethianism holds that sin, greed, and corruption are means to attain temporary pleasure, but that they inherently harm the person's soul as well as other human beings.

In *The Canterbury Tales,* we observe more instances of bad actions punished than goodness being rewarded. This would appear to be some reflection of Boethianism. In the "Pardoner's Tale," all three thieves wind up dead, which is a result of their desire for wealth. Each wrong doer pays with their life, and they are unable to enjoy the wealth they worked to steal. Within his tales, Chaucer gives reprieve to people undergoing struggle, but also interweaves stories of contemptible individuals being cosmically punished for their wickedness. The thieves idolize physical wealth, which leads to their downfall. This same theme and ideological principle of Boethianism is repeated in the "Friar's Tale," whose summoner character attempts to gain further wealth by partnering with a demon. The summoner's refusal to repent for his avarice and corruption leads to the demon dragging his soul to Hell. Again, we see the theme of the individual who puts faith and morality aside in favor for a physical prize. The result, of course, is that the summoner loses everything.

The examples of the righteous being rewarded tend to appear in a spiritual context within the *Canterbury Tales*. However, there are a few instances where we see goodness resulting in physical reward. In the Prioress' Tale, we see corporal punishment for barbarism *and* a reward for goodness. The Jews are punished for their murder of the child, giving a sense of law and order (though racist) to the plot. While the boy does die, he is granted a lasting reward by being able to sing even after his death, a miracle that marks that the murdered youth led a pure life. Here, the miracle represents eternal favor with God.

Again, we see the theological philosophy of Boethianism in Chaucer's *The Canterbury Tales* through acts of sin and righteousness and the consequences that follow. When pleasures of the world are sought instead of God's favor, we see characters being punished in tragic ways. However, the absence of worldly lust has its own set of consequences for the characters seeking to obtain God's favor.

15. What would be a potential reward for living a good life, as described in Boethianism?
    a. A long life sustained by the good deeds one has done over a lifetime
    b. Wealth and fertility for oneself and the extension of one's family line
    c. Vengeance for those who have been persecuted by others who have a capacity for committing wrongdoing
    d. God's divine favor for one's righteousness

16. What might be the main reason why the author chose to discuss Boethianism through examining The Canterbury Tales?
    a. *The Canterbury Tales* is a well-known text.
    b. *The Canterbury Tales* is the only known fictional text that contains use of Boethianism.
    c. *The Canterbury Tales* presents a manuscript written in the medieval period that can help illustrate Boethianism through stories and show how people of the time might have responded to the idea.
    d. Within each individual tale in *The Canterbury Tales*, the reader can read about different levels of Boethianism and how each level leads to greater enlightenment.

17. What "ideological principle" is the author referring to in the middle of the second paragraph when talking about the "Friar's Tale"?
    a. The principle that the act of ravaging another's possessions is the same as ravaging one's soul.
    b. The principle that thieves who idolize physical wealth will be punished in an earthly sense as well as eternally.
    c. The principle that fraternization with a demon will result in one losing everything, including his or her life.
    d. The principle that a desire for material goods leads to moral malfeasance punishable by a higher being.

18. Which of the following words, if substituted for the word *avarice* in paragraph two, would LEAST change the meaning of the sentence?
    a. Perniciousness
    b. Pithiness
    c. Covetousness
    d. Precariousness

19. Based on the passage, what view does Boethianism take on desire?
    a. Desire does not exist in the context of Boethianism
    b. Desire is a virtue and should be welcomed
    c. Having desire is evidence of demonic possession
    d. Desire for pleasure can lead toward sin

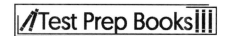

Questions 20-27 are based on the following passages:

Passage I

Lethal force, or deadly force, is defined as the physical means to cause death or serious harm to another individual. The law holds that lethal force is only accepted when you or another person are in immediate and unavoidable danger of death or severe bodily harm. For example, a person could be beating a weaker person in such a way that they are suffering severe enough trauma that could result in death or serious harm. This would be an instance where lethal force would be acceptable and possibly the only way to save that person from irrevocable damage.

Another example of when to use lethal force would be when someone enters your home with a deadly weapon. The intruder's presence and possession of the weapon indicate mal-intent and the ability to inflict death or severe injury to you and your loved ones. Again, lethal force can be used in this situation. Lethal force can also be applied to prevent the harm of another individual. If a woman is being brutally assaulted and is unable to fend off an attacker, lethal force can be used to defend her as a last-ditch effort. If she is in immediate jeopardy of rape, harm, and/or death, lethal force could be the only response that could effectively deter the assailant.

The key to understanding the concept of lethal force is the term *last resort*. Deadly force cannot be taken back; it should be used only to prevent severe harm or death. The law does distinguish whether the means of one's self-defense is fully warranted, or if the individual goes out of control in the process. If you continually attack the assailant after they are rendered incapacitated, this would be causing unnecessary harm, and the law can bring charges against you. Likewise, if you kill an attacker unnecessarily after defending yourself, you can be charged with murder. This would move lethal force beyond necessary defense, making it no longer a last resort but rather a use of excessive force.

Passage II

Assault is the unlawful attempt of one person to apply apprehension on another individual by an imminent threat or by initiating offensive contact. Assaults can vary, encompassing physical strikes, threatening body language, and even provocative language. In the case of the latter, even if a hand has not been laid, it is still considered an assault because of its threatening nature.

Let's look at an example: A homeowner is angered because his neighbor blows fallen leaves into his freshly mowed lawn. Irate, the homeowner gestures a fist to his fellow neighbor and threatens to bash his head in for littering on his lawn. The homeowner's physical motions and verbal threat heralds a physical threat against the other neighbor. These factors classify the homeowner's reaction as an assault. If the angry neighbor hits the threatening homeowner in retaliation, that would constitute an assault as well because he physically hit the homeowner.

Assault also centers on the involvement of weapons in a conflict. If someone fires a gun at another person, this could be interpreted as an assault unless the shooter acted in self-defense. If an individual drew a gun or a knife on someone with the intent to harm them, that would be considered assault. However, it's also considered an assault if someone simply aimed a weapon, loaded or not, at another person in a threatening manner.

20. What is the purpose of the second passage?
    a. To inform the reader about what assault is and how it is committed
    b. To inform the reader about how assault is a minor example of lethal force
    c. To disprove the previous passage concerning lethal force
    d. The author is recounting an incident in which they were assaulted

21. Which of the following situations, according to the passages, would not constitute an illegal use of lethal force?
    a. A disgruntled cashier yells obscenities at a customer.
    b. A thief is seen running away with stolen cash.
    c. A man is attacked in an alley by another man with a knife.
    d. A woman punches another woman in a bar.

22. Given the information in the passages, which of the following must be true about assault?
    a. Assault charges are more severe than unnecessary use of force charges.
    b. There are various forms of assault.
    c. Smaller, weaker people cannot commit assaults.
    d. Assault is justified only as a last resort.

23. Which of the following, if true, would most seriously undermine the explanation proposed by the author of Passage I in the third paragraph?
    a. An instance of lethal force in self-defense is not absolutely absolved from blame. The law considers the necessary use of force at the time it is committed.
    b. An individual who uses lethal force under necessary defense is in direct compliance of the law under most circumstances.
    c. Lethal force in self-defense should be forgiven in all cases for the peace of mind of the primary victim.
    d. The use of lethal force is not evaluated on the intent of the user but rather the severity of the primary attack that warranted self-defense.

24. Based on the passages, what can be inferred about the relationship between assault and lethal force?
    a. An act of lethal force always leads to a type of assault.
    b. An assault will result in someone using lethal force.
    c. An assault with deadly intent can lead to an individual using lethal force to preserve their well-being.
    d. If someone uses self-defense in a conflict, it is called deadly force; if actions or threats are intended, it is called assault.

25. Which of the following best describes the way the passages are structured?
    a. Both passages open by defining a legal concept and then continue to describe situations that further explain the concept.
    b. Both passages begin with situations, introduce accepted definitions, and then cite legal ramifications.
    c. Passage I presents a long definition while the Passage II begins by showing an example of assault.
    d. Both cite specific legal doctrines, then proceed to explain the rulings.

26. What can be inferred about the role of intent in lethal force and assault?
    a. Intent is irrelevant. The law does not take intent into account.
    b. Intent is vital for determining the lawfulness of using lethal force.
    c. Intent is very important for determining both lethal force and assault; intent is examined in both parties and helps determine the severity of the issue.
    d. The intent of the assailant is the main focus for determining legal ramifications; it is used to determine if the defender was justified in using force to respond.

27. The author uses the example in the second paragraph of Passage II in order to do what?
    a. To demonstrate two different types of assault by showing how each specifically relates to the other
    b. To demonstrate a single example of two different types of assault, then adding in the third type of assault in the example's conclusion
    c. To prove that the definition of lethal force is altered when the victim in question is a homeowner and his property is threatened
    d. To suggest that verbal assault can be an exaggerated crime by the law and does not necessarily lead to physical violence

Questions 28-33 are based upon the following passage:

> This excerpt is adapted from "What to the Slave is the Fourth of July?" Rochester, New York July 5, 1852.
>
> Fellow citizens—Pardon me, and allow me to ask, why am I called upon to speak here today? What have I, or those I represent, to do with your national independence? Are the great principles of political freedom and of natural justice, embodied in that Declaration of Independence, extended to us? And am I therefore called upon to bring our humble offering to the national altar, and to confess the benefits, and express devout gratitude for the blessings, resulting from your independence to us?
>
> Would to God, both for your sakes and ours, ours that an affirmative answer could be truthfully returned to these questions! Then would my task be light, and my burden easy and delightful. For who is there so cold that a nation's sympathy could not warm him? Who so obdurate and dead to the claims of gratitude that would not thankfully acknowledge such priceless benefits? Who so stolid and selfish, that would not give his voice to swell the hallelujahs of a nation's jubilee, when the chains of servitude had been torn from his limbs? I am not that man. In a case like that, the dumb my eloquently speak, and the lame man leap as an hart.
>
> But, such is not the state of the case. I say it with a sad sense of the disparity between us. I am not included within the pale of this glorious and anniversary. Oh pity! Your high independence only reveals the immeasurable distance between us. The blessings in

which you this day rejoice, I do not enjoy in common. The rich inheritance of justice, liberty, prosperity, and independence, bequeathed by your fathers, is shared by *you*, not by *me*. This Fourth of July is *yours,* not *mine*. You may rejoice, *I* must mourn. To drag a man in fetters into the grand illuminated temple of liberty, and call upon him to join you in joyous anthems, were inhuman mockery and sacrilegious irony. Do you mean, citizens, to mock me, by asking me to speak today? If so there is a parallel to your conduct. And let me warn you that it is dangerous to copy the example of a nation whose crimes, towering up to heaven, were thrown down by the breath of the Almighty, burying that nation and irrecoverable ruin! I can today take up the plaintive lament of a peeled and woe-smitten people.

By the rivers of Babylon, there we sat down. Yea! We wept when we remembered Zion. We hanged our harps upon the willows in the midst thereof. For there, they that carried us away captive, required of us a song; and they who wasted us required of us mirth, saying, "Sing us one of the songs of Zion." How can we sing the Lord's song in a strange land? If I forget thee, O Jerusalem, let my right hand forget her cunning. If I do not remember thee, let my tongue cleave to the roof of my mouth.

28. What is the tone of the first paragraph of this passage?
    a. Exasperated
    b. Inclusive
    c. Contemplative
    d. Nonchalant

29. Which word CANNOT be used synonymously with the term *obdurate* as it is conveyed in the text below?

    Who so obdurate and dead to the claims of gratitude, that would not thankfully acknowledge such priceless benefits?

    a. Steadfast
    b. Stubborn
    c. Contented
    d. Unwavering

30. What is the central purpose of this text?
    a. To demonstrate the author's extensive knowledge of the Bible
    b. To address the hypocrisy of the Fourth of July holiday
    c. To convince wealthy landowners to adopt new holiday rituals
    d. To explain why minorities often relished the notion of segregation in government institutions

31. Which statement serves as evidence for the question above?
    a. By the rivers of Babylon...down.
    b. Fellow citizens...today.
    c. I can...woe-smitten people.
    d. The rich inheritance of justice...*not by me*.

32. The statement below features an example of which of the following literary devices?
    Oh pity! Your high independence only reveals the immeasurable distance between us.

    a. Assonance
    b. Parallelism
    c. Amplification
    d. Hyperbole

33. The speaker's use of biblical references, such as "rivers of Babylon" and the "songs of Zion," helps the reader to do all EXCEPT which of the following?
    a. Identify with the speaker using common text
    b. Convince the audience that injustices have been committed by referencing another group of people who have been previously affected by slavery
    c. Display the equivocation of the speaker and those that he represents
    d. Appeal to the listener's sense of humanity

Questions 34-39 are based upon the following passage:

> This excerpt is an adaptation from Abraham Lincoln's Address Delivered at the Dedication of the Cemetery at Gettysburg, November 19, 1863.
>
> Four score and seven years ago our fathers brought forth on this continent, a new nation, conceived in liberty, and dedicated to the proposition that all men are created equal.
>
> Now we are engaged in a great civil war, testing whether that nation, or any nation so conceived and so dedicated, can long endure. We are met on a great battlefield of that war. We have come to dedicate a portion of that field, as a final resting place for those who here gave their lives that this nation might live. It is altogether fitting and proper that we should do this.
>
> But, in a larger sense, we cannot dedicate—we cannot consecrate that we cannot hallow—this ground. The brave men, living and dead, who struggled here, have consecrated it, far above our poor power to add or detract. The world will little note, nor long remember what we say here, but it can never forget what they did here. It is for us the living, rather, to be dedicated here to the unfinished work which they who fought here have thus far so nobly advanced. It is rather for us to be here and dedicated to the great task remaining before us—that from these honored dead we take increased devotion to that cause for which they gave the last full measure of devotion—that we here highly resolve that these dead shall not have died in vain—that this nation, under God, shall have a new birth of freedom—and that government of people, by the people, for the people, shall not perish from the earth.

34. The best description for the phrase *four score and seven years ago* is which of the following?
    a. A unit of measurement
    b. A period of time
    c. A literary movement
    d. A statement of political reform

35. What is the setting of this text?
    a. A battleship off of the coast of France
    b. A desert plain on the Sahara Desert
    c. A battlefield in North America
    d. The residence of Abraham Lincoln

36. Which war is Abraham Lincoln referring to in the following passage?
    Now we are engaged in a great civil war, testing whether that nation, or any nation so conceived and so dedicated, can long endure.

    a. World War I
    b. The War of the Spanish Succession
    c. World War II
    d. The American Civil War

37. What message is the author trying to convey through this address?
    a. The audience should perpetuate the ideals of freedom that the soldiers died fighting for.
    b. The audience should honor the dead by establishing an annual memorial service.
    c. The audience should form a militia that would overturn the current political structure.
    d. The audience should forget the lives that were lost and discredit the soldiers.

38. Which rhetorical device is being used in the following passage?
    ...we here highly resolve that these dead shall not have died in vain—that this nation, under God, shall have a new birth of freedom—and that government of people, by the people, for the people, shall not perish from the earth.

    a. Antimetabole
    b. Antiphrasis
    c. Anaphora
    d. Epiphora

39. What is the effect of Lincoln's statement in the following passage?
    But, in a larger sense, we cannot dedicate—we cannot consecrate that we cannot hallow—this ground. The brave men, living and dead, who struggled here, have consecrated it, far above our poor power to add or detract.

    a. His comparison emphasizes the great sacrifice of the soldiers who fought in the war.
    b. His comparison serves as a reminder of the inadequacies of his audience.
    c. His comparison serves as a catalyst for guilt and shame among audience members.
    d. His comparison attempts to illuminate the great differences between soldiers and civilians.

Questions 40-45 are based upon the following passage:

This excerpt is adapted from Charles Dickens' speech in Birmingham in England on December 30, 1853 on behalf of the Birmingham and Midland Institute.

My Good Friends,—When I first imparted to the committee of the projected Institute my particular wish that on one of the evenings of my readings here the main body of my audience should be composed of working men and their families, I was animated by two desires; first, by the wish to have the great pleasure of meeting you face to face at this

Christmas time, and accompany you myself through one of my little Christmas books; and second, by the wish to have an opportunity of stating publicly in your presence, and in the presence of the committee, my earnest hope that the Institute will, from the beginning, recognise one great principle—strong in reason and justice—which I believe to be essential to the very life of such an Institution. It is, that the working man shall, from the first unto the last, have a share in the management of an Institution which is designed for his benefit, and which calls itself by his name.

I have no fear here of being misunderstood—of being supposed to mean too much in this. If there ever was a time when any one class could of itself do much for its own good, and for the welfare of society—which I greatly doubt—that time is unquestionably past. It is in the fusion of different classes, without confusion; in the bringing together of employers and employed; in the creating of a better common understanding among those whose interests are identical, who depend upon each other, who are vitally essential to each other, and who never can be in unnatural antagonism without deplorable results, that one of the chief principles of a Mechanics' Institution should consist. In this world, a great deal of the bitterness among us arises from an imperfect understanding of one another. Erect in Birmingham a great Educational Institution, properly educational; educational of the feelings as well as of the reason; to which all orders of Birmingham men contribute; in which all orders of Birmingham men meet; wherein all orders of Birmingham men are faithfully represented—and you will erect a Temple of Concord here which will be a model edifice to the whole of England.

Contemplating as I do the existence of the Artisans' Committee, which not long ago considered the establishment of the Institute so sensibly, and supported it so heartily, I earnestly entreat the gentlemen—earnest I know in the good work, and who are now among us—by all means to avoid the great shortcoming of similar institutions; and in asking the working man for his confidence, to set him the great example and give him theirs in return. You will judge for yourselves if I promise too much for the working man, when I say that he will stand by such an enterprise with the utmost of his patience, his perseverance, sense, and support; that I am sure he will need no charitable aid or condescending patronage; but will readily and cheerfully pay for the advantages which it confers; that he will prepare himself in individual cases where he feels that the adverse circumstances around him have rendered it necessary; in a word, that he will feel his responsibility like an honest man, and will most honestly and manfully discharge it. I now proceed to the pleasant task to which I assure you I have looked forward for a long time.

40. Which word is most closely synonymous with the word *patronage* as it appears in the following statement?

...that I am sure he will need no charitable aid or condescending patronage

a. Auspices
b. Aberration
c. Acerbic
d. Adulation

41. Which term is most closely aligned with the definition of the term *working man* as it is defined in the following passage?

> You will judge for yourselves if I promise too much for the working man, when I say that he will stand by such an enterprise with the utmost of his patience, his perseverance, sense, and support...

    a. Plebeian
    b. Viscount
    c. Entrepreneur
    d. Bourgeois

42. Which of the following statements most closely correlates with the definition of the term *working man* as it is defined in Question 41?

    a. A working man is not someone who works for institutions or corporations, but someone who is well-versed in the workings of the soul.
    b. A working man is someone who is probably not involved in social activities because the physical demand for work is too high.
    c. A working man is someone who works for wages among the middle class.
    d. The working man has historically taken to the field, to the factory, and now to the screen.

43. Based upon the contextual evidence provided in the passage above, what is the meaning of the term *enterprise* in the third paragraph?
    a. Company
    b. Courage
    c. Game
    d. Cause

44. The speaker addresses his audience as *My Good Friends.* What kind of credibility does this salutation give to the speaker?

    a. The speaker is an employer addressing his employees, so the salutation is a way for the boss to bridge the gap between himself and his employees.
    b. The speaker's salutation is one from an entertainer to his audience and uses the friendly language to connect to his audience before a serious speech.
    c. The salutation is used ironically to give a somber tone to the serious speech that follows.
    d. The speech is one from a politician to the public, so the salutation is used to grab the audience's attention.

45. According to the passage, what is the speaker's second desire for his time in front of the audience?
    a. To read a Christmas story
    b. For the working man to have a say in his institution, which is designed for his benefit.
    c. To have an opportunity to stand in their presence
    d. For the life of the institution to be essential to the audience as a whole

Questions 46-51 are based upon the following passage:

> "MANKIND being originally equals in the order of creation, the equality could only be destroyed by some subsequent circumstance; the distinctions of rich, and poor, may in a great measure be accounted for, and that without having recourse to the harsh ill sounding names of oppression and avarice. Oppression is often the consequence, but

seldom or never the means of riches; and though avarice will preserve a man from being necessitously poor, it generally makes him too timorous to be wealthy.

But there is another and greater distinction for which no truly natural or religious reason can be assigned, and that is, the distinction of men into KINGS and SUBJECTS. Male and female are the distinctions of nature, good and bad the distinctions of heaven; but how a race of men came into the world so exalted above the rest, and distinguished like some new species, is worth enquiring into, and whether they are the means of happiness or of misery to mankind.

In the early ages of the world, according to the scripture chronology, there were no kings; the consequence of which was there were no wars; it is the pride of kings which throw mankind into confusion. Holland without a king hath enjoyed more peace for this last century than any of the monarchical governments in Europe. Antiquity favors the same remark; for the quiet and rural lives of the first patriarchs hath a happy something in them, which vanishes away when we come to the history of Jewish royalty.

Government by kings was first introduced into the world by the Heathens, from whom the children of Israel copied the custom. It was the most prosperous invention the Devil ever set on foot for the promotion of idolatry. The Heathens paid divine honors to their deceased kings, and the Christian world hath improved on the plan by doing the same to their living ones. How impious is the title of sacred majesty applied to a worm, who in the midst of his splendor is crumbling into dust!

As the exalting one man so greatly above the rest cannot be justified on the equal rights of nature, so neither can it be defended on the authority of scripture; for the will of the Almighty, as declared by Gideon and the prophet Samuel, expressly disapproves of government by kings. All anti-monarchical parts of scripture have been very smoothly glossed over in monarchical governments, but they undoubtedly merit the attention of countries, which have their governments yet to form. "Render unto Caesar the things which are Caesar's" is the scripture doctrine of courts, yet it is no support of monarchical government, for the Jews at that time were without a king, and in a state of vassalage to the Romans.

Near three thousand years passed away from the Mosaic account of the creation, till the Jews under a national delusion requested a king. Till then their form of government (except in extraordinary cases, where the Almighty interposed) was a kind of republic administered by a judge and the elders of the tribes. Kings they had none, and it was held sinful to acknowledge any being under that title but the Lord of Hosts. And when a man seriously reflects on the idolatrous homage which is paid to the persons of Kings, he need not wonder, that the Almighty ever jealous of his honor, should disapprove of a form of government which so impiously invades the prerogative of heaven.

Excerpt From: Thomas Paine. "Common Sense."

46. According to passage, what role does avarice, or greed, play in poverty?
    a. It can make a man very wealthy.
    b. It is the consequence of wealth.
    c. Avarice can prevent a man from being poor, but too fearful to be very wealthy.
    d. Avarice is what drives a person to be very wealthy

47. Of these distinctions, which does the author believe to be beyond natural or religious reason?
    a. Good and bad
    b. Male and female
    c. Human and animal
    d. King and subjects

48. According to the passage, what are the Heathens responsible for?
    a. Government by kings
    b. Quiet and rural lives of patriarchs
    c. Paying divine honors to their living kings
    d. Equal rights of nature

49. Which of the following best states Paine's rationale for the denouncement of monarchy?
    a. It is against the laws of nature.
    b. It is against the equal rights of nature and is denounced in scripture.
    c. Despite scripture, a monarchal government is unlawful.
    d. Neither the law nor scripture denounce monarchy.

50. Based on the passage, what is the best definition of the word *idolatrous*?
    a. Worshipping heroes
    b. Being deceitful
    c. Sinfulness
    d. Engaging in illegal activities

Answer Explanations

1. C: The author contrasts two different viewpoints, then builds a case showing preference for one over the other. Choice *A* is incorrect because the introduction does not contain an impartial definition, but rather, an opinion. Choice *B* is incorrect. There is no puzzling phenomenon given, as the author doesn't mention any peculiar cause or effect that is in question regarding poetry. Choice *D* does contain another's viewpoint at the beginning of the passage; however, to say that the author has no stake in this argument is incorrect; the author uses personal experiences to build their case.

2. B: Choice *B* accurately describes the author's argument in the text: that poetry is not irrelevant. While the author does praise, and even value, Buddy Wakefield as a poet, the author never heralds him as a genius. Eliminate Choice *A*, as it is an exaggeration. Not only is Choice *C* an exaggerated statement, but the author never mentions spoken word poetry in the text. Choice *D* is wrong because this statement contradicts the writer's argument.

3. D: *Exiguously* means not occurring often, or occurring rarely, so Choice *D* would LEAST change the meaning of the sentence. Choice *A*, *indolently*, means unhurriedly, or slow, and does not fit the context of the sentence. Choice *B*, *inaudibly*, means quietly or silently. Choice *C*, *interminably*, means endlessly, or all the time, and is the opposite of the word *exiguously*.

4. D: A student's insistence that psychoanalysis is a subset of modern psychology is the most analogous option. The author of the passage tries to insist that performance poetry is a subset of modern poetry, and therefore, tries to prove that modern poetry is not "dying," but thriving on social media for the masses. Choice *A* is incorrect, as the author is not refusing any kind of validation. Choice *B* is incorrect; the author's insistence is that poetry will *not* lose popularity. Choice *C* mimics the topic but compares two different genres, while the author does no comparison in this passage.

5. B: The author's purpose is to disprove Gioia's article claiming that poetry is a dying art form that only survives in academic settings. In order to prove his argument, the author educates the reader about new developments in poetry (Choice *A*) and describes the brilliance of a specific modern poet (Choice *C*), but these serve as examples of a growing poetry trend that counters Gioia's argument. Choice *D* is incorrect because it contradicts the author's argument.

6. D: This question is difficult because the choices offer real reasons as to why the author includes the quote. However, the question specifically asks for the *main reason* for including the quote. The quote from a recently written poem shows that people are indeed writing, publishing, and performing poetry (Choice *B*). The quote also shows that people are still listening to poetry (Choice *C*). These things are true, and by their nature, serve to disprove Gioia's views (Choice *A*), which is the author's goal. However, Choice *D* is the most direct reason for including the quote, because the article analyzes the quote for its "complex themes" that "draws listeners and appreciation" right after it's given.

7. C: *Extraneous* most nearly means *superfluous*, or *trivial*. Choice *A*, *indispensable*, is incorrect because it means the opposite of *extraneous*. Choice *B*, *bewildering*, means *confusing* and is not relevant to the context of the sentence. Finally, Choice *D* is wrong because although the prefix of the word is the same, *ex-*, the word *exuberant* means *elated* or *enthusiastic*, and is irrelevant to the context of the sentence.

8. A: The author's purpose is to bring to light an alternative view on human perception by examining the role of technology in human understanding. This is a challenging question because the author's purpose is somewhat open-ended. The author concludes by stating that the questions regarding human perception and observation can be approached from many angles. Thus, the author does not seem to be attempting to prove one thing or another. Choice B is incorrect because we cannot know for certain whether the electron experiment is the latest discovery in astroparticle physics because no date is given. Choice *C* is a broad generalization that does not reflect accurately on the writer's views. While the author does appear to reflect on opposing views of human understanding (Choice *D*), the best answer is Choice *A*.

9. C: It presents a problem, explains the details of that problem, and then ends with more inquiry. The beginning of this paragraph literally "presents a conundrum," explains the problem of partial understanding, and then ends with more questions, or inquiry. There is no solution offered in this paragraph, making Choices *A and B* incorrect. Choice *D* is incorrect because the paragraph does not begin with a definition.

10. D: Looking back in the text, the author describes that classical philosophy holds that understanding can be reached by careful observation. This will not work if they are overly invested or biased in their pursuit. Choices *A*, *B*, and *C* are in no way related and are completely unnecessary. A specific theory is not necessary to understanding, according to classical philosophy mentioned by the author.

11. B: The electrons passed through both holes and then onto the plate. Choices *A* and *C* are wrong because such movement is not mentioned at all in the text. In the passage the author says that electrons that were physically observed appeared to pass through one hole or another. Remember, the electrons that were observed doing this were described as acting like particles. Therefore, Choice *D* is wrong. Recall that the plate actually recorded electrons passing through both holes simultaneously and hitting the plate. This behavior, the electron activity that wasn't seen by humans, was characteristic of waves. Thus, Choice *B* is the right answer.

12. C: The author mentions "gravity" to demonstrate an example of natural phenomena humans discovered and understood without the use of tools or machines. Choice *A* mirrors the language in the beginning of the paragraph but is incorrect in its intent. Choice *B* is incorrect; the paragraph mentions nothing of "not knowing the true nature of gravity." Choice *D* is incorrect as well. There is no mention of an "alternative solution" in this paragraph.

13. A: The important thing to keep in mind is that we must choose a scenario that best parallels, or is most similar to, the discovery of the experiment mentioned in the passage. The important aspects of the experiment can be summed up like so: humans directly observed one behavior of electrons and then through analyzing a tool (the plate that recorded electron hits), discovered that there was another electron behavior that could not be physically seen by human eyes. This summary best parallels the scenario in Choice *A*. Like Feynman, the colorblind person can observe one aspect of the world but through the special goggles (a tool), he is able to see a natural phenomenon that he could not physically see on his own. While Choice *D* is compelling, the x-ray helps humans see the broken bone, but it does not necessarily reveal that the bone is broken in the first place. The other choices do not parallel the scenario in question. Therefore, Choice *A* is the best choice.

14. B: The author would not agree that technology renders human observation irrelevant. Choice *A* is incorrect because much of the passage discusses how technology helps humans observe what cannot be seen with the naked eye; therefore, the author would agree with this statement. This line of reasoning is also why the author would agree with Choice *D*, making it incorrect as well. As indicated in the second paragraph, the author seems to think that humans create inventions and tools with the goal of studying phenomena more precisely. This indicates increased understanding as people recognize limitations and

develop items to help bypass the limitations and learn. Therefore, Choice *C* is incorrect as well. Again, the author doesn't attempt to disprove or dismiss classical philosophy.

15. D: The author explains that Boethianism is a Medieval theological philosophy that attributes sin to temporary pleasure and righteousness with virtue and God's providence. Besides Choice *D,* the choices listed are all physical things. While these could still be divine rewards, Boethianism holds that the true reward for being virtuous is in God's favor. It is also stressed in the article that physical pleasures cannot be taken into the afterlife. Therefore, the best choice is *D,* God's favor.

16. C: *The Canterbury Tales* presents a manuscript written in the medieval period that can help illustrate Boethianism through stories and show how people of the time might have responded to the idea. Choices *A* and *B* are generalized statements, and we have no evidence to support Choice *B*. Choice *D* is very compelling, but it looks at Boethianism in a way that the author does not. The author does not mention "different levels of Boethianism" when discussing the tales, only that the concept appears differently in different tales. Boethianism also doesn't focus on enlightenment.

17. D: The author is referring to the principle that a desire for material goods leads to moral malfeasance punishable by a higher being. Choice *A* is incorrect; while the text does mention thieves ravaging others' possessions, it is only meant as an example and not as the principle itself. Choice *B* is incorrect for the same reason as *A*. Choice *C* is mentioned in the text and is part of the example that proves the principle, and also not the principle itself.

18. C: The word *avarice* most nearly means *covetousness*, or extremely desirous of money or wealth. Choice *A* means *evil* or *mischief* and does not relate to the context of the sentence. Choice *B* is also incorrect, because *pithiness* means *shortness* or *conciseness*. Choice *D* is close because *precariousness* means dangerous or instability, which goes well with the context. However, we are told of the summoner's specific characteristic of greed, which makes Choice *C* the best answer.

19. D: Desire for pleasure can lead toward sin. Boethianism acknowledges desire as something that leads out of holiness, so Choice *A* is incorrect. Choice *B* is incorrect because in the passage, Boethianism is depicted as being wary of desire and anything that binds people to the physical world. Choice *C* can be eliminated because the author never says that desire indicates demonic.

20. A: The purpose is to inform the reader about what assault is and how it is committed. Choice *B* is incorrect because the passage does not state that assault is a lesser form of lethal force, only that an assault can use lethal force, or alternatively, lethal force can be utilized to counter a dangerous assault. Choice *C* is incorrect because the passage is informative and does not have a set agenda. Finally, Choice *D* is incorrect because although the author uses an example in order to explain assault, it is not indicated that this is the author's personal account.

21. C: If the man being attacked in an alley by another man with a knife used self-defense by lethal force, it would not be considered illegal. The presence of a deadly weapon indicates mal-intent and because the individual is isolated in an alley, lethal force in self-defense may be the only way to preserve his life. Choices *A* and *B* can be ruled out because in these situations, no one is in danger of immediate death or bodily harm by someone else. Choice *D* is an assault and does exhibit intent to harm, but this situation isn't severe enough to merit lethal force; there is no intent to kill.

22. B: As discussed in the second passage, there are several forms of assault, like assault with a deadly weapon, verbal assault, or threatening posture or language. Choice *A* is incorrect because the author does mention what the charges are on assaults; therefore, we cannot assume that they are more or less than unnecessary use of force charges. Choice *C* is incorrect because anyone is capable of assault; the author does not state that one group of people cannot commit assault. Choice *D* is incorrect because assault is never justified. Self-defense resulting in lethal force can be justified.

23. D: The use of lethal force is not evaluated on the intent of the user, but rather on the severity of the primary attack that warranted self-defense. This statement most undermines the last part of the passage because it directly contradicts how the law evaluates the use of lethal force. Choices *A* and *B* are stated in the paragraph, so they do not undermine the explanation from the author. Choice *C* does not necessarily undermine the passage, but it does not support the passage either. It is more of an opinion that does not offer strength or weakness to the explanation.

24. C: An assault with deadly intent can lead to an individual using lethal force to preserve their well-being. Choice *C* is correct because it clearly establishes what both assault and lethal force are and gives the specific way in which the two concepts meet. Choice *A* is incorrect because lethal force doesn't necessarily result in assault. This is also why Choice *B* is incorrect. Not all assaults would necessarily be life-threatening to the point where lethal force is needed for self-defense. Choice *D* is compelling but ultimately too vague; the statement touches on aspects of the two ideas but fails to present the concrete way in which the two are connected to each other.

25. A: Both passages open by defining a legal concept and then continue to describe situations in order to further explain the concept. Choice *D* is incorrect because while the passages utilize examples to help explain the concepts discussed, the author doesn't indicate that they are specific court cases. It's also clear that the passages don't open with examples, but instead, they begin by defining the terms addressed in each passage. This eliminates Choice *B,* and ultimately reveals Choice *A* to be the correct answer. Choice *A* accurately outlines the way both passages are structured. Because the passages follow a nearly identical structure, the Choice *C* can easily be ruled out.

26. C: Intent is very important for determining both lethal force and assault; intent is examined in both parties and helps determine the severity of the issue. Choices *A* and *B* are incorrect because it is clear in both passages that intent is a prevailing theme in both lethal force and assault. Choice *D* is compelling, but if a person uses lethal force to defend himself or herself, the intent of the defender is also examined in order to help determine if there was excessive force used. Choice *C* is correct because it states that intent is important for determining both lethal force and assault, and that intent is used to gauge the severity of the issues. Remember, just as lethal force can escalate to excessive use of force, there are different kinds of assault. Intent dictates several different forms of assault.

27. B: The example is used to demonstrate a single example of two different types of assault, then adding in a third type of assault to the example's conclusion. The example mainly provides an instance of "threatening body language" and "provocative language" with the homeowner gesturing threats to his neighbor. It ends the example by adding a third type of assault: physical strikes. This example is used to show the variant nature of assaults. Choice *A* is incorrect because it doesn't mention the "physical strike" assault at the end and is not specific enough. Choice *C* is incorrect because the example does not say anything about the definition of lethal force or how it might be altered. Choice *D* is incorrect, as the example mentions nothing about cause and effect.

28. A: The tone is exasperated. While contemplative is an option because of the inquisitive nature of the text, Choice *A* is correct because the speaker is frustrated by the thought of being included when he felt that the fellow members of his race were being excluded. The speaker is not nonchalant, nor accepting of the circumstances which he describes.

29. C: Choice *C*, *contented*, is the only word that has different meaning. Furthermore, the speaker expresses objection and disdain throughout the entire text.

30. B: To address the hypocrisy of the Fourth of July holiday. While the speaker makes biblical references, it is not the main focus of the passage, thus eliminating Choice *A* as an answer. The passage also makes no mention of wealthy landowners and doesn't speak of any positive response to the historical events, so Choices *C* and *D* are not correct.

31: D: Choice *D* is the correct answer because it clearly makes reference to justice being denied.

32: D: It is an example of hyperbole. Choices *A* and *B* are unrelated. Assonance is the repetition of sounds and commonly occurs in poetry. Parallelism refers to two statements that correlate in some manner. Choice *C* is incorrect because amplification normally refers to clarification of meaning by broadening the sentence structure, while hyperbole refers to a phrase or statement that is being exaggerated.

33: C: Display the equivocation of the speaker and those that he represents. Choice *C* is correct because the speaker is clear about his intention and stance throughout the text. Choice *A* could be true, but the words "common text" is arguable. Choice *B* is also partially true, as another group of people affected by slavery are being referenced. However, the speaker is not trying to convince the audience that injustices have been committed, as it is already understood there have been injustices committed. Choice *D* is also close to the correct answer, but it is not the *best* answer choice possible.

34. B: It denotes a period of time. It is apparent that Lincoln is referring to a period of time within the context of the passage because of how the sentence is structured with the word *ago*.

35. C: Lincoln's reference to *the brave men, living and dead, who struggled here,* proves that he is referring to a battlefield. Choices *A* and *B* are incorrect, as a *civil war* is mentioned and not a war with France or a war in the Sahara Desert. Choice *D* is incorrect because it does not make sense to consecrate a President's ground instead of a battlefield ground for soldiers who died during the American Civil War.

36. D: Abraham Lincoln is a former president of the United States, and he referenced a "civil war" during his address.

37. A: The audience should perpetuate the ideals of freedom that the soldiers died fighting for. Lincoln doesn't address any of the topics outlined in Choices *B*, *C*, or *D*. Therefore, Choice *A* is the correct answer.

38. D: Choice *D* is the correct answer because of the repetition of the word *people* at the end of the passage. Choice *A*, *antimetatabole*, is the repetition of words in a succession. Choice *B*, *antiphrasis*, is a form of denial of an assertion in a text. Choice *C*, *anaphora*, is the repetition that occurs at the beginning of sentences.

39. A: Choice *A* is correct because Lincoln's intention was to memorialize the soldiers who had fallen as a result of war as well as celebrate those who had put their lives in danger for the sake of their country. Choices *B* and *D* are incorrect because Lincoln's speech was supposed to foster a sense of pride among the members of the audience while connecting them to the soldiers' experiences.

40. A: The word *patronage* most nearly means *auspices*, which means *protection* or *support*. Choice *B*, *aberration*, means *deformity* and does not make sense within the context of the sentence. Choice *C*, *acerbic,* means *bitter* and also does not make sense in the sentence. Choice *D, adulation,* is a positive word meaning *praise,* and thus does not fit with the word *condescending* in the sentence.

41. D: *Working man* is most closely aligned with Choice *D, bourgeois.* In the context of the speech, the word *bourgeois* means *working* or *middle class.* Choice *A, Plebeian,* does suggest *common people;* however, this is a term that is specific to ancient Rome. Choice *B, viscount,* is a European title used to describe a specific degree of nobility. Choice *C, entrepreneur,* is a person who operates their own business.

42. C: In the context of the speech, the term *working man* most closely correlates with Choice *C, working man is someone who works for wages among the middle class.* Choice *A* is not mentioned in the passage and is off-topic. Choice *B* may be true in some cases, but it does not reflect the sentiment described for the term *working man* in the passage. Choice *D* may also be arguably true. However, it is not given as a definition but as *acts* of the working man, and the topics of *field, factory,* and *screen* are not mentioned in the passage.

43. D: *Enterprise* most closely means *cause.* Choices *A, B,* and *C* are all related to the term *enterprise.* However, Dickens speaks of a *cause* here, not a company, courage, or a game. *He will stand by such an enterprise* is a call to stand by a cause to enable the working man to have a certain autonomy over his own economic standing. The very first paragraph ends with the statement that the working man *shall . . . have a share in the management of an institution which is designed for his benefit.*

44. B: The speaker's salutation is one from an entertainer to his audience and uses the friendly language to connect to his audience before a serious speech. Recall in the first paragraph that the speaker is there to "accompany [the audience] . . . through one of my little Christmas books," making him an author there to entertain the crowd with his own writing. The speech preceding the reading is the passage itself, and, as the tone indicates, a serious speech addressing the "working man." Although the passage speaks of employers and employees, the speaker himself is not an employer of the audience, so Choice *A* is incorrect. Choice *C* is also incorrect, as the salutation is not used ironically, but sincerely, as the speech addresses the well-being of the crowd. Choice *D* is incorrect because the speech is not given by a politician, but by a writer.

45: B: Choice *A* is incorrect because that is the speaker's *first* desire, not his second. Choices *C* and *D* are tricky because the language of both of these is mentioned after the word *second.* However, the speaker doesn't get to the second wish until the next sentence. Choices *C* and *D* are merely prepositions preparing for the statement of the main clause, Choice *B,* for the working man to have a say in his institution, which is designed for his benefit.

46. C: In lines 6 and 7, it is stated that avarice can prevent a man from being "necessitously poor", but too timorous, or fearful, to achieve real wealth. According to the passage, avarice does not tend to make a person very wealthy. The passage states that oppression, not avarice, is the consequence of wealth. The passage does not state that avarice drives a person's desire to be wealthy.

47. D: Paine believes that the distinction that is beyond a natural or religious reason is between king and subjects. He states that the distinction between good and bad is made in heaven. The distinction between male and female is natural. He does not mention anything about the distinction between humans and animals.

48. A: The passage states that the Heathens were the first to introduce government by kings into the world. The quiet lives of patriarchs came before the Heathens introduced this type of government. It

was Christians, not Heathens, who paid divine honors to living kings. Heathens honored deceased kings. Equal rights of nature are mentioned in the paragraph, but not in relation to the Heathens.

49. B: Paine asserts that a monarchy is against the equal rights of nature and cites several parts of scripture that also denounce it. He doesn't say it is against the laws of nature. Because he uses scripture to further his argument, it is not despite scripture that he denounces the monarchy. Paine addresses the law by saying the courts also do not support a monarchical government.

50. A: To be *idolatrous* is to worship idols or heroes, in this case, kings. It is not defined as being deceitful. While idolatry is considered a sin, it is an example of a sin, not a synonym for it. Idolatry may have been considered illegal in some cultures, but it is not a definition for the term.

Writing

Brainstorming

One of the most important steps in writing an essay is prewriting. Before drafting an essay, it's helpful to think about the topic for a moment or two, in order to gain a more solid understanding of what the task is. Then, spending about five minutes jotting down the immediate ideas that could work for the essay is recommended. It is a way to get some words on the page and offer a reference for ideas when drafting. Scratch paper is provided for writers to use any prewriting techniques such as webbing, free writing, or listing. The goal is to get ideas out of the mind and onto the page.

Considering Opposing Viewpoints

In the planning stage, it's important to consider all aspects of the topic, including different viewpoints on the subject. There are more than two ways to look at a topic, and a strong argument considers those opposing viewpoints. Considering opposing viewpoints can help writers present a fair, balanced, and informed essay that shows consideration for all readers. This approach can also strengthen an argument by recognizing and potentially refuting the opposing viewpoint(s).

Drawing from personal experience may help to support ideas. For example, if the goal for writing is a personal narrative, then the story should be from the writer's own life. Many writers find it helpful to draw from personal experience, even in an essay that is not strictly narrative. Personal anecdotes or short stories can help to illustrate a point in other types of essays as well.

Moving from Brainstorming to Planning

Once the ideas are on the page, it's time to turn them into a solid plan for the essay. The best ideas from the brainstorming results can then be developed into a more formal outline. An outline typically has one main point (the thesis) and at least three sub-points that support the main point. Here's an example:

Main Idea

- Point #1
- Point #2
- Point #3

Of course, there will be details under each point, but this approach is the best for dealing with timed writing.

Staying on Track

Basing the essay on the outline aids in both organization and coherence. The goal is to ensure that there is enough time to develop each sub-point in the essay, roughly spending an equal amount of time on each idea. Keeping an eye on the time will help. If there are fifteen minutes left to draft the essay, then it makes sense to spend about 5 minutes on each of the ideas. Staying on task is critical to success, and timing out the parts of the essay can help writers avoid feeling overwhelmed.

Parts of the Essay

The **introduction** has to do a few important things:

- Establish the **topic** of the essay in original wording (i.e., not just repeating the prompt)
- Clarify the significance/importance of the topic or purpose for writing (not too many details, a brief overview)
- Offer a **thesis statement** that identifies the writer's own viewpoint on the topic (typically one-two brief sentences as a clear, concise explanation of the main point on the topic)

Body paragraphs reflect the ideas developed in the outline. Three-four points is probably sufficient for a short essay, and they should include the following:

- A **topic sentence** that identifies the sub-point (e.g., a reason why, a way how, a cause or effect)
- A detailed **explanation** of the point, explaining why the writer thinks this point is valid
- Illustrative examples, such as personal examples or real-world examples, that support and validate the point (i.e., "prove" the point)
- A **concluding sentence** that connects the examples, reasoning, and analysis to the point being made

The **conclusion,** or final paragraph, should be brief and should reiterate the focus, clarifying why the discussion is significant or important. It is important to avoid adding specific details or new ideas to this paragraph. The purpose of the conclusion is to sum up what has been said to bring the discussion to a close.

Don't Panic!

Writing an essay can be overwhelming, and performance panic is a natural response. The outline serves as a basis for the writing and helps to keep writers focused. Getting stuck can also happen, and it's helpful to remember that brainstorming can be done at any time during the writing process. Following the steps of the writing process is the best defense against writer's block.

Timed essays can be particularly stressful, but assessors are trained to recognize the necessary planning and thinking for these timed efforts. Using the plan above and sticking to it helps with time management. Timing each part of the process helps writers stay on track. Sometimes writers try to cover too much in their essays. If time seems to be running out, this is an opportunity to determine whether all of the ideas in the outline are necessary. Three body paragraphs are sufficient, and more than that is probably too much to cover in a short essay.

More isn't always *better* in writing. A strong essay will be clear and concise. It will avoid unnecessary or repetitive details. It is better to have a concise, five-paragraph essay that makes a clear point, than a ten-paragraph essay that doesn't. The goal is to write one-two pages of quality writing. Paragraphs should also reflect balance; if the introduction goes to the bottom of the first page, the writing may be going off-track or be repetitive. It's best to fall into the one-two page range, but a complete, well-developed essay is the ultimate goal.

The Final Steps

Leaving a few minutes at the end to revise and proofread offers an opportunity for writers to polish things up. Putting one's self in the reader's shoes and focusing on what the essay actually says helps writers identify problems—it's a movement from the mindset of writer to the mindset of editor. The goal is to have a clean, clear copy of the essay. The following areas should be considered when proofreading:

- Sentence fragments
- Awkward sentence structure
- Run-on sentences
- Incorrect word choice
- Grammatical agreement errors
- Spelling errors
- Punctuation errors
- Capitalization errors

The Short Overview

The essay may seem challenging, but following these steps can help writers focus:

- Take one-two minutes to think about the topic.
- Generate some ideas through brainstorming (three-four minutes).
- Organize ideas into a brief outline, selecting just three-four main points to cover in the essay (eventually the body paragraphs).
- Develop essay in parts:
- Introduction paragraph, with intro to topic and main points
- Viewpoint on the subject at the end of the introduction
- Body paragraphs, based on outline
- Each paragraph: makes a main point, explains the viewpoint, uses examples to support the point
- Brief conclusion highlighting the main points and closing
- Read over the essay (last five minutes).
- Look for any obvious errors, making sure that the writing makes sense.

Authors utilize a wide range of techniques to tell a story or communicate information. Readers should be familiar with the most common of these techniques. Techniques of writing are also commonly known as rhetorical devices.

Grammar/Usage

Nouns

A **noun** is a person, place, thing, or idea. All nouns fit into one of two types, common or proper.

A **common noun** is a word that identifies any of a class of people, places, or things. Examples include numbers, objects, animals, feelings, concepts, qualities, and actions. *A, an,* or *the* usually precedes the

common noun. These parts of speech are called *articles*. Here are some examples of sentences using nouns preceded by articles.

A building is under construction.
The girl would like to move to *the* city.

A **proper noun** (also called a **proper name**) is used for the specific name of an individual person, place, or organization. The first letter in a proper noun is capitalized. "My name is *Mary.*" "I work for *Walmart.*"

Nouns sometimes serve as **adjectives** (which themselves describe nouns), such as "hockey player" and "state government."

Pronouns

A word used in place of a noun is known as a **pronoun**. Pronouns are words like *I, mine, hers,* and *us.*

Pronouns can be split into different classifications (as shown below) which make them easier to learn; however, it's not important to memorize the classifications.

- **Personal pronouns:** refer to people

- **First person pronouns:** we, I, our, mine

- **Second person pronouns:** you, yours

- **Third person pronouns:** he, she, they, them, it

- **Possessive pronouns:** demonstrate ownership (mine, his, hers, its, ours, theirs, yours)

- **Interrogative pronouns:** ask questions (what, which, who, whom, whose)

- **Relative pronouns:** include the five interrogative pronouns and others that are relative (whoever, whomever, that, when, where)

- **Demonstrative pronouns:** replace something specific (this, that, those, these)

- **Reciprocal pronouns:** indicate something was done or given in return (each other, one another)

- **Indefinite pronouns:** have a nonspecific status (anybody, whoever, someone, everybody, somebody)

Indefinite pronouns such as *anybody, whoever, someone, everybody*, and *somebody* command a singular verb form, but others such as *all, none,* and *some* could require a singular or plural verb form.

Antecedents
An **antecedent** is the noun to which a pronoun refers; it needs to be written or spoken before the pronoun is used. For many pronouns, antecedents are imperative for clarity. In particular, a lot of the personal, possessive, and demonstrative pronouns need antecedents. Otherwise, it would be unclear who or what someone is referring to when they use a pronoun like *he* or *this*.

Pronoun reference means that the pronoun should refer clearly to one, clear, unmistakable noun (the antecedent).

Pronoun-antecedent agreement refers to the need for the antecedent and the corresponding pronoun to agree in gender, person, and number. Here are some examples:

The *kidneys* (plural antecedent) are part of the urinary system. *They* (plural pronoun) serve several roles.

The kidneys are part of the *urinary system* (singular antecedent). *It* (singular pronoun) is also known as the renal system.

Pronoun Cases

The **subjective pronouns** —*I, you, he/she/it, we, they,* and *who*—are the subjects of the sentence.

Example: *They* have a new house.

The **objective pronouns**—*me, you* (*singular*), *him/her, us, them,* and *whom*—are used when something is being done for or given to someone; they are objects of the action.

Example: The teacher has an apple for *us*.

The **possessive pronouns**—*mine, my, your, yours, his, hers, its, their, theirs, our,* and *ours*—are used to denote that something (or someone) belongs to someone (or something).

Example: It's *their* chocolate cake.

Even Better Example: It's *my* chocolate cake!

One of the greatest challenges and worst abuses of pronouns concerns *who* and *whom*. Just knowing the following rule can eliminate confusion. *Who* is a subjective-case pronoun used only as a subject or subject complement. *Whom* is only objective-case and, therefore, the object of the verb or preposition.

Who is going to the concert?

You are going to the concert with *whom*?

Hint: When using *who* or *whom*, think of whether someone would say *he* or *him*. If the answer is *he*, use *who*. If the answer is *him*, use *whom*. This trick is easy to remember because *he* and *who* both end in vowels, and *him* and *whom* both end in the letter *M*.

Many possessive pronouns sound like contractions. For example, many people get *it's* and *its* confused. The word *it's* is the contraction for *it is*. The word *its* without an apostrophe is the possessive form of *it*.

I love that wooden desk. It's beautiful. (contraction)

I love that wooden desk. Its glossy finish is beautiful. (possessive)

If you are not sure which version to use, replace *it's/its* with *it is* and see if that sounds correct. If so, use the contraction (*it's*). That trick also works for *who's/whose, you're/your,* and *they're/their*.

Adjectives

"The *extraordinary* brain is the *main* organ of the central nervous system." The adjective *extraordinary* describes the brain in a way that causes one to realize it is more exceptional than some of the other organs while the adjective *main* defines the brain's importance in its system.

An **adjective** is a word or phrase that names an attribute that describes or clarifies a noun or pronoun. This helps the reader visualize and understand the characteristics—size, shape, age, color, origin, etc.— of a person, place, or thing that otherwise might not be known. Adjectives breathe life, color, and depth into the subjects they define. Life would be *drab* and *colorless* without adjectives!

Adjectives often precede the nouns they describe.

> S*he drove her <u>new</u> car.*

However, adjectives can also come later in the sentence.

> *Her car is <u>new</u>.*

Adjectives using the prefix *a–* can only be used after a verb.

> Correct: The dog was *alive* until the car ran up on the curb and hit him.

> Incorrect: The *alive* dog was hit by a car that ran up on the curb.

Other examples of this rule include *awake, ablaze, ajar, alike,* and *asleep.*

Other adjectives used after verbs concern states of health.

> The girl was finally *well* after a long bout of pneumonia.

> The boy was *fine* after the accident.

An adjective phrase is not a bunch of adjectives strung together, but a group of words that describes a noun or pronoun and, thus, functions as an adjective. Very happy is an adjective phrase; so are way too hungry and passionate about traveling.

Possessives

In grammar, **possessive nouns** show ownership, which was seen in previous examples like *mine, yours,* and *theirs*.

Singular nouns are generally made possessive with an apostrophe and an *s* (*'s*).

> My *uncle's* new car is silver.

> The *dog's* bowl is empty.

> *James's* ties are becoming outdated.

Plural nouns ending in *s* are generally made possessive by just adding an apostrophe ('):

> The pistachio nuts' saltiness is added during roasting. (The saltiness of pistachio nuts is added during roasting.)

> The students' achievement tests are difficult. (The achievement tests of the students are difficult.)

If the plural noun does not end in an *s* such as *women,* then it is made possessive by adding an *apostrophe s ('s)—women's.*

Indefinite possessive pronouns such as *nobody* or *someone* become possessive by adding an *apostrophe s— nobody's* or *someone's.*

Verbs

A verb is the part of speech that describes an action, state of being, or occurrence.

A verb forms the main part of a predicate of a sentence. This means that the verb explains what the noun (which will be discussed shortly) is doing. A simple example is *time <u>flies</u>*. The verb *flies* explains what the action of the noun, *time,* is doing. This example is a *main* verb.

Helping (auxiliary) verbs are words like *have, do, be, can, may, should, must,* and *will.* "I *should* go to the store." Helping verbs assist main verbs in expressing tense, ability, possibility, permission, or obligation.

Particles are minor function words like *not, in, out, up,* or *down* that become part of the verb itself. "I might *not.*"

Participles are words formed from verbs that are often used to modify a noun, noun phrase, verb, or verb phrase.

> The *running* teenager collided with the cyclist.

Participles can also create compound verb forms.

> He is *speaking.*

Verbs have five basic forms: the **base** form, the **-s** form, the **-ing** form, the **past** form, and the **past participle** form.

The past forms are either **regular** (*love/loved; hate/hated*) or **irregular** because they don't end by adding the common past tense suffix "-ed" (*go/went; fall/fell; set/set*).

Adverbs

Adverbs have more functions than adjectives because they modify or qualify verbs, adjectives, or other adverbs as well as word groups that express a relation of place, time, circumstance, or cause. Therefore, adverbs answer any of the following questions: *How, when, where, why, in what way, how often, how much, in what condition,* and/or *to what degree. How good looking is he? He is <u>very</u> handsome.*

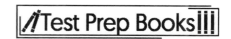

Here are some examples of adverbs for different situations:

- how: quickly
- when: daily
- where: there
- in what way: easily
- how often: often
- how much: much
- in what condition: badly
- what degree: hardly

As one can see, for some reason, many adverbs end in *-ly*.

Adverbs do things like emphasize (*really, simply,* and *so*), amplify (*heartily, completely,* and *positively*), and tone down (*almost, somewhat,* and *mildly*).

Adverbs also come in phrases.

The dog ran as <u>though his life depended on it.</u>

Prepositions

Prepositions are connecting words and, while there are only about 150 of them, they are used more often than any other individual groups of words. They describe relationships between other words. They are placed before a noun or pronoun, forming a phrase that modifies another word in the sentence. **Prepositional phrases** begin with a preposition and end with a noun or pronoun, the **object of the preposition.** *A pristine lake is <u>near the store</u> and <u>behind the bank</u>.*

Some commonly used prepositions are *about, after, anti, around, as, at, behind, beside, by, for, from, in, into, of, off, on, to,* and *with.*

Complex prepositions, which also come before a noun or pronoun, consist of two or three words such as *according to, in regards to,* and *because of.*

Interjections

Interjections are words used to express emotion. Examples include *wow, ouch,* and *hooray*. Interjections are often separate from sentences; in those cases, the interjection is directly followed by an exclamation point. In other cases, the interjection is included in a sentence and followed by a comma. The punctuation plays a big role in the intensity of the emotion that the interjection is expressing. Using a comma or semicolon indicates less excitement than using an exclamation mark.

Conjunctions

Conjunctions are vital words that connect words, phrases, thoughts, and ideas. Conjunctions show relationships between components. There are two types:

Coordinating conjunctions are the primary class of conjunctions placed between words, phrases, clauses, and sentences that are of equal grammatical rank; the coordinating conjunctions are *for, and,*

nor, but, or, yet, and *so.* A useful memorization trick is to remember that all the first letters of these conjunctions collectively spell the word fanboys.

I need to go shopping, *but* I must be careful to leave enough money in the bank.

She wore a black, red, *and* white shirt.

Subordinating conjunctions are the secondary class of conjunctions. They connect two unequal parts, one **main** (or **independent**) and the other **subordinate** (or **dependent**). I must go to the store *even though* I do not have enough money in the bank.

Because I read the review, I do not want to go to the movie.

Notice that the presence of subordinating conjunctions makes clauses dependent. *I read the review* is an independent clause, but *because* makes the clause dependent. Thus, it needs an independent clause to complete the sentence.

Subject-Verb Agreement

The subject of a sentence and its verb must agree. The cornerstone rule of subject-verb agreement is that subject and verb must agree in number. Whether the subject is singular or plural, the verb must follow suit.

    Incorrect: The houses is new.
    Correct: The houses are new.
    Also Correct: The house is new.

In other words, a singular subject requires a singular verb; a plural subject requires a plural verb. The words or phrases that come between the subject and verb do not alter this rule.

    Incorrect: The houses built of brick is new.
    Correct: The houses built of brick are new.

    Incorrect: The houses with the sturdy porches is new.
    Correct: The houses with the sturdy porches are new.

The subject will always follow the verb when a sentence begins with *here* or *there.* Identify these with care.

    Incorrect: Here *is* the *houses* with sturdy porches.
    Correct: Here *are* the *houses* with sturdy porches.

The subject in the sentences above is not *here*, it is *houses*. Remember, *here* and *there* are never subjects. Be careful that contractions such as *here's* or *there're* do not cause confusion!

Two subjects joined by *and* require a plural verb form, except when the two combine to make one thing:

    Incorrect: Garrett and Jonathan is over there.
    Correct: Garrett and Jonathan are over there.

Incorrect: Spaghetti and meatballs are a delicious meal!
Correct: Spaghetti and meatballs is a delicious meal!

In the example above, *spaghetti and meatballs* is a compound noun. However, *Garrett and Jonathan* is not a compound noun.

Two singular subjects joined by *or, either/or,* or *neither/nor* call for a singular verb form.

Incorrect: Butter or syrup are acceptable.
Correct: Butter or syrup is acceptable.

Plural subjects joined by *or, either/or*, or *neither/nor* are, indeed, plural.

The chairs or the boxes are being moved next.

If one subject is singular and the other is plural, the verb should agree with the closest noun.

Correct: The chair or the boxes are being moved next.
Correct: The chairs or the box is being moved next.

Some plurals of money, distance, and time call for a singular verb.

Incorrect: Three dollars *are* enough to buy that.
Correct: Three dollars *is* enough to buy that.

For words declaring degrees of quantity such as *many of, some of,* or *most of,* let the noun that follows *of* be the guide:

Incorrect: Many of the books is in the shelf.
Correct: Many of the books are in the shelf.

Incorrect: Most of the pie *are* on the table.
Correct: Most of the pie *is* on the table.

For indefinite pronouns like anybody or everybody, use singular verbs.

Everybody *is* going to the store.

However, the pronouns *few, many, several, all, some,* and *both* have their own rules and use plural forms.

Some *are* ready.

Some nouns like *crowd* and *congress* are called *collective nouns* and they require a singular verb form.

Congress *is* in session.
The news *is* over.

Books and movie titles, though, including plural nouns such as *Great Expectations*, also require a singular verb. Remember that only the subject affects the verb. While writing tricky subject-verb arrangements, say them aloud. Listen to them. Once the rules have been learned, one's ear will become sensitive to them, making it easier to pick out what's right and what's wrong.

Sentence Structure and Formation

Sentence Structure

Simple sentence: composed of one independent clause

Many people watch hummingbirds.

Note that it has one subject and one verb; however, a simple sentence can have a compound subject and/or a compound verb.

Adults and children often enjoy watching and photographing hummingbirds.

Compound sentence: composed of two independent clauses

The wind knocked down lots of trees, but no trees in my yard were affected.

Complex sentence: composed of one independent clause and one dependent clause

Although the wind knocked down lots of trees, no trees in my yard were affected.

Sentence Fluency

Learning and utilizing the mechanics of structure will encourage effective, professional results, and adding some creativity will elevate one's writing to a higher level.

First, the basic elements of sentences will be reviewed.

A **sentence** is a set of words that make up a grammatical unit. The words must have certain elements and be spoken or written in a specific order to constitute a complete sentence that makes sense.

1. A sentence must have a **subject** (a noun or noun phrase). The subject tells whom or what the sentence is addressing (i.e. what it is about).

2. A sentence must have an **action** or **state of being** (*a* verb). To reiterate: A verb forms the main part of the predicate of a sentence. This means that it explains what the noun is doing.

3. A sentence must convey a complete thought.

When examining writing, readers should be mindful of grammar, structure, spelling, and patterns. Sentences can come in varying sizes and shapes, so the point of grammatical correctness is not to stamp out creativity or diversity in writing. Rather, grammatical correctness ensures that writing will be enjoyable and clear. One of the most common methods successful test takers employ to catch errors is to mouth the words as they read them. Many typos are fixed automatically by the brain, but mouthing the words often circumvents this instinct and helps one read what's actually on the page. Often, grammar errors are caught not by memorization of grammar rules but by the training of one's mind to know whether something *sounds* right or not.

Types of Sentences

There isn't an overabundance of absolutes in grammar, but here is one: every sentence in the English language falls into one of four categories.

> **Declarative:** a simple statement that ends with a period
>
> > The price of milk per gallon is the same as the price of gasoline.
>
> **Imperative:** a command, instruction, or request that ends with a period
>
> > Buy milk when you stop to fill up your car with gas.
>
> **Interrogative:** a question that ends with a question mark
>
> > Will you buy the milk?
>
> **Exclamatory:** a statement or command that expresses emotions like anger, urgency, or surprise and ends with an exclamation mark
>
> > Buy the milk now!

Declarative sentences are the most common type, probably because they are comprised of the most general content, without any of the bells and whistles that the other three types contain. They are, simply, declarations or statements of any degree of seriousness, importance, or information.

Imperative sentences often seem to be missing a subject. The subject is there, though; it is just not visible or audible because it is *implied*. For example:

> Buy the milk when you fill up your car with gas.

In this sentence, *you* is the implied subject, the one to whom the command is issued. This is sometimes called *the understood you* because it is understood that *you* is the subject of the sentence.

Interrogative sentences—those that ask questions—are defined as such from the idea of the word **interrogation**, the action of questions being asked of suspects by investigators. Although that is serious business, interrogative sentences apply to all kinds of questions.

To exclaim is at the root of **exclamatory** sentences. These are made with strong emotions behind them. The only technical difference between a declarative or imperative sentence and an exclamatory one is the exclamation mark at the end. The example declarative and imperative sentences can both become an exclamatory one simply by putting an exclamation mark at the end of the sentences.

> The price of milk per gallon is the same as the price of gasoline!
>
> Buy milk when you stop to fill up your car with gas!

After all, someone might be really excited by the price of gas or milk, or they could be mad at the person that will be buying the milk! However, as stated before, exclamation marks in abundance defeat their own purpose! After a while, they begin to cause fatigue! When used only for their intended purpose, they can have their expected and desired effect.

Transitions

Transitions are the glue used to make organized thoughts adhere to one another. Transitions are the glue that helps put ideas together seamlessly, within sentences and paragraphs, between them, and (in longer documents) even between sections. Transitions may be single words, sentences, or whole paragraphs (as in the prior example). Transitions help readers to digest and understand what to feel about what has gone on and clue readers in on what is going on, what will be, and how they might react to all these factors. Transitions are like good clues left at a crime scene.

Parallel Structure in a Sentence

Parallel structure, also known as **parallelism**, refers to using the same grammatical form within a sentence. This is important in lists and for other components of sentences.

Incorrect: At the recital, the boys and girls were dancing, singing, and played musical instruments.

Correct: At the recital, the boys and girls were dancing, singing, and playing musical instruments.

Notice that in the first example, *played* is not in the same verb tense as the other verbs nor is it compatible with the helping verb *were*. To test for parallel structure in lists, try reading each item as if it were the only item in the list.

The boys and girls were dancing.

The boys and girls were singing.

The boys and girls were played musical instruments.

Suddenly, the error in the sentence becomes very clear. Here's another example:

Incorrect: After the accident, I informed the police *that Mrs. Holmes backed* into my car, *that Mrs. Holmes got out* of her car to look at the damage, and *she was driving* off without leaving a note.

Correct: After the accident, I informed the police *that Mrs. Holmes backed* into my car, *that Mrs. Holmes got out* of her car to look at the damage, and *that Mrs. Holmes drove off* without leaving a note.

Correct: After the accident, I informed the police that Mrs. Holmes *backed* into my car, *got out* of her car to look at the damage, and *drove off* without leaving a note.

Note that there are two ways to fix the nonparallel structure of the first sentence. The key to parallelism is consistent structure.

Capitalization/Punctuation/Spelling

Capitalization

Here's a non-exhaustive list of things that should be capitalized.

- The first word of every sentence
- The first word of every line of poetry
- The first letter of proper nouns (World War II)
- Holidays (Valentine's Day)

- The days of the week and months of the year (Tuesday, March)
- The first word, last word, and all major words in the titles of books, movies, songs, and other creative works (In the novel, *To Kill a Mockingbird*, note that *a* is lowercase since it's not a major word, but *to* is capitalized since it's the first word of the title.)
- Titles when preceding a proper noun (President Roberto Gonzales, Aunt Judy)

When simply using a word such as president or secretary, though, the word is not capitalized.

Officers of the new business must include a *president* and *treasurer*.

Seasons—spring, fall, etc.—are not capitalized.

North, *south*, *east*, and *west* are capitalized when referring to regions but are not when being used for directions. In general, if it's preceded by *the* it should be capitalized.

I'm from the South.
I drove south.

Punctuation

Ellipses

An **ellipsis** (. . .) consists of three handy little dots that can speak volumes on behalf of irrelevant material. Writers use them in place of words, lines, phrases, list content, or paragraphs that might just as easily have been omitted from a passage of writing. This can be done to save space or to focus only on the specifically relevant material.

Exercise is good for some unexpected reasons. Watkins writes, "Exercise has many benefits such as . . . reducing cancer risk."

In the example above, the ellipsis takes the place of the other benefits of exercise that are more expected.

The ellipsis may also be used to show a pause in sentence flow.

"I'm wondering . . . how this could happen," Dylan said in a soft voice.

Commas

A **comma** (,) is the punctuation mark that signifies a pause—breath—between parts of a sentence. It denotes a break of flow. As with so many aspects of writing structure, authors will benefit by reading their writing aloud or mouthing the words. This can be particularly helpful if one is uncertain about whether the comma is needed.

In a complex sentence—one that contains a **subordinate** (**dependent**) clause or clauses—the use of a comma is dictated by where the subordinate clause is located. If the subordinate clause is located before the main clause, a comma is needed between the two clauses.

I will not pay for the steak, *because I don't have that much money.*

Generally, if the subordinate clause is placed after the main clause, no punctuation is needed.

I did well on my exam because I studied two hours the night before.

Notice how the last clause is dependent because it requires the earlier independent clauses to make sense.

Use a comma on both sides of an interrupting phrase.

> I will pay for the ice cream, *chocolate and vanilla*, and then will eat it all myself.

The words forming the phrase in italics are nonessential (extra) information. To determine if a phrase is nonessential, try reading the sentence without the phrase and see if it's still coherent.

A comma is not necessary in this next sentence because no interruption—nonessential or extra information—has occurred. Read sentences aloud when uncertain.

I will pay for his chocolate and vanilla ice cream and then will eat it all myself.

If the nonessential phrase comes at the beginning of a sentence, a comma should only go at the end of the phrase. If the phrase comes at the end of a sentence, a comma should only go at the beginning of the phrase.

Other types of interruptions include the following:

- interjections: Oh no, I am not going.
- abbreviations: Barry Potter, M.D., specializes in heart disorders.
- direct addresses: Yes, Claudia, I am tired and going to bed.
- parenthetical phrases: His wife, lovely as she was, was not helpful.
- transitional phrases: Also, it is not possible.

The second comma in the following sentence is called an Oxford comma.

> I will pay for ice cream, syrup, and pop.

It is a comma used after the second-to-last item in a series of three or more items. It comes before the word *or* or *and*. Not everyone uses the Oxford comma; it is optional, but many believe it is needed. The comma functions as a tool to reduce confusion in writing. So, if omitting the Oxford comma would cause confusion, then it's best to include it.

Commas are used in math to mark the place of thousands in numerals, breaking them up so they are easier to read. Other uses for commas are in dates (*March 19, 2016*), letter greetings (*Dear Sally,*), and in between cities and states (*Louisville, KY*).

Semicolons

The **semicolon** (;) might be described as a heavy-handed comma. Take a look at these two examples:

> I will pay for the ice cream, but I will not pay for the steak.
> I will pay for the ice cream; I will not pay for the steak.

What's the difference? The first example has a comma and a conjunction separating the two independent clauses. The second example does not have a conjunction, but there are two independent clauses in the sentence, so something more than a comma is required. In this case, a semicolon is used.

Two independent clauses can only be joined in a sentence by either a comma and conjunction or a semicolon. If one of those tools is not used, the sentence will be a run-on. Remember that while the clauses are independent, they need to be closely related in order to be contained in one sentence.

Another use for the semicolon is to separate items in a list when the items themselves require commas.

> The family lived in Phoenix, Arizona; Oklahoma City, Oklahoma; and Raleigh, North Carolina.

Colons

Colons (:) have many miscellaneous functions. Colons can be used to precede further information or a list. In these cases, a colon should only follow an independent clause.

> Humans take in sensory information through five basic senses: sight, hearing, smell, touch, and taste.

The meal includes the following components:

- Caesar salad
- spaghetti
- garlic bread
- cake

The family got what they needed: a reliable vehicle.

While a comma is more common, a colon can also proceed a formal quotation.

> He said to the crowd: "Let's begin!"

The colon is used after the greeting in a formal letter.

> Dear Sir:
> To Whom It May Concern:

In the writing of time, the colon separates the minutes from the hour (*4:45 p.m.*). The colon can also be used to indicate a ratio between two numbers (*50:1*).

Hyphens

The **hyphen** (-) is a little hash mark that can be used to join words to show that they are linked.

Hyphenate two words that work together as a single adjective (a compound adjective).

> honey-covered biscuits

Some words always require hyphens, even if not serving as an adjective.

> merry-go-round

Hyphens always go after certain prefixes like *anti-* & *all-*.

Hyphens should also be used when the absence of the hyphen would cause a strange vowel combination (*semi-engineer*) or confusion. For example, *re-collect* should be used to describe something being gathered twice rather than being written as *recollect*, which means to remember.

Parentheses and Dashes

Parentheses are half-round brackets that look like this: *( )*. They set off a word, phrase, or sentence that is an afterthought, explanation, or side note relevant to the surrounding text but not essential. A pair of commas is often used to set off this sort of information, but parentheses are generally used for information that would not fit well within a sentence or that the writer deems not important enough to be structurally part of the sentence.

> The picture of the heart (see above) shows the major parts you should memorize.
> Mount Everest is one of three mountains in the world that are over 28,000 feet high (K2 and Kanchenjunga are the other two).

See how the sentences above are complete without the parenthetical statements? In the first example, *see above* would not have fit well within the flow of the sentence. The second parenthetical statement could have been a separate sentence, but the writer deemed the information not pertinent to the topic.

The **em-dash** (—) is a mark longer than a hyphen used as a punctuation mark in sentences and to set apart a relevant thought. Even after plucking out the line separated by the dash marks, the sentence will be intact and make sense.

> Looking out the airplane window at the landmarks—Lake Clarke, Thompson Community College, and the bridge—she couldn't help but feel excited to be home.

The dashes use is similar to that of parentheses or a pair of commas. So, what's the difference? Many believe that using dashes makes the clause within them stand out while using parentheses is subtler. It's advised to not use dashes when commas could be used instead.

Quotation Marks

Here are some instances where **quotation marks** should be used:

- Dialogue for characters in narratives. When characters speak, the first word should always be capitalized, and the punctuation goes inside the quotes. For example:

  > Janie said, "The tree fell on my car during the hurricane."

- Around titles of songs, short stories, essays, and chapter in books
- To emphasize a certain word
- To refer to a word as the word itself

Apostrophes

This punctuation mark, the apostrophe ('), is a versatile little mark. It has a few different functions:

- Quotes: Apostrophes are used when a second quote is needed within a quote.

  > In my letter to my friend, I wrote, "The girl had to get a new purse, and guess what Mary did? She said, 'I'd like to go with you to the store.' I knew Mary would buy it for her."

- Contractions: Another use for an apostrophe in the quote above is a contraction. *I'd* is used for *I would.*

- Possession: An apostrophe followed by the letter *s* shows possession (*Mary's* purse). If the possessive word is plural, the apostrophe generally just follows the word.

    The trees' leaves are all over the ground.

Spelling

Spelling might or might not be important to some, or maybe it just doesn't come naturally, but those who are willing to discover some new ideas and consider their benefits can learn to spell better and improve their writing. Misspellings reduce a writer's credibility and can create misunderstandings. Spell checkers built into word processors are not a substitute for accuracy. They are neither foolproof nor without error. In addition, a writer's misspelling of one word may also be a valid (but incorrect) word. For example, a writer intending to spell *herd* might accidentally type *s* instead of *d* and unintentionally spell *hers*. Since *hers* is a word, it would not be marked as a misspelling by a spell checker. In short, writers should use spell check, but not rely on them.

Guidelines for Spelling

Saying and listening to a word serves as the beginning of knowing how to spell it. Writers should keep these subsequent guidelines in mind, remembering there are often exceptions because the English language is replete with them.

Guideline #1: Syllables must have at least one vowel. In fact, every syllable in every English word has a vowel.

- d*o*g
- h*a*yst*a*ck
- *a*nsw*e*r*i*ng
- *a*bst*e*nt*i*o*u*s
- s*i*mpl*e*

Guideline #2: The long and short of it. When the vowel has a short vowel sound as in *mad* or *bed,* only the single vowel is needed. If the word has a long vowel sound, add another vowel, either alongside it or separated by a consonant: bed/*bead*; mad/*made.* When the second vowel is separated by two consonants— *madder*—it does not affect the first vowel's sound.

Guideline #3: Suffixes. Refer to the examples listed above.

Guideline #4: Which comes first; the *i* or the *e*? Remember the saying, "*I* before *e* except after *c* or when sounding as *a* as in *neighbor* or *weigh*." Keep in mind that these are only guidelines and that there are always exceptions to every rule.

Guideline #5: Vowels in the right order. Another helpful rhyme is, "When two vowels go walking, the first one does the talking." When two vowels are in a row, the first one often has a long vowel sound and the other is silent. An example is *team.*

If one has difficulty spelling words, he or she can determine a strategy to help. Some people work on spelling by playing word games like Scrabble or Words with Friends. Others use phonics, which is sounding words out by slowly and surely stating each syllable. People try repeating and memorizing spellings as well as picturing words in their head, or they may try making up silly memory aids. Each person should experiment and see what works best.

Homophones

Homophones are two or more words that have no particular relationship to one another except their identical pronunciations. Homophones make spelling English words fun and challenging. Examples include:

Common Homophones
affect, effect
allot, a lot
barbecue, barbeque
bite, byte
brake, break
capital, capitol
cash, cache
cell, sell
colonel, kernel
do, due, dew
dual, duel
eminent, imminent
flew, flu, flue
gauge, gage
holy, wholly
it's, its
knew, new
libel, liable
principal, principle
their, there, they're
to, too, two
yoke, yolk

Irregular Plurals

Irregular plurals are words that aren't made plural the usual way.

- Most nouns are made plural by adding –s (books, televisions, skyscrapers).

- Most nouns ending in *ch, sh, s, x,* or *z* are made plural by adding –es (churches, marshes).

- Most nouns ending in a vowel + *y* are made plural by adding –s (days, toys).

- Most nouns ending in a consonant + *y,* are made plural by the -y becoming -ies (baby becomes *babies*).

- Most nouns ending in an *o* are made plural by adding –s (pianos, photos).

- Some nouns ending in an *o,* though, may be made plural by adding –es (example: potatoes, volcanoes), and, of note, there is no known rhyme or reason for this!

82

- Most nouns ending in an *f* or *fe* are made plural by the *-f* or *-fe* becoming *-ves*! (example: wolf becomes *wolves*).

- Some words function as both the singular and plural form of the word (fish, deer).

- Other exceptions include *man* becomes *men, mouse* becomes *mice, goose* becomes *geese,* and *foot* becomes *feet.*

Contractions

The basic rule for making **contractions** is one area of spelling that is pretty straightforward: combine the two words by inserting an apostrophe (') in the space where a letter is omitted. For example, to combine *you* and *are*, drop the *a* and put the apostrophe in its place: *you're*.

> he + is = he's

> you + all = y'all (informal, but often misspelled)

Note that *it's*, when spelled with an apostrophe, is always the contraction for *it is*. The possessive form of the word is written without an apostrophe as *its*.

Knowledge of Language

Precision

People often think of precision in terms of math, but precise word choice is another key to successful writing. Since language itself is imprecise, it's important for the writer to find the exact word or words to convey the full, intended meaning of a given situation. For example:

> The number of deaths has gone down since seat belt laws started.

There are several problems with this sentence. First, the word *deaths* is too general. From the context, it's assumed that the writer is referring only to deaths caused by car accidents. However, without clarification, the sentence lacks impact and is probably untrue. The phrase *gone down* might be accurate, but a more precise word would provide more information and greater accuracy. Did the numbers show a slow and steady decrease in highway fatalities or a sudden drop? If the latter is true, the writer is missing a chance to make his or her point more dramatically. Instead of *gone down* the author could substitute *plummeted, fallen drastically,* or *rapidly diminished* to bring the information to life. Also, the phrase *seat belt laws* is unclear. Does it refer to laws requiring cars to include seat belts or to laws requiring drivers and passengers to use them? Finally, *started* is not a strong verb. Words like *enacted* or *adopted* are more direct and make the content more real. When put together, these changes create a far more powerful sentence:

> The number of highway fatalities has plummeted since laws requiring seat belt usage were enacted.

However, it's important to note that precise word choice can sometimes be taken too far. If the writer of the sentence above takes precision to an extreme, it might result in the following:

> The incidence of high-speed, automobile accident-related fatalities has decreased 75% and continued to remain at historical lows since the initial set of federal legislations requiring seat belt use were enacted in 1992.

This sentence is extremely precise, but it takes so long to achieve that precision that it suffers from a lack of clarity. Precise writing is about finding the right balance between information and flow. This is also an issue of conciseness (discussed in the next section).

The last thing for writers to consider with precision is a word choice that's not only unclear or uninteresting, but also confusing or misleading. For example:

> The number of highway fatalities has become hugely lower since laws requiring seat belt use were enacted.

In this case, the reader might be confused by the word *hugely*. Huge means large, but here the writer uses *hugely* in an incorrect and awkward manner. Although most readers can decipher this, doing so disconnects them from the flow of the writing and makes the writer's point less effective.

Concision

"Less is more" is a good rule for writers to follow when composing a sentence. Unfortunately, writers often include extra words and phrases that seem necessary at the time but add nothing to the main idea. This confuses the reader and creates unnecessary repetition. Writing that lacks conciseness is usually guilty of excessive wordiness and redundant phrases. Here's an example containing both of these issues:

> When legislators decided to begin creating legislation making it mandatory for automobile drivers and passengers to make use of seat belts while in cars, a large number of them made those laws for reasons that were political reasons.

There are several empty or "fluff" words here that take up too much space. These can be eliminated while still maintaining the writer's meaning. For example:

- *decided to begin* could be shortened to *began*
- *making it mandatory for* could be shortened to *requiring*
- *make use of* could be shortened to *use*
- *a large number* could be shortened to *many*

In addition, there are several examples of redundancy that can be eliminated:

- legislators decided to begin creating legislation and made those laws
- automobile drivers and passengers and while in cars
- reasons that were political reasons

These changes are incorporated as follows:

> When legislators began requiring drivers and passengers to use seat belts, many of them did so for political reasons.

There are many general examples of redundant phrases, such as *add an additional, complete and total, time schedule*, and *transportation vehicle*. If asked to identify a redundant phrase on the exam, test takers should look for words that are close together with the same (or similar) meanings.

Consistency in Style and Tone

Style and tone are often thought to be the same thing. Though they're closely related, there are important differences to keep in mind. The easiest way to do this is to remember that **style** creates and affects **tone**. More specifically, style is *how the writer uses words* to create the desired tone for his or her writing.

Style

Style can include any number of technical writing choices, and some may have to be analyzed on the test. A few examples of style choices include:

- Sentence Construction: When presenting facts, does the writer use shorter sentences to create a quicker sense of the supporting evidence, or does he or she use longer sentences to elaborate and explain the information?

- Technical Language: Does the writer use jargon to demonstrate his or her expertise in the subject, or do the writer use ordinary language to help the reader understand things in simple terms?

- Formal Language: Does the writer refrain from using contractions such as *won't* or *can't* to create a more formal tone, or does he or she use a colloquial, conversational style to connect to the reader?

- Formatting: Does the writer use a series of shorter paragraphs to help the reader follow a line of argument, or does he or she use longer paragraphs to examine an issue in great detail and demonstrate his or her knowledge of the topic?

On the exam, test takers should examine the writer's style and how his or her writing choices affect the way the text comes across.

Tone

Tone refers to the writer's attitude toward the subject matter. Tone is usually explained in terms of a work of fiction. For example, the tone conveys how the writer feels about the characters and the situations in which they're involved. Nonfiction writing is sometimes thought to have no tone at all; however, this is incorrect.

A lot of nonfiction writing has a neutral tone, which is an important one for the writer to use. A neutral tone demonstrates that the writer is presenting a topic impartially and letting the information speak for itself. On the other hand, nonfiction writing can be just as effective and appropriate if the tone isn't neutral. The following short passage provides an example of tone in nonfiction writing:

> Seat belts save more lives than any other automobile safety feature. Many studies show that airbags save lives as well; however, not all cars have airbags. For instance, some older cars don't. Furthermore, air bags aren't entirely reliable. For example, studies show that in 15% of accidents airbags don't deploy as designed, but, on the other hand, seat belt malfunctions are extremely rare. The number of highway fatalities has plummeted since laws requiring seat belt usage were enacted.

In this passage, the writer mostly chooses to retain a neutral tone when presenting information. If instead, the author chose to include his or her own personal experience of losing a friend or family member in a car accident, the tone would change dramatically. The tone would no longer be neutral and would show that the writer has a personal stake in the content, allowing him or her to interpret the information in a different way. When analyzing tone, the reader should consider what the writer is trying to achieve in the text and how they *create* the tone using style.

Practice Questions

1. Which of the following sentences has an error in capitalization?
   a. The East Coast has experienced very unpredictable weather this year.
   b. My Uncle owns a home in Florida, where he lives in the winter.
   c. I am taking English Composition II on campus this fall.
   d. There are several nice beaches we can visit on our trip to the Jersey Shore this summer.

2. Julia Robinson, an avid photographer in her spare time, was able to capture stunning shots of the local wildlife on her last business trip to Australia.

Which of the following is an adjective in the preceding sentence?

   a. Time
   b. Capture
   c. Avid
   d. Photographer

3. Which of the following sentences uses correct punctuation?
   a. Carole is not currently working; her focus is on her children at the moment.
   b. Carole is not currently working and her focus is on her children at the moment.
   c. Carole is not currently working, her focus is on her children at the moment.
   d. Carole is not currently working her focus is on her children at the moment.

4. Which of these examples is a compound sentence?
   a. Alex and Shane spent the morning coloring and later took a walk down to the park.
   b. After coloring all morning, Alex and Shane spent the afternoon at the park.
   c. Alex and Shane spent the morning coloring, and then they took a walk down to the park.
   d. After coloring all morning and spending part of the day at the park, Alex and Shane took a nap.

5. Which of these examples shows incorrect use of subject-verb agreement?
   a. Neither of the cars are parked on the street.
   b. Both of my kids are going to camp this summer.
   c. Any of your friends are welcome to join us on the trip in November.
   d. Each of the clothing options is appropriate for the job interview.

6. When it gets warm in the spring, _____ and _____ like to go fishing at Cobbs Creek.

Which of the following word pairs should be used in the blanks above?

   a. me, him
   b. he, I
   c. him, I
   d. he, me

7. Which example shows correct comma usage for dates?
    a. The due date for the final paper in the course is Monday, May 16, 2016.
    b. The due date for the final paper in the course is Monday, May 16 2016.
    c. The due date for the final project in the course is Monday, May, 16, 2016.
    d. The due date for the final project in the course is Monday May 16, 2016.

8. At last night's company function, in honor of Mr. Robertson's retirement, several employees spoke kindly about his career achievements.

In the preceding sentence, what part of speech is the word *function*?

    a. Adjective
    b. Adverb
    c. Verb
    d. Noun

9. Which of the examples uses the correct plural form?
    a. Tomatos
    b. Analysis
    c. Cacti
    d. Criterion

10. Which of the following examples uses correct punctuation?
    a. The moderator asked the candidates, "Is each of you prepared to discuss your position on global warming?".
    b. The moderator asked the candidates, "Is each of you prepared to discuss your position on global warming?"
    c. The moderator asked the candidates, 'Is each of you prepared to discuss your position on global warming?'
    d. The moderator asked the candidates, "Is each of you prepared to discuss your position on global warming"?

11. In which of the following sentences does the word *part* function as an adjective?
    a. The part Brian was asked to play required many hours of research.
    b. She parts ways with the woodsman at the end of the book.
    c. The entire team played a part in the success of the project.
    d. Ronaldo is part Irish on his mother's side of the family.

12. All of Shannon's family and friends helped her to celebrate her 50th birthday at Café Sorrento.

Which of the following is the complete subject of the preceding sentence?

    a. Family and friends
    b. All
    c. All of Shannon's family and friends
    d. Shannon's family and friends

13. Which of the following examples correctly uses quotation marks?
    a. "Where the Red Fern Grows" was one of my favorite novels as a child.
    b. Though he is famous for his roles in films like "The Great Gatsby" and "Titanic," Leonardo DiCaprio has never won an Oscar.
    c. Sylvia Plath's poem, "Daddy" will be the subject of this week's group discussion.
    d. "The New York Times" reported that many fans are disappointed in some of the trades made by the Yankees this off-season.

14. Which of the following sentences shows correct word usage?
    a. It's often been said that work is better then rest.
    b. Its often been said that work is better then rest.
    c. It's often been said that work is better than rest.
    d. Its often been said that work is better than rest.

15. Which of the following is an imperative sentence?
    a. Pennsylvania's state flag includes two draft horses and an eagle.
    b. Go down to the basement and check the hot water heater for signs of a leak.
    c. You must be so excited to have a new baby on the way!
    d. How many countries speak Spanish?

16. Which of the following examples is a compound sentence?
    a. Shawn and Jerome played soccer in the backyard for two hours.
    b. Marissa last saw Elena and talked to her this morning.
    c. The baby was sick, so I decided to stay home from work.
    d. Denise, Kurt, and Eric went for a run after dinner.

17. Which of the following sentences uses correct subject-verb agreement?
    a. There is two constellations that can be seen from the back of the house.
    b. At least four of the sheep needs to be sheared before the end of summer.
    c. Lots of people were auditioning for the singing competition on Saturday.
    d. Everyone in the group have completed the assignment on time.

18. Philadelphia is home to some excellent walking tours where visitors can learn more about the culture and rich history of the city of brotherly love.

What are the adjectives in the preceding sentence?

    a. Philadelphia, tours, visitors, culture, history, city, love
    b. Excellent, walking, rich, brotherly
    c. Is, can, learn
    d. To, about, of

19. The realtor showed _____ and _____ a house on Wednesday afternoon.

Which of the following pronoun pairs should be used in the blanks above?

    a. She, I
    b. She, me
    c. Me, her
    d. Her, me

20. Which of the following examples uses correct punctuation?
    a. Recommended supplies for the hunting trip include the following: rain gear, large backpack, hiking boots, flashlight, and non-perishable foods.
    b. I left the store, because I forgot my wallet.
    c. As soon as the team checked into the hotel; they met in the lobby for a group photo.
    d. None of the furniture came in on time: so they weren't able to move in to the new apartment.

21. Which of the following sentences shows correct word usage?
    a. Your going to have to put you're jacket over their.
    b. You're going to have to put your jacket over there.
    c. Your going to have to put you're jacket over they're.
    d. You're going to have to put your jacket over their.

22. A teacher notices that, when students are talking to each other between classes, they are using their own unique vocabulary words and expressions to talk about their daily lives. When the teacher hears these non-standard words that are specific to one age or cultural group, what type of language is she listening to?
    a. Slang
    b. Jargon
    c. Dialect
    d. Vernacular

23. A teacher wants to counsel a student about using the word *ain't* in a research paper for a high school English class. What advice should the teacher give?
    a. *Ain't* is not in the dictionary, so it isn't a word.
    b. Because the student isn't in college yet, *ain't* is an appropriate expression for a high school writer.
    c. *Ain't* is incorrect English and should not be part of a serious student's vocabulary because it sounds uneducated.
    d. *Ain't* is a colloquial expression, and while it may be appropriate in a conversational setting, it is not standard in academic writing.

24. What is the structure of the following sentence?

    The restaurant is unconventional because it serves both Chicago style pizza and New York style pizza.

    a. Simple
    b. Compound
    c. Complex
    d. Compound-complex

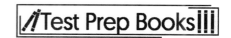

25. The following sentence contains what kind of error?

    This summer, I'm planning to travel to Italy, take a Mediterranean cruise, going to Pompeii, and eat a lot of Italian food.

    a. Parallelism
    b. Sentence fragment
    c. Misplaced modifier
    d. Subject-verb agreement

26. The following sentence contains what kind of error?

    Forgetting that he was supposed to meet his girlfriend for dinner, Anita was mad when Fred showed up late.

    a. Parallelism
    b. Run-on sentence
    c. Misplaced modifier
    d. Subject-verb agreement

27. The following sentence contains what kind of error?

    Some workers use all their sick leave, other workers cash out their leave.

    a. Parallelism
    b. Comma splice
    c. Sentence fragment
    d. Subject-verb agreement

28. A student writes the following in an essay:

    Protestors filled the streets of the city. Because they were dissatisfied with the government's leadership.

Which of the following is an appropriately-punctuated correction for this sentence?

    a. Protestors filled the streets of the city, because they were dissatisfied with the government's leadership.
    b. Protesters, filled the streets of the city, because they were dissatisfied with the government's leadership.
    c. Because they were dissatisfied with the government's leadership protestors filled the streets of the city.
    d. Protestors filled the streets of the city because they were dissatisfied with the government's leadership.

29. What is the part of speech of the underlined word in the sentence?

We need to come up with a fresh <u>approach</u> to this problem.

a. Noun
b. Verb
c. Adverb
d. Adjective

30. What is the part of speech of the underlined word in the sentence?

Investigators conducted an <u>exhaustive</u> inquiry into the accusations of corruption.

a. Noun
b. Verb
c. Adverb
d. Adjective

31. The underlined portion of the sentence is an example of which sentence component?

New students should report <u>to the student center</u>.

a. Dependent clause
b. Adverbial phrase
c. Adjective clause
d. Noun phrase

32. Which word choices will correctly complete the sentence?

Increasing the price of bus fares has had a greater [affect / effect] on ridership [then / than] expected.

a. affect; then
b. affect; than
c. effect; then
d. effect; than

33. The following is an example of what type of sentence?

Although I wished it were summer, I accepted the change of seasons, and I started to appreciate the fall.

a. Compound
b. Simple
c. Complex
d. Compound-Complex

34. A student reads the following sentence:

> A hundred years ago, automobiles were rare, but now cars are ubiquitous.

However, she doesn't know what the word *ubiquitous* means. Which key context clue is essential to decipher the word's meaning?

    a. Ago
    b. Cars
    c. Now
    d. Rare

35. Which word in the following sentence is a proper noun?

> People think the Statue of Liberty is an awesome sight.

    a. People
    b. Statue of Liberty
    c. Awesome
    d. Sight

36. Which word in the following sentence is a plural noun?

> The black kitten was the girl's choice from the litter of kittens.

    a. Kitten
    b. Girl's
    c. Choice
    d. Kittens

37. Which pronoun makes the following sentence grammatically correct?

> _____ ordered the flowers?

    a. Whose
    b. Whom
    c. Who
    d. Who've

38. Which pronoun makes the following sentence grammatically correct?

> The giraffe nudged _____ baby.

    a. it's
    b. hers
    c. their
    d. its

39. What is the word *several* in the following sentence called?

    Several are laughing loudly on the bus.

    a. Singular indefinite pronoun
    b. Plural indefinite pronoun
    c. Singular objective pronoun
    d. Indefinite adjective

40. Which word in the following sentence is an adjective?

    The connoisseur slowly enjoyed the delectable meal.

    a. Delectable
    b. Connoisseur
    c. Slowly
    d. Enjoyed

41. Which choice identifies all of the prepositions in the following sentence?

    We went down by the water, near the lake, before dawn, to see the pretty sunrise.

    a. Went, to see, pretty
    b. By, near, before
    c. Water, lake, dawn, sunrise
    d. We, down, the, pretty

42. Which sentence has an interjection?
    a. The cookie was full of chocolaty goodness.
    b. Well, Carrie didn't like the cookie.
    c. Can't you see that cookie is broken?
    d. That's too bad, but I'll still eat it!

43. Identify the complete subject in the following sentence.

    The heaviest green bike is mine.

    a. bike
    b. green bike
    c. The heaviest green bike
    d. is mine

44. Identify the complete predicate in the following sentence.

My house is the yellow one at the end of the street.

a. My house
b. is the yellow one
c. at the end of the street.
d. is the yellow one at the end of the street.

45. Which sentence shows incorrect subject/verb agreement?
a. All of the kittens in the litter show their courage.
b. The black kitten pounce on the ball of yarn.
c. The calico kitten eats voraciously.
d. My favorite kitten snuggles with its mother.

46. What is the indirect object in the following sentence?

Calysta brought her mother the beautiful stained-glass lamp.

a. Stained-glass lamp
b. Brought
c. Her mother
d. Beautiful

47. Which sentence is grammatically correct?
a. They're on their way to New Jersey but there not there yet.
b. Their on their way to New Jersey but they're not there yet.
c. They're on their way to New Jersey but they're not there yet.
d. They're on their way to New Jersey but there not their yet.

48. Identify the prepositional phrase in the following sentence.

For the longest time, I have wanted to learn to roller skate.

a. I have wanted
b. wanted to learn
c. learn to roller skate
d. For the longest time

49. Identify the sentence structure of the following sentence.

The weight of the world was on his shoulders, so he took a long walk.

a. Simple sentence
b. Compound sentence
c. Complex sentence
d. Compound-complex sentence

50. Identify the sentence structure of the following sentence.

The last thing she wanted to do was see the Eiffel Tower before the flight.

a. Simple sentence
b. Compound sentence
c. Complex sentence
d. Compound-complex sentence

Answer Explanations

1. B: In Choice *B* the word *Uncle* should not be capitalized, because it is not functioning as a proper noun. If the word named a specific uncle, such as *Uncle Jerry*, then it would be considered a proper noun and should be capitalized. Choice *A* correctly capitalizes the proper noun *East Coast*, and does not capitalize *winter*, which functions as a common noun in the sentence. Choice *C* correctly capitalizes the name of a specific college course, which is considered a proper noun. Choice *D* correctly capitalizes the proper noun *Jersey Shore*.

2. C: In Choice *C*, *avid* is functioning as an adjective that modifies the word photographer. *Avid* describes the photographer Julia Robinson's style. The words *time* and *photographer* are functioning as nouns, and the word *capture* is functioning as a verb in the sentence. Other words functioning as adjectives in the sentence include, *local*, *business*, and *spare*, as they all describe the nouns they precede.

3. A: Choice *A* is correctly punctuated because it uses a semicolon to join two independent clauses that are related in meaning. Each of these clauses could function as an independent sentence. Choice *B* is incorrect because the conjunction is not preceded by a comma. A comma and conjunction should be used together to join independent clauses. Choice *C* is incorrect because a comma should only be used to join independent sentences when it also includes a coordinating conjunction such as *and* or *so*. Choice *D* does not use punctuation to join the independent clauses, so it is considered a fused (same as a run-on) sentence.

4. C: Choice *C* is a compound sentence because it joins two independent clauses with a comma and the coordinating conjunction *and*. The sentences in Choices *B* and *D* include one independent clause and one dependent clause, so they are complex sentences, not compound sentences. The sentence in Choice *A* has both a compound subject, *Alex and Shane*, and a compound verb, *spent and took*, but the entire sentence itself is one independent clause.

5. A: Choice *A* uses incorrect subject-verb agreement because the indefinite pronoun *neither* is singular and must use the singular verb form *is*. The pronoun *both* is plural and uses the plural verb form of *are*. The pronoun *any* can be either singular or plural. In this example, it is used as a plural, so the plural verb form *are* is used. The pronoun *each* is singular and uses the singular verb form *is*.

6. B: Choice *B* is correct because the pronouns *he* and *I* are in the subjective case. *He* and *I* are the subjects of the verb *like* in the independent clause of the sentence. Choice *A*, *C*, and *D* are incorrect because they all contain at least one objective pronoun (*me* and *him*). Objective pronouns should not be used as the subject of the sentence, but rather, they should come as an object of a verb. To test for correct pronoun usage, try reading the pronouns as if they were the only pronoun in the sentence. For example, *he* and *me* may appear to be the correct answer choices, but try reading them as the only pronoun.

He like[s] to go fishing...

Me like to go fishing...

When looked at that way, *me* is an obviously incorrect choice.

7. A: It is necessary to put a comma between the date and the year. It is also required to put a comma between the day of the week and the month. Choice *B* is incorrect because it is missing the comma between the day and year. Choice *C* is incorrect because it adds an unnecessary comma between the month and date. Choice *D* is missing the necessary comma between day of the week and the month.

8. D: In Choice *D*, the word function is a noun. While the word *function* can also act as a verb, in this particular sentence it is acting as a noun as the object of the preposition *at*. Choices *A* and *B* are incorrect because the word *function* cannot be used as an adjective or adverb.

9. C: Cacti is the correct plural form of the word *cactus*. Choice *A* (*tomatos*) includes an incorrect spelling of the plural of *tomato*. Both Choice *B* (*analysis*) and Choice *D* (*criterion*) are incorrect because they are in singular form. The correct plural form for these choices would be *criteria* and analyses.

10. B: Quotation marks are used to indicate something someone said. The example sentences feature a direct quotation that requires the use of double quotation marks. Also, the end punctuation, in this case a question mark, should always be contained within the quotation marks. Choice *A* is incorrect because there is an unnecessary period after the quotation mark. Choice *C* is incorrect because it uses single quotation marks, which are used for a quote within a quote. Choice *D* is incorrect because it places the punctuation outside of the quotation marks.

11. D: In Choice *D*, the word *part* functions as an adjective that modifies the word *Irish*. Choices *A* and *C* are incorrect because the word *part* functions as a noun in these sentences. Choice *B* is incorrect because the word *part* functions as a verb.

12. C: *All of Shannon's family and friends* is the complete subject because it includes who or what is doing the action in the sentence as well as the modifiers that go with it. Choice *A* is incorrect because it only includes the simple subject of the sentence. Choices *B* and *D* are incorrect because they only include part of the complete subject.

13. C: Choice *C* is correct because quotation marks should be used for the title of a short work such as a poem. Choices *A*, *B*, and *D* are incorrect because the titles of novels, films, and newspapers should be placed in italics, not quotation marks.

14. C: This question focuses on the correct usage of the commonly confused word pairs of *it's/its* and *then/than*. *It's* is a contraction for *it is* or *it has*. *Its* is a possessive pronoun. The word *than* shows comparison between two things. *Then* is an adverb that conveys time. Choice *C* correctly uses *it's* and *than*. *It's* is a contraction for *it has* in this sentence, and *than* shows comparison between *work* and *rest*. None of the other answers choices use both of the correct words.

15. B: Choice *B* is an imperative sentence because it issues a command. In addition, it ends with a period, and an imperative sentence must end in a period or exclamation mark. Choice *A* is a declarative sentence that states a fact and ends with a period. Choice *C* is an exclamatory sentence that shows strong emotion and ends with an exclamation point. Choice *D* is an interrogative sentence that asks a question and ends with a question mark.

16. C: Choice *C* is a compound sentence because it joins two independent clauses—*The baby was sick* and *I decided to stay home from work*—with a comma and the coordinating conjunction *so*. Choices *A, B,* and *D*, are all simple sentences, each containing one independent clause with a complete subject and predicate. Choices *A* and *D* each contain a compound subject, or more than one subject, but they are still simple sentences that only contain one independent clause. Choice *B* contains a compound verb (more than one verb), but it's still a simple sentence.

17. C: The simple subject of this sentence, the word *lots*, is plural. It agrees with the plural verb form *were*. Choice *A* is incorrect, because the simple subject *there*, referring to the two constellations, is considered plural. It does not agree with the singular verb form *is*. In Choice *B*, the plural subject *four*, does not agree with the singular verb form *needs*. In Choice *D* the singular subject *everyone* does not agree with the third person plural verb form *have*.

18. B: *Excellent* and *walking* are adjectives modifying the noun *tours*. *Rich* is an adjective modifying the noun *history*, and *brotherly* is an adjective modifying the noun *love*. Choice *A* is incorrect because all of these words are functioning as nouns in the sentence. Choice *C* is incorrect because all of these words are functioning as verbs in the sentence. Choice *D* is incorrect because all of these words are considered prepositions, not adjectives.

19. D: The object pronouns *her* and *me* act as the indirect objects of the sentence. If *me* is in a series of object pronouns, it should always come last in the series. Choice *A* is incorrect because it uses subject pronouns *she* and *I*. Choice *B* is incorrect because it uses the subject pronoun *she*. Choice *C* uses the correct object pronouns, but they are in the wrong order.

20. A: In this example, a colon is correctly used to introduce a series of items. Choice *B* places an unnecessary comma before the word *because*. A comma is not needed before the word *because* when it introduces a dependent clause at the end of a sentence and provides necessary information to understand the sentence. Choice *C* is incorrect because it uses a semi-colon instead of a comma to join a dependent clause and an independent clause. Choice *D* is incorrect because it uses a colon in place of a comma and coordinating conjunction to join two independent clauses.

21. B: Choice *B* correctly uses the contraction for *you are* as the subject of the sentence, and it correctly uses the possessive pronoun *your* to indicate ownership of the jacket. It also correctly uses the adverb *there*, indicating place. Choice *A* is incorrect because it reverses the possessive pronoun *your* and the contraction for *you are*. It also uses the possessive pronoun *their* instead of the adverb *there*. Choice *C* is incorrect because it reverses *your* and *you're* and uses the contraction for *they are* in place of the adverb *there*. Choice *D* incorrectly uses the possessive pronoun *their* instead of the adverb *there*.

22. A: Slang refers to non-standard expressions that are not used in elevated speech and writing. Slang tends to be specific to one group or time period and is commonly used within groups of young people during their conversations with each other. Jargon refers to the language used in a specialized field. The vernacular is the native language of a local area, and a dialect is one form of a language in a certain region. Thus, *B, C,* and *D* are incorrect.

23. D: Colloquial language is that which is used conversationally or informally, in contrast to professional or academic language. While *ain't* is common in conversational English, it is a non-standard expression in academic writing. For college-bound students, high school should introduce them to the expectations of a college classroom, so *B* is not the best answer. Teachers should also avoid placing moral or social value on certain patterns of speech. Rather than teaching students that their familiar speech patterns are bad, teachers should help students learn when and how to use appropriate forms of expression, so *C* is wrong. *Ain't* is in the dictionary, so *A* is incorrect, both in the reason for counseling and in the factual sense.

24. C: A complex sentence joins an independent or main clause with a dependent or subordinate clause. In this case, the main clause is "The restaurant is unconventional." This is a clause with one subject-verb combination that can stand alone as a grammatically-complete sentence. The dependent clause is "because it serves both Chicago style pizza and New York style pizza." This clause begins with the subordinating conjunction *because* and also consists of only one subject-verb combination. *A* is incorrect because a simple sentence consists of only one verb-subject combination—one independent clause. *B* is incorrect because a compound sentence contains two independent clauses connected by a conjunction. *D* is incorrect because a complex-compound sentence consists of two or more independent clauses and one or more dependent clauses.

25. A: Parallelism refers to consistent use of sentence structure or word form. In this case, the list within the sentence does not utilize parallelism; three of the verbs appear in their base form—*travel, take,* and *eat*—but one appears as a gerund—*going*. A parallel version of this sentence would be "This summer, I'm planning to travel to Italy, take a Mediterranean cruise, go to Pompeii, and eat a lot of Italian food." *B* is incorrect because this description is a complete sentence. *C* is incorrect as a misplaced modifier is a modifier that is not located appropriately in relation to the word or words they modify. *D* is incorrect because subject-verb agreement refers to the appropriate conjugation of a verb in relation to its subject.

26. C: In this sentence, the modifier is the phrase "Forgetting that he was supposed to meet his girlfriend for dinner." This phrase offers information about Fred's actions, but the noun that immediately follows it is Anita, creating some confusion about the "do-er" of the phrase. A more appropriate sentence arrangement would be "Forgetting that he was supposed to meet his girlfriend for dinner, Fred made Anita mad when he showed up late." *A* is incorrect as parallelism refers to the consistent use of sentence structure and verb tense, and this sentence is appropriately consistent. *B* is incorrect as a run-on sentence does not contain appropriate punctuation for the number of independent clauses presented, which is not true of this description. *D* is incorrect because subject-verb agreement refers to the appropriate conjugation of a verb relative to the subject, and all verbs have been properly conjugated.

27. B: A comma splice occurs when a comma is used to join two independent clauses together without the additional use of an appropriate conjunction. One way to remedy this problem is to replace the comma with a semicolon. Another solution is to add a conjunction: "Some workers use all their sick leave, but other workers cash out their leave." *A* is incorrect as parallelism refers to the consistent use of sentence structure and verb tense; all tenses and structures in this sentence are consistent. *C* is incorrect because a sentence fragment is a phrase or clause that cannot stand alone—this sentence contains two independent clauses. *D* is incorrect because subject-verb agreement refers to the proper conjugation of a verb relative to the subject, and all verbs have been properly conjugated.

28. D: The problem in the original passage is that the second sentence is a dependent clause that cannot stand alone as a sentence; it must be attached to the main clause found in the first sentence. Because the main clause comes first, it does not need to be separated by a comma. However, if the dependent clause came first, then a comma would be necessary, which is why Choice *C* is incorrect. *A* and *B* also insert unnecessary commas into the sentence.

29. A: A noun refers to a person, place, thing, or idea. Although the word *approach* can also be used as a verb, in the sentence it functions as a noun within the noun phrase "a fresh approach," so *B* is incorrect. An adverb is a word or phrase that provides additional information of the verb, but because the verb is *need* and not *approach*, then *C* is false. An adjective is a word that describes a noun, used here as the word *fresh*, but it is not the noun itself. Thus, *D* is also incorrect.

30. D: An adjective modifies a noun, answering the question "Which one?" or "What kind?" In this sentence, the word *exhaustive* is an adjective that modifies the noun *investigation*. Another clue that this word is an adjective is the suffix *–ive*, which means "having the quality of." The nouns in this sentence are *investigators, inquiry, accusations,* and *corruption*; therefore, *A* is incorrect. The verb in this sentence is *conducted* because this was the action taken by the subject *the investigators*; therefore, *B* is incorrect. *C* is incorrect because an adverb is a word or phrase that provides additional information about the verb, expressing how, when, where, or in what manner.

31. B: In this case, the phrase functions as an adverb modifying the verb *report*, so *B* is the correct answer. "To the student center" does not consist of a subject-verb combination, so it is not a clause; thus, Choices *A* and *C* can be eliminated. This group of words is a phrase. Phrases are classified by either the controlling word in the phrase or its function in the sentence. *D* is incorrect because a noun phrase is a series of words that describe or modify a noun.

32. D: In this sentence, the first answer choice requires a noun meaning *impact* or *influence*, so *effect* is the correct answer. For the second answer choice, the sentence is drawing a comparison. *Than* shows a comparative relationship whereas *then* shows sequence or consequence. *A* and *C* can be eliminated because they contain the choice *then*. *B* is incorrect because *affect* is a verb while this sentence requires a noun.

33. D: Since the sentence contains two independent clauses and a dependent clause, the sentence is categorized as compound-complex:

> Independent clause: I accepted the change of seasons
> Independent clause: I started to appreciate the fall
> Dependent clause: Although I wished it were summer

34. D: Students can use context clues to make a careful guess about the meaning of unfamiliar words. Although all of the words in a sentence can help contribute to the overall sentence, in this case, the adjective that pairs with *ubiquitous* gives the most important hint to the student—cars were first *rare*, but now they are *ubiquitous*. The inversion of *rare* is what gives meaning to the rest of the sentence and *ubiquitous* means "existing everywhere" or "not rare." *A* is incorrect because *ago* only indicates a time frame. *B* is incorrect because *cars* does not indicate a contrasting relationship to the word *ubiquitous* to provide a good context clue. *C* is incorrect because it also only indicates a time frame, but used together with *rare*, it provides the contrasting relationship needed to identify the meaning of the unknown word.

35. B: Proper nouns are specific. *Statue of Liberty* is a proper noun and specifies exactly which statue is being discussed. Choice *A* is incorrect because the word *people* is a common noun and it is only capitalized because it is at the beginning of the sentence. Choice *C* is incorrect. The word *awesome* is an adjective describing the sight. Choice *D* is incorrect. The word *sight* is a common noun. A clue to eliminate answer Choices *C* and *D* is that they were not capitalized. Proper nouns are always capitalized.

36. D: The word *kittens* is plural, meaning more than one kitten. Choice *A* is incorrect. The word *kitten* is singular. Choice *B* is incorrect. The word *girl's* is a singular possessive form. The girl is making the choice.

There is only one girl involved. Choice *C* is incorrect. The word *choice* is a singular noun. The girl has only one choice to make. The word *litter* in this sentence is a collective plural noun meaning a group of kittens.

37. C: The word *who* in the sentence is a subjective interrogative pronoun and the sentence needed a subject that begins a question. Choice *A* is incorrect. The word *whose* is a possessive pronoun and it is not being asked who owns the flowers. Choice *B* is incorrect. The word *whom* is always an objective pronoun—never a subjective one; a subjective pronoun is needed in this sentence. Choice *D* is incorrect. The word *who've* is a contraction of the words *who* and *have*. We would not say, *"Who have ordered the flowers?"*

38. D: The word *its* in the sentence is the singular possessive form of the pronoun that stands in place for the word *giraffe's*. There is one baby that belongs to one giraffe. Choice *A* is incorrect. It is a contraction of the words *it* and *is*. You would not say, *"The giraffe nudged it is baby."* Choice *B* is incorrect. We do not know the gender of the giraffe and if it was female the proper word would be *her* baby not *hers* baby. Choice *C* is incorrect. The word *their* is a plural possessive pronoun and we need a singular possessive pronoun because there is only one giraffe doing the nudging.

39. B: The word *several* stands in for a plural noun at the beginning of the sentence, such as the noun *people*. It is also an indefinite pronoun because the number of people, for example, is not defined. Choice *A* is incorrect. The pronoun is plural not singular. It indicates more than one person. We can tell because the sentence works with the plural word *are* for the verb; substituting the singular word *is* would not make sense. We wouldn't say, *"Several is laughing loudly on the bus."* Choice *C* is incorrect. *Several* is the subject of the sentence. Therefore, it is a subjective pronoun not an objective one. Choice *D* is incorrect. The word *several* does not modify a noun in the sentence. If the sentence said, *"Several people are laughing loudly on the bus,"* then the word *several* would be an indefinite adjective modifying the word *people*.

40. A: The word *delectable* is an adjective modifying the noun *meal* in the sentence. It answers the question: *"What kind of meal?"* Choice *B* is incorrect. The word *connoisseur* is a noun that is the subject of the sentence. Choice *C* is incorrect. The word *slowly* is an adverb telling how the subject enjoyed the meal. Choice *D* is incorrect. The word *enjoyed* is the past-tense verb in the sentence telling us what action the subject had taken.

41. B: The word *by* is a positional preposition telling where we are in relation to the water. The word *near* is also a positional preposition telling where *we* are in relation to the lake. The word *before* is a time preposition telling when in relation to the time of day, dawn. Choice *A* is incorrect because *went* and *to see* are both verbs and *pretty* is an adjective modifying the word *sunrise*. Choice *C* is incorrect because *water, lake, dawn,* and *sunrise* are all nouns in the sentence. Choice *D* is incorrect because the word *we* is a pronoun, the word *down* is an adverb modifying the verb *went*, the word *the* is an article, and the word *pretty* is an adjective modifying *sunrise*.

42. B: The word *well* at the beginning of the sentence is set apart from the rest of the sentence with a comma and is a mild interjection. Choice *A* is incorrect. The word *goodness* at the end of the sentence is a noun. It is the idea/state of being for the cookie. Choice *C* is incorrect. It is an interrogative sentence and all of the words in the sentence can be identified as other parts of speech. *Can't* is a contraction of the word cannot and it works with the word *see* as the verb in the sentence. The word *you* is a pronoun; the word *that* is an adjective modifying the word *cookie*; *cookie* is a noun; *is* is another verb; and *broken* is an adjective modifying the word *cookie*. Choice *D* is incorrect because the exclamation mark at the

end of the sentence is not there to set apart an interjection. Rather, it is there to punctuate the exclamatory sentence.

43. C: The simple subject is *bike* and its modifiers *the heaviest* and *green* are included to form the complete subject. Choice *A* is incorrect because *bike* is the simple subject. Choice *B* is incorrect because it includes only one of the modifiers (*green*) of the word *bike*. Choice *D* is incorrect because *is mine* is the predicate of the sentence, not the subject.

44. D: The subject of the sentence is *my house;* therefore, the rest of the sentence is the predicate. Choice *A* is incorrect because *my house* is the subject, not the predicate. Choice *B* is incorrect because *is the yellow one* is only part of the predicate, but the sentence does not end there. Choice *C* is incorrect because *at the end of the street* is only a portion of the predicate.

45. B: The *kitten* is a singular subject and so the singular verb *pounces* should be used instead of *pounce*. Choice *A* is incorrect because the plural subject *kittens* agrees with the plural verb *show*. Choice *C* is incorrect because the singular subject *kitten* agrees with the singular verb *eats*. Choice *D* is incorrect because the singular subject *kitten* agrees with the singular verb *snuggles.*

46. C: *Her mother* is to whom Calysta brought the lamp. Choice *A* is incorrect because *stained-glass lamp* is the direct object of the sentence. Choice *B* is incorrect because *brought* is the verb. Choice *D* is incorrect because *beautiful* is an adjective modifying the noun *lamp*.

47. C: *They're* (they are) on *their* (possessive) way to New Jersey but *they're* (they are) not *there* (location) yet. This sentence makes sense. Choice *A* is incorrect because after the word *but* should be the word *they're* (they are). Choice *B* is incorrect because the sentence should begin with *they're* (they are) instead of *their* (possessive). Choice *D* is incorrect because after the word *not* should be *there* (location) instead of *their* (possessive).

48. D: *For the longest time* is an introductory prepositional phrase beginning with the preposition *for* and all the modifiers for the word *time*. Choice *A* is incorrect because it includes the past-tense verb *have wanted* creating a clause. Choice *B* is incorrect because it includes the verb *wanted* and the infinitive *to learn*. Choice *C* is incorrect because it includes the verb *learn* and the infinitive *to skate*.

49. B: *The weight of the world was on his shoulders* and *he took a long walk* are both independent clauses connected with a comma and a coordinating conjunction. Choice *A* is incorrect because there are two independent clauses and a simple sentence has only one independent clause. Choice *C* is incorrect because the sentence has no dependent clauses and a complex sentence needs at least one dependent clause. Choice *D* is incorrect because, although there are two independent clauses, there are no dependent clauses and a compound-complex sentence will have at least two independent clauses and at least one dependent clause.

50. A: *The last thing she wanted to do was see the Eiffel Tower before the flight* has only one independent clause and no dependent clauses therefore it is a simple sentence. Choice *B* is incorrect because the sentence has only one independent clause. *The last thing she wanted to do* is a gerund phrase serving as the noun subject of the sentence, therefore it is not an independent clause. Choice *C* is incorrect because the sentence has no dependent clauses therefore it cannot be a complex sentence. Choice *D* is incorrect because the sentence has only one independent clause and no dependent clauses. A compound-complex sentence needs at least two independent clauses and at least one dependent clause.

Writing Prompt

Informational Essay Sample Writing Prompt

Below there are two passages. The first passage explains the concept of lethal force, while the second one depicts situations for assault. For your essay, read both texts and then write an informational essay on the consequences of both lethal force and assault, and detail the main differences. Use evidence from the passages.

Before you begin planning and writing, read the two texts:

1. Lethal Force
2. Assault

As you read the texts, think about evidence you can use in your essay. You can take notes on your own paper while you read.

After you read the passages, organize your essay. Think about the examples, facts, and details in the passages. Think about how you will effectively introduce your topic to catch the audience's attention, and how you will separate each paragraph with its own main ideas.

Now write your informational essay. Make sure to:

- Use information from the two texts so that you include details from both.
- Introduce your topic in a clear way and organize your essay in a way that makes sense.
- Include topic development using evidence from the two passages or separate details and facts about the topics.
- Use effective transitions between paragraphs.
- Make the relationship between ideas and concepts clear.
- Make sure your language and vocabulary are clear and correct.
- Write a conclusion that follows the information presented.

Passages

Passage I: Lethal Force

Lethal force, or deadly force, is defined as the physical means to cause death or serious harm to another individual. The law holds that lethal force is only accepted when you or another person are in immediate and unavoidable danger of death or severe bodily harm. For example, a person could be beating a weaker person in such a way that they are suffering severe enough trauma that could result in death or serious harm. This would be an instance where lethal force would be acceptable and possibly the only way to save that person from irrevocable damage.

Another example of when to use lethal force would be when someone enters your home with a deadly weapon. The intruder's presence and possession of the weapon indicate mal-intent and the ability to inflict death or severe injury to you and your loved ones. Again, lethal force can be used in this situation. Lethal force can also be applied to prevent the harm of another individual. If a woman is being brutally assaulted and is

unable to fend off an attacker, lethal force can be used to defend her as a last-ditch effort. If she is in immediate jeopardy of rape, harm, and/or death, lethal force could be the only response that could effectively deter the assailant.

The key to understanding the concept of lethal force is the term *last resort*. Deadly force cannot be taken back; it should be used only to prevent severe harm or death. The law does distinguish whether the means of one's self-defense is fully warranted, or if the individual goes out of control in the process. If you continually attack the assailant after they are rendered incapacitated, this would be causing unnecessary harm, and the law can bring charges against you. Likewise, if you kill an attacker unnecessarily after defending yourself, you can be charged with murder. This would move lethal force beyond necessary defense, making it no longer a last resort but rather a use of excessive force.

Passage II: Assault

Assault is the unlawful attempt of one person to apply apprehension on another individual by an imminent threat or by initiating offensive contact. Assaults can vary, encompassing physical strikes, threatening body language, and even provocative language. In the case of the latter, even if a hand has not been laid, it is still considered an assault because of its threatening nature.

Let's look at an example: A homeowner is angered because his neighbor blows fallen leaves into his freshly mowed lawn. Irate, the homeowner gestures a fist to his fellow neighbor and threatens to bash his head in for littering on his lawn. The homeowner's physical motions and verbal threat heralds a physical threat against the other neighbor. These factors classify the homeowner's reaction as an assault. If the angry neighbor hits the threatening homeowner in retaliation, that would constitute an assault as well because he physically hit the homeowner.

Assault also centers on the involvement of weapons in a conflict. If someone fires a gun at another person, this could be interpreted as an assault unless the shooter acted in self-defense. If an individual drew a gun or a knife on someone with the intent to harm them, that would be considered assault. However, it's also considered an assault if someone simply aimed a weapon, loaded or not, at another person in a threatening manner.

Mathematics

Numbers and Quantity

Whole Number Operations

Addition

Addition is the combination of two numbers so their quantities are added together cumulatively. The sign for an addition operation is the + symbol. For example, 9 + 6 = 15. The 9 and 6 combine to achieve a cumulative value, called a **sum**.

Addition holds the commutative property, which means that the order of the numbers in an addition equation can be switched without altering the result. The formula for the commutative property is a + b = b + a. The following examples can demonstrate how the commutative property works:

$$7 = 3 + 4 = 4 + 3 = 7$$

$$20 = 12 + 8 = 8 + 12 = 20$$

Addition also holds the **associative property**, which means that the grouping of numbers does not matter in an addition problem. In other words, the presence or absence of parentheses is irrelevant. The formula for the associative property is (a + b) + c = a + (b + c). Here are some examples of the associative property at work:

$$30 = (6 + 14) + 10 = 6 + (14 + 10) = 30$$

$$35 = 8 + (2 + 25) = (8 + 2) + 25 = 35$$

Subtraction

Subtraction is taking away one number from another, so their quantities are reduced. The sign designating a subtraction operation is the − symbol, and the result is called the **difference.** For example, 9 - 6 = 3. The number *6* detracts from the number *9* to reach the difference *3*.

Unlike addition, subtraction follows neither the commutative nor associative properties. The order and grouping in subtraction impact the result.

$$15 = 22 - 7 \neq 7 - 22 = -15$$

$$3 = (10 - 5) - 2 \neq 10 - (5 - 2) = 7$$

When working through subtraction problems involving larger numbers, it's necessary to regroup the numbers. The following practice problem uses regrouping:

$$\begin{array}{r} 3\ 2\ 5 \\ -\ \ 7\ 7 \\ \hline \end{array}$$

Here, it is clear that the ones and tens columns for 77 are greater than the ones and tens columns for 325. To subtract this number, one needs to borrow from the tens and hundreds columns. When borrowing from a column, subtracting 1 from the lender column will add 10 to the borrower column:

$$
\begin{array}{ccc}
3\text{-}1 & 10\text{+}2\text{-}1 & 10\text{+}5 \\
- & 7 & 7 \\
\hline
\end{array}
=
\begin{array}{ccc}
2 & 11 & 15 \\
- & 7 & 7 \\
\hline
2 & 4 & 8
\end{array}
$$

After ensuring that each digit in the top row is greater than the digit in the corresponding bottom row, subtraction can proceed as normal, and the answer is found to be 248.

Multiplication

Multiplication involves adding together multiple copies of a number. It is indicated by an × symbol or a number immediately outside of a parenthesis. For example:

$$5(8 - 2)$$

The two numbers being multiplied together are called **factors**, and their result is called a **product**. For example, $9 \times 6 = 54$. This can be shown alternatively by expansion of either the 9 or the 6:

$$9 \times 6 = 9 + 9 + 9 + 9 + 9 + 9 = 54$$

$$9 \times 6 = 6 + 6 + 6 + 6 + 6 + 6 + 6 + 6 + 6 = 54$$

Like addition, multiplication holds the commutative and associative properties:

$$115 = 23 \times 5 = 5 \times 23 = 115$$

$$84 = 3 \times (7 \times 4) = (3 \times 7) \times 4 = 84$$

Multiplication also follows the **distributive property**, which allows the multiplication to be distributed through parentheses. The formula for distribution is $a \times (b + c) = ab + ac$. This is clear after the examples:

$$45 = 5 \times 9 = 5(3 + 6) = (5 \times 3) + (5 \times 6) = 15 + 30 = 45$$

$$20 = 4 \times 5 = 4(10 - 5) = (4 \times 10) - (4 \times 5) = 40 - 20 = 20$$

Multiplication becomes slightly more complicated when multiplying numbers with decimals. The easiest way to answer these problems is to ignore the decimals and multiply as if they were whole numbers. After multiplying the factors, a decimal gets placed in the product. The placement of the decimal is determined by taking the cumulative number of decimal places in the factors.

For example:

$$0.7$$
$$\times 3$$
$$\overline{2.1}$$

$$2.6$$
$$\times 4.2$$
$$\overline{10.92}$$

$$1.5$$
$$\times 6.4$$
$$\overline{9.60}$$

Starting with the first example, the first step is to ignore the decimal and multiply the numbers as though they were whole numbers, which results in a product of 21. The next step is to count the number of digits that follow a decimal (one, in this case). Finally, the decimal place gets moved that many positions to the left, because the factors have only one decimal place. The second example works the same way, except that there are two total decimal places in the factors, so the product's decimal is moved two places over. In the third example, the decimal should be moved over two digits, but the digit zero is no longer needed, so it is erased, and the final answer is 9.6.

Division

Division and multiplication are inverses of each other in the same way that addition and subtraction are opposites. The signs designating the division operation are the ÷ and / symbols. In division, the second number divides into the first.

The number before the division sign is called the **dividend** or, if expressed as a fraction, the **numerator.** For example, in $a \div b$, a is the dividend, while in $\frac{a}{b}$, a is the numerator.

The number after the division sign is called the **divisor** or, if expressed as a fraction, the **denominator.** For example, in $a \div b$, b is the divisor, while in $\frac{a}{b}$, b is the denominator.

Like subtraction, division doesn't follow the commutative property, as it matters which number comes before the division sign, and division doesn't follow the associative or distributive properties for the same reason. For example:

$$\frac{3}{2} = 9 \div 6 \neq 6 \div 9 = \frac{2}{3}$$

$$2 = 10 \div 5 = (30 \div 3) \div 5 \neq 30 \div (3 \div 5) = 30 \div \frac{3}{5} = 50$$

$$25 = 20 + 5 = (40 \div 2) + (40 \div 8) \neq 40 \div (2 + 8) = 40 \div 10 = 4$$

If a divisor doesn't divide into a dividend an integer number of times, whatever is left over is termed the **remainder**. The remainder can be further divided out into decimal form by using long division; however, this doesn't always give a **quotient** with a finite number of decimal places, so the remainder can also be expressed as a fraction over the original divisor.

Division with decimals is similar to multiplication with decimals in that when dividing a decimal by a whole number, one should ignore the decimal and divide as if it was a whole number.

Upon finding the answer, or quotient, the decimal point is inserted at the decimal place equal to that in the dividend.

$$15.75 \div 3 = 5.25$$

When the divisor is a decimal number, both the divisor and dividend get multiplied by 10. This process is repeated until the divisor is a whole number, then one needs to complete the division operation as described above.

$$17.5 \div 2.5 = 175 \div 25 = 7$$

Order of Operations

When solving equations with multiple operations, special rules apply. These rules are known as the **Order of Operations**. The order is as follows: Parentheses, Exponents, Multiplication and Division from left to right, and Addition and Subtraction from left to right. A popular mnemonic device to help remember the order is Please Excuse My Dear Aunt Sally (PEMDAS).

Evaluate the following two problems to understand the Order of Operations:

1) $4 + (3 \times 2)^2 \div 4$

First, solve the operation within the parentheses: $4 + 6^2 \div 4$.
Second, solve the exponent: $4 + 36 \div 4$.
Third, solve the division operation: $4 + 9$.
Fourth, finish the operation with addition for the answer, 13.

2) $2 \times (6 + 3) \div (2 + 1)^2$

$2 \times 9 \div (3)^2$
$2 \times 9 \div 9$
$18 \div 9$
2

Estimation and Rounding

Estimation is finding a value that is close to a solution but is not the exact answer. For example, if there are values in the thousands to be multiplied, then each value can be estimated to the nearest thousand and the calculation performed. This value provides an approximate solution that can be determined very quickly.

When estimating, it's often convenient to **round** a number, which means to give an approximate figure to make it easier to compare amounts or perform mental math. Round up when the digit is 5 or more. The digit used to determine the rounding, and all subsequent digits, become 0, and the selected place value is increased by 1. Here are some examples:

75 rounded to the nearest ten is 80
380 rounded to the nearest hundred is 400
22.697 rounded to the nearest hundredth is 22.70

Round down when rounding on any digit that is below 5. The rounded digit, and all subsequent digits, becomes 0, and the preceding digit stays the same. Here are some examples:

92 rounded to the nearest ten is 90
839 rounded to the nearest hundred is 800
22.643 rounded to the nearest hundredth is 22.64

The same estimation strategies and techniques used when working with standard math problems can be employed when working with real-life situations. Estimation is frequently used in calculations involving money, such as for determining if one has enough money for a purchase, how much one needs to save weekly to buy a desired product, or how much a restaurant bill will sum to.

Applying Operations to Real-World Contexts

Addition and subtraction are **inverse operations**. Adding a number and then subtracting the same number will cancel each other out, resulting in the original number, and vice versa. For example, $8 + 7 - 7 = 8$ and $137 - 100 + 100 = 137$. Similarly, multiplication and division are inverse operations. Therefore, multiplying by a number and then dividing by the same number results in the original number, and vice versa. For example, $8 \times 2 \div 2 = 8$ and $12 \div 4 \times 4 = 12$. Inverse operations are used to work backwards to solve problems. In the case that 7 and a number add to 18, the inverse operation of subtraction is used to find the unknown value ($18 - 7 = 11$). If a school's entire 4th grade was divided evenly into 3 classes each with 22 students, the inverse operation of multiplication is used to determine the total students in the grade ($22 \times 3 = 66$). Additional scenarios involving inverse operations are included in the tables below.

There are a variety of real-world situations in which one or more of the operators is used to solve a problem. The tables below display the most common scenarios.

Addition & Subtraction

	Unknown Result	Unknown Change	Unknown Start
Adding to	5 students were in class. 4 more students arrived. How many students are in class? $5 + 4 = ?$	8 students were in class. More students arrived late. There are now 18 students in class. How many students arrived late? $8 + ? = 18$ Solved by inverse operations $18 - 8 = ?$	Some students were in class early. 11 more students arrived. There are now 17 students in class. How many students were in class early? $? + 11 = 17$ Solved by inverse operations $17 - 11 = ?$
Taking from	15 students were in class. 5 students left class. How many students are in class now? $15 - 5 = ?$	12 students were in class. Some students left class. There are now 8 students in class. How many students left class? $12 - ? = 8$ Solved by inverse operations $8 + ? = 12 \rightarrow 12 - 8 = ?$	Some students were in class. 3 students left class. Then there were 13 students in class. How many students were in class before? $? - 3 = 13$ Solved by inverse operations $13 + 3 = ?$

	Unknown Total	Unknown Addends (Both)	Unknown Addends (One)
Putting together/ taking apart	The homework assignment is 10 addition problems and 8 subtraction problems. How many problems are in the homework assignment? $10 + 8 = ?$	Bobby has $9. How much can Bobby spend on candy and how much can Bobby spend on toys? $9 = ? + ?$	Bobby has 12 pairs of pants. 5 pairs of pants are shorts, and the rest are long. How many pairs of long pants does he have? $12 = 5 + ?$ Solved by inverse operations $12 - 5 = ?$

	Unknown Difference	Unknown Larger Value	Unknown Smaller Value
Comparing	Bobby has 5 toys. Tommy has 8 toys. How many more toys does Tommy have than Bobby? $5+?=8$ Solved by inverse operations $8-5=?$ Bobby has \$6. Tommy has \$10. How many fewer dollars does Bobby have than Tommy? $10-6=?$	Tommy has 2 more toys than Bobby. Bobby has 4 toys. How many toys does Tommy have? $2+4=?$ Bobby has 3 fewer dollars than Tommy. Bobby has \$8. How many dollars does Tommy have? $?-3=8$ Solved by inverse operations $8+3=?$	Tommy has 6 more toys than Bobby. Tommy has 10 toys. How many toys does Bobby have? $?+6=10$ Solved by inverse operations $10-6=?$ Bobby has \$5 less than Tommy. Tommy has \$9. How many dollars does Bobby have? $9-5=?$

Multiplication and Division

	Unknown Product	Unknown Group Size	Unknown Number of Groups
Equal groups	There are 5 students, and each student has 4 pieces of candy. How many pieces of candy are there in all? $5\times4=?$	14 pieces of candy are shared equally by 7 students. How many pieces of candy does each student have? $7\times?=14$ Solved by inverse operations $14\div7=?$	If 18 pieces of candy are to be given out 3 to each student, how many students will get candy? $?\times3=18$ Solved by inverse operations $18\div3=?$

	Unknown Product	**Unknown Factor**	**Unknown Factor**
Arrays	There are 5 rows of students with 3 students in each row. How many students are there? $5 \times 3 = ?$	If 16 students are arranged into 4 equal rows, how many students will be in each row? $4 \times ? = 16$ Solved by inverse operations $16 \div 4 = ?$	If 24 students are arranged into an array with 6 columns, how many rows are there? $? \times 6 = 24$ Solved by inverse operations $24 \div 6 = ?$

	Larger Unknown	**Smaller Unknown**	**Multiplier Unknown**
Comparing	A small popcorn costs $1.50. A large popcorn costs 3 times as much as a small popcorn. How much does a large popcorn cost? $1.50 \times 3 = ?$	A large soda costs $6 and that is 2 times as much as a small soda costs. How much does a small soda cost? $2 \times ? = 6$ Solved by inverse operations $6 \div 2 = ?$	A large pretzel costs $3 and a small pretzel costs $2. How many times as much does the large pretzel cost as the small pretzel? $? \times 2 = 3$ Solved by inverse operations $3 \div 2 = ?$

Fraction Operations

A **fraction** is a part of something that is whole. Items such as apples can be cut into parts to help visualize fractions. If an apple is cut into 2 equal parts, each part represents ½ of the apple. If each half is cut into two parts, the apple now is cut into quarters. Each piece now represents ¼ of the apple. In this example, each part is equal because they all have the same size. Geometric shapes, such as circles and squares, can also be utilized to help visualize the idea of fractions. For example, a circle can be drawn and divided into 6 equal parts:

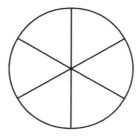

Shading can be used to represent parts of the circle that can be translated into fractions. The top of the fraction, the **numerator**, can represent how many segments are shaded. The bottom of the fraction, the **denominator**, can represent the number of segments that the circle is broken into. A pie is a good analogy to use in this example. If one piece of the circle is shaded, or one piece of pie is cut out, $^1/_6$ of the object is being referred to. An apple, a pie, or a circle can be utilized in order to compare simple fractions. For example, showing that ½ is larger than ¼ and that ¼ is smaller than $1/_3$ can be accomplished through shading. A **unit fraction** is a fraction in which the numerator is 1, and the denominator is a positive whole number. It represents one part of a whole—one piece of pie.

Imagine that an apple pie has been baked for a holiday party, and the full pie has eight slices. After the party, there are five slices left. How could the amount of the pie that remains be expressed as a fraction? The numerator is 5 since there are 5 pieces left, and the denominator is 8 since there were eight total slices in the whole pie. Thus, expressed as a fraction, the leftover pie totals $\frac{5}{8}$ of the original amount.

Fractions come in three different varieties: proper fractions, improper fractions, and mixed numbers. **Proper fractions** have a numerator less than the denominator, such as $\frac{3}{8}$, but **improper fractions** have a numerator greater than the denominator, such as $\frac{15}{8}$. **Mixed numbers** combine a whole number with a proper fraction, such as $3\frac{1}{2}$. Any mixed number can be written as an improper fraction by multiplying the integer by the denominator, adding the product to the value of the numerator, and dividing the sum by the original denominator. For example:

$$3\frac{1}{2} = \frac{3 \times 2 + 1}{2} = \frac{7}{2}$$

Whole numbers can also be converted into fractions by placing the whole number as the numerator and making the denominator 1. For example, $3 = \frac{3}{1}$.

The bar in a fraction represents division. Therefore $^6/_5$ is the same as $6 \div 5$. In order to rewrite it as a mixed number, division is performed to obtain $6 \div 5 = 1\ R1$. The remainder is then converted into fraction form. The actual remainder becomes the numerator of a fraction, and the divisor becomes the denominator. Therefore $1\ R1$ is written as $1\frac{1}{5}$, a mixed number. A mixed number can also decompose into the addition of a whole number and a fraction. For example,

$$1\frac{1}{5} = 1 + \frac{1}{5} \text{ and } 4\frac{5}{6} = 4 + \frac{1}{6} + \frac{1}{6} + \frac{1}{6} + \frac{1}{6} + \frac{1}{6}$$

Every fraction can be built from a combination of unit fractions.

One of the most fundamental concepts of fractions is their ability to be manipulated by multiplication or division. This is possible since $\frac{n}{n} = 1$ for any non-zero integer. As a result, multiplying or dividing by $\frac{n}{n}$ will not alter the original fraction since any number multiplied or divided by 1 doesn't change the value of that number. Fractions of the same value are known as equivalent fractions. For example, $\frac{2}{8}, \frac{25}{100},$ and $\frac{40}{160}$ are equivalent, as they all equal $\frac{1}{4}$.

Like fractions, or **equivalent fractions**, are the terms used to describe these fractions that are made up of different numbers but represent the same quantity. For example, the given fractions are $^4/_8$ and $^3/_6$. If a pie was cut into 8 pieces and 4 pieces were removed, half of the pie would remain. Also, if a pie was split into 6 pieces and 3 pieces were eaten, half of the pie would also remain. Therefore, both of the fractions represent half of a pie. These two fractions are referred to as like fractions. **Unlike fractions** are fractions that are different and cannot be thought of as representing equal quantities. When working with fractions in mathematical expressions, like fractions should be simplified. Both $^4/_8$ and $^3/_6$ can be simplified into $^1/_2$.

Comparing fractions can be completed through the use of a number line. For example, if $^3/_5$ and $^6/_{10}$ need to be compared, each fraction should be plotted on a number line. To plot $^3/_5$, the area from 0 to 1 should be broken into 5 equal segments, and the fraction represents 3 of them. To plot $^6/_{10}$, the area from 0 to 1 should be broken into 10 equal segments, and the fraction represents 6 of them.

It can be seen that $\dfrac{3}{5} = \dfrac{6}{10}$

Like fractions are plotted at the same point on a number line. Unit fractions can also be used to compare fractions. For example, if it is known that

$$\frac{4}{5} > \frac{1}{2}$$

and

$$\frac{1}{2} > \frac{4}{10}$$

then it is also known that

$$\frac{4}{5} > \frac{4}{10}$$

Also, converting improper fractions to mixed numbers can be helpful in comparing fractions because the whole number portion of the number is more visible.

Adding and subtracting mixed numbers and fractions can be completed by decomposing fractions into a sum of whole numbers and unit fractions. For example, the given problem is

$$5\frac{3}{7} + 2\frac{1}{7}$$

Decomposing into

$$5 + \frac{1}{7} + \frac{1}{7} + \frac{1}{7} + 2 + \frac{1}{7}$$

This shows that the whole numbers can be added separately from the unit fractions. The answer is:

$$5 + 2 + \frac{1}{7} + \frac{1}{7} + \frac{1}{7} + \frac{1}{7} = 7 + \frac{4}{7} = 7\frac{4}{7}$$

Although many equivalent fractions exist, they are easier to compare and interpret when reduced or simplified. The numerator and denominator of a simple fraction will have no factors in common other than 1. When reducing or simplifying fractions, divide the numerator and denominator by the greatest common factor. A simple strategy is to divide the numerator and denominator by low numbers, like 2, 3, or 5 until arriving at a simple fraction, but the same thing could be achieved by determining the greatest common factor for both the numerator and denominator and dividing each by it. Using the first method is preferable when both the numerator and denominator are even, end in 5, or are obviously a multiple of another number. However, if no numbers seem to work, it will be necessary to factor the numerator and denominator to find the GCF. Let's look at examples:

1) Simplify the fraction $\frac{6}{8}$:

Dividing the numerator and denominator by 2 results in $\frac{3}{4}$, which is a simple fraction.

2) Simplify the fraction $\frac{12}{36}$:

Dividing the numerator and denominator by 2 leaves $\frac{6}{18}$. This isn't a simple fraction, as both the numerator and denominator have factors in common. Diving each by 3 results in $\frac{2}{6}$, but this can be further simplified by dividing by 2 to get $\frac{1}{3}$. This is the simplest fraction, as the numerator is 1. In cases like this, multiple division operations can be avoided by determining the greatest common factor between the numerator and denominator.

3) Simplify the fraction $\frac{18}{54}$ by dividing by the greatest common factor:

First, determine the factors for the numerator and denominator. The factors of 18 are 1, 2, 3, 6, 9, and 18. The factors of 54 are 1, 2, 3, 6, 9, 18, 27, and 54. Thus, the greatest common factor is 18. Dividing $\frac{18}{54}$ by 18 leaves $\frac{1}{3}$, which is the simplest fraction. This method takes slightly more work, but it definitively arrives at the simplest fraction.

Adding and Subtracting Fractions

Adding and subtracting fractions that have the same denominators involves adding or subtracting the numerators. The denominator will stay the same. Therefore, the decomposition process can be made simpler, and the fractions do not have to be broken into unit fractions.

For example, the given problem is:

$$4\frac{7}{8} - 2\frac{6}{8}$$

The answer is found by adding the answers to both

$$4 - 2 \text{ and } \frac{7}{8} - \frac{6}{8}$$

$$2 + \frac{1}{8} = 2\frac{1}{8}$$

A common mistake would be to add the denominators so that

$$\frac{1}{4} + \frac{1}{4} = \frac{1}{8} \text{ or } \frac{2}{8}$$

However, conceptually, it is known that two quarters make a half, so neither one of these are correct.

If two fractions have different denominators, equivalent fractions must be used to add or subtract them. The fractions must be converted into fractions that have common denominators. A **least common denominator** or the product of the two denominators can be used as the common denominator. For example, in the problem $\frac{5}{6} + \frac{2}{3}$, both 6, which is the least common denominator, and 18, which is the product of the denominators, can be used. In order to use 6, $\frac{2}{3}$ must be converted to sixths. A number line can be used to show the equivalent fraction is $\frac{4}{6}$. What happens is that $\frac{2}{3}$ is multiplied times a fractional form of 1 to obtain a denominator of 6. Hence, $\frac{2}{3} \times \frac{2}{2} = \frac{4}{6}$. Therefore, the problem is now $\frac{5}{6} + \frac{4}{6} = \frac{9}{6}$, which can be simplified into $\frac{3}{2}$. In order to use 18, both fractions must be converted into having 18 as their denominator. $\frac{5}{6}$ would have to be multiplied times $\frac{3}{3}$, and $\frac{2}{3}$ would need to be multiplied times $\frac{6}{6}$. The addition problem would be $\frac{15}{18} + \frac{12}{18} = \frac{27}{18}$, which reduces into $\frac{3}{2}$.

It is always possible to find a common denominator by multiplying the denominators. However, when the denominators are large numbers, this method is unwieldy, especially if the answer must be provided in its simplest form. Thus, it's beneficial to find the **least common denominator** of the fractions—the least common denominator is incidentally also the **least common multiple**.

Once equivalent fractions have been found with common denominators, simply add or subtract the numerators to arrive at the answer:

1) $\frac{1}{2} + \frac{3}{4} = \frac{2}{4} + \frac{3}{4} = \frac{5}{4}$

2) $\frac{3}{12} + \frac{11}{20} = \frac{15}{60} + \frac{33}{60} = \frac{48}{60} = \frac{4}{5}$

3) $\frac{7}{9} - \frac{4}{15} = \frac{35}{45} - \frac{12}{45} = \frac{23}{45}$

4) $\frac{5}{6} - \frac{7}{18} = \frac{15}{18} - \frac{7}{18} = \frac{8}{18} = \frac{4}{9}$

Multiplying and Dividing Fractions

Of the four basic operations that can be performed on fractions, the one that involves the least amount of work is multiplication. To multiply two fractions, simply multiply the numerators together, multiply the denominators together, and place the products of each as a fraction. Whole numbers and mixed numbers can also be expressed as a fraction, as described above, to multiply with a fraction.

Because multiplication is commutative, multiplying a fraction times a whole number is the same as multiplying a whole number times a fraction. The problem involves adding a fraction a specific number of times. The problem $3 \times \frac{1}{4}$ can be translated into adding the unit fraction 3 times:

$$\frac{1}{4} + \frac{1}{4} + \frac{1}{4} = \frac{3}{4}$$

In the problem $4 \times \frac{2}{5}$, the fraction can be decomposed into $\frac{1}{5} + \frac{1}{5}$ and then added 4 times to obtain $\frac{8}{5}$. Also, both of these answers can be found by just multiplying the whole number times the numerator of the fraction being multiplied.

The whole numbers can be written in fraction form as:

$$\frac{3}{1} \times \frac{1}{4} = \frac{3}{4}$$

$$\frac{4}{1} \times \frac{2}{5} = \frac{8}{5}$$

Multiplying a fraction times a fraction involves multiplying the numerators together separately and the denominators together separately. For example,

$$\frac{3}{8} \times \frac{2}{3} = \frac{3 \times 2}{8 \times 3} = \frac{6}{24}$$

This can then be reduced to $^1/_4$.

Dividing a fraction by a fraction is actually a multiplication problem. It involves flipping the divisor and then multiplying normally. For example,

$$\frac{22}{5} \div \frac{1}{2} = \frac{22}{5} \times \frac{2}{1} = \frac{44}{5}$$

The same procedure can be implemented for division problems involving fractions and whole numbers. The whole number can be rewritten as a fraction over a denominator of 1, and then division can be completed.

A common denominator approach can also be used in dividing fractions. Considering the same problem, $\frac{22}{5} \div \frac{1}{2}$, a common denominator between the two fractions is 10. $\frac{22}{5}$ would be rewritten as $\frac{22}{5} \times \frac{2}{2} = \frac{44}{10}$, and $\frac{1}{2}$ would be rewritten as $\frac{1}{2} \times \frac{5}{5} = \frac{5}{10}$. Dividing both numbers straight across results in:

$$\frac{44}{10} \div \frac{5}{10} = \frac{^{44}/_5}{^{10}/_{10}} = \frac{^{44}/_5}{1} = ^{44}/_5$$

Many real-world problems will involve the use of fractions. Key words include actual fraction values, such as half, quarter, third, fourth, etc. The best approach to solving word problems involving fractions is to draw a picture or diagram that represents the scenario being discussed, while deciding which type of operation is necessary in order to solve the problem. A phrase such as "one fourth of 60 pounds of

coal" creates a scenario in which multiplication should be used, and the mathematical form of the phrase is $\frac{1}{4} \times 60$.

Decimal Operations

The **decimal system** is a way of writing out numbers that uses ten different numerals: 0, 1, 2, 3, 4, 5, 6, 7, 8, and 9. This is also called a "base ten" or "base 10" system. Other bases are also used. For example, computers work with a base of 2. This means they only use the numerals 0 and 1.

The **decimal place** denotes how far to the right of the decimal point a numeral is. The first digit to the right of the decimal point is in the **tenths** place. The next is the **hundredths**. The third is the *thousandths*.

So, 3.142 has a 1 in the tenths place, a 4 in the hundredths place, and a 2 in the thousandths place.

The **decimal point** is a period used to separate the *ones* place from the *tenths* place when writing out a number as a decimal.

A **decimal number** is a number written out with a decimal point instead of as a fraction, for example, 1.25 instead of $\frac{5}{4}$. Depending on the situation, it can sometimes be easier to work with fractions and sometimes easier to work with decimal numbers.

A decimal number is **terminating** if it stops at some point. It is called **repeating** if it never stops but repeats a pattern over and over. It is important to note that every rational number can be written as a terminating decimal or as a repeating decimal.

Addition with Decimals

To add decimal numbers, each number needs to be lined up by the decimal point in vertical columns. For each number being added, the zeros to the right of the last number need to be filled in so that each of the numbers has the same number of places to the right of the decimal. Then, the columns can be added together. Here is an example of 2.45 + 1.3 + 8.891 written in column form:

$$2.450$$

$$1.300$$

$$+\ 8.891$$

Zeros have been added in the columns so that each number has the same number of places to the right of the decimal.

Added together, the correct answer is 12.641:

$$2.450$$

$$1.300$$

$$+\ 8.891$$

$$12.641$$

Subtraction with Decimals

Subtracting decimal numbers is the same process as adding decimals. Here is $7.89 - 4.235$ written in column form:

$$7.890$$

$$- \underline{4.235}$$

$$3.655$$

A zero has been added in the column so that each number has the same number of places to the right of the decimal.

Multiplication with Decimals

The simplest way to multiply decimals is to calculate the product as if the decimals are not there, then count the number of decimal places in the original problem. Use that total to place the decimal the same number of places over in your answer, counting from right to left. For example, 0.5 x 1.25 can be rewritten and multiplied as 5 x 125, which equals 625. Then the decimal is added three places from the right for .625.

The final answer will have the same number of decimal *points* as the total number of decimal *places* in the problem. The first number has one decimal place, and the second number has two decimal places. Therefore, the final answer will contain three decimal places:

$$0.5 \times 1.25 = 0.625$$

Division with Decimals

Dividing a decimal by a whole number entails using long division first by ignoring the decimal point. Then, the decimal point is moved the number of places given in the problem.

For example, $6.8 \div 4$ can be rewritten as $68 \div 4$, which is 17. There is one non-zero integer to the right of the decimal point, so the final solution would have one decimal place to the right of the solution. In this case, the solution is 1.7.

Dividing a decimal by another decimal requires changing the divisor to a whole number by moving its decimal point. The decimal place of the dividend should be moved by the same number of places as the divisor. Then, the problem is the same as dividing a decimal by a whole number.

For example, $5.72 \div 1.1$ has a divisor with one decimal point in the denominator. The expression can be rewritten as $57.2 \div 11$ by moving each number one decimal place to the right to eliminate the decimal. The long division can be completed as $572 \div 11$ with a result of 52. Since there is one non-zero integer to the right of the decimal point in the problem, the final solution is 5.2.

In another example, $8 \div 0.16$ has a divisor with two decimal points in the denominator. The expression can be rewritten as $800 \div 16$ by moving each number two decimal places to the right to eliminate the decimal in the divisor. The long division can be completed with a result of 50.

Estimation and Rounding with Decimals

Prior to performing operations and calculating the answer to a problem involving addition, subtraction, multiplication, or division, it is helpful to estimate the result. Doing so will enable the test taker to

determine whether his or her computed answer is logical within the context of a given problem and prevent careless errors. For example, it is unfortunately common under the pressure of a testing situation for test takers to inadvertently perform the incorrect operation or make a simple calculation error on an otherwise easy math problem. By quickly estimating the answer by eyeballing the numbers, rounding if needed, and performing some simple mental math, test takers can establish an approximate expected outcome before calculating the specific answer. The derived result after computation can then be evaluated by its nearness to the expected answer. This is performed by approximating given values to perform mental math. Numbers should be rounded to the nearest value possible to check the initial results.

As mentioned, sometimes when performing operations such as multiplying numbers, the result can be estimated by rounding. For example, to estimate the value of 11.2×2.01, each number can be rounded to the nearest integer. This will yield a result of 22.

Rounding numbers helps with estimation because it changes the given number to a simpler, although less accurate, number than the exact given number. Rounding allows for easier calculations, which estimate the results of using the exact given number. The accuracy of the estimate and ease of use depends on the place value to which the number is rounded. First, the place value is specified. Then, the digit to its right is looked at. For example, if rounding to the nearest hundreds place, the digit in the tens place is used. If it is a zero, one, 2, 3, or 4, the digit being rounded to is left alone. If it is a 5, 6, 7, 8 or 9, the digit being rounded to is increased by one. All other digits before the decimal point are then changed to zeros, and the digits in decimal places are dropped. If a decimal place is being rounded to, all digits that come after are just dropped.

For example, if 845,231.45 was to be rounded to the nearest thousands place, the answer would be 845,000. The 5 would remain the same due to the 2 in the hundreds place. Also, if 4.567 were to be rounded to the nearest tenths place, the answer would be 4.6. The 5 increased to 6 due to the 6 in the hundredths place, and the rest of the decimal is dropped.

Percent Problems

Think of percentages as fractions with a denominator of 100. In fact, percentage means "per hundred." The basic percent equation is the following:

$$\frac{is}{of} = \frac{\%}{100}$$

The placement of numbers in the equation depends on what the question asks.

Example 1
Find 40% of 80.

Basically, the problem is asking, "What is 40% of 80?" The 40% is the percent, and 80 is the number to find the percent "of." The equation is:

$$\frac{x}{80} = \frac{40}{100}$$

After cross-multiplying, the problem becomes 100x = 80(40). Solving for x gives the answer: x = 32.

Example 2
What percent of 100 is 20?

The 20 fills in the "is" portion, while 100 fills in the "of." The question asks for the percent, so that will be x, the unknown. The following equation is set up:

$$\frac{20}{100} = \frac{x}{100}$$

Cross-multiplying yields the equation 100x = 20(100). Solving for x gives the answer of 20%.

Example 3
30% of what number is 30?

The following equation uses the clues and numbers in the problem:

$$\frac{30}{x} = \frac{30}{100}$$

Cross-multiplying results in the equation 30(100) = 30x. Solving for x gives the answer x = 100.

Relationships in Numerical Data

In some cases, it is useful to compare numerical data and determine the relationship between values. One of the best ways to mathematically compare two values is to compute the percentage difference between the two values. For example, consider a given music shop that had a net profit of $120,000 in the first year of operation and $185,000 over the second year. Rather than simply finding the net difference between the two years (using subtraction), the business owner may want to know by what percentage his profit increased; in other words, how much his profit in the second year increased relative to his first year. In such cases, the percentage change is desired. The following sections provide some guidance for this process.

Percent Increase/Decrease
Problems dealing with percentages may involve an original value, a change in that value, and a percentage change. A problem will provide two pieces of information and ask to find the third. To do so, this formula is used: $\frac{change}{original\ value}$ x 100 = percent change. Here's a sample problem:

Attendance at a baseball stadium has dropped 16% from last year. Last year's average attendance was 40,000. What is this year's average attendance?

Using the formula and information, the change is unknown (x), the original value is 40,000, and the percent change is 16%. The formula can be written as: $\frac{x}{40,000}$ x 100 = 16. When solving for x, it is determined the change was 6,400. The problem asked for this year's average attendance, so to calculate, the change (6,400) is subtracted from last year's attendance (40,000) to determine this year's average attendance is 33,600.

Percent More Than/Less Than

Percentage problems may give a value and what percent that given value is more than or less than an original unknown value. Here's a sample problem:

A store advertises that all its merchandise has been reduced by 25%. The new price of a pair of shoes is $60. What was the original price?

This problem can be solved by writing a proportion. Two ratios should be written comparing the cost and the percent of the original cost. The new cost is 75% of the original cost (100% - 25%); and the original cost is 100% of the original cost. The unknown original cost can be represented by *x*. The proportion would be set up as: $\frac{60}{75} = \frac{x}{100}$. Solving the proportion, it is determined the original cost was $80.

Real-World Rate, Percent, and Measurement Problems

A **ratio** compares the size of one group to the size of another. For example, there may be a room with 4 tables and 24 chairs. The ratio of tables to chairs is 4: 24. Such ratios behave like fractions in that both sides of the ratio by the same number can be multiplied or divided. Thus, the ratio 4:24 is the same as the ratio 2:12 and 1:6.

One quantity is **proportional** to another quantity if the first quantity is always some multiple of the second. For instance, the distance travelled in five hours is always five times to the speed as travelled. The distance is proportional to speed in this case.

One quantity is **inversely proportional** to another quantity if the first quantity is equal to some number divided by the second quantity. The time it takes to travel one hundred miles will be given by 100 divided by the speed travelled. The time is inversely proportional to the speed.

When dealing with word problems, there is no fixed series of steps to follow, but there are some general guidelines to use. It is important that the quantity to be found is identified. Then, it can be determined how the given values can be used and manipulated to find the final answer.

Example 1

Jana wants to travel to visit Alice, who lives one hundred and fifty miles away. If she can drive at fifty miles per hour, how long will her trip take?

The quantity to find is the *time* of the trip. The time of a trip is given by the distance to travel divided by the speed to be traveled. The problem determines that the distance is one hundred and fifty miles, while the speed is fifty miles per hour. Thus, 150 divided by 50 is $150 \div 50 = 3$. Because *miles* and *miles per hour* are the units being divided, the miles cancel out. The result is 3 hours.

Example 2

Bernard wishes to paint a wall that measures twenty feet wide by eight feet high. It costs ten cents to paint one square foot. How much money will Bernard need for paint?

The final quantity to compute is the *cost* to paint the wall. This will be ten cents ($0.10) for each square foot of area needed to paint. The area to be painted is unknown, but the dimensions of the wall are given; thus, it can be calculated.

The dimensions of the wall are 20 feet wide and 8 feet high. Since the area of a rectangle is length multiplied by width, the area of the wall is 8 x 20 = 160 square feet. Multiplying 0.1 x 160 yields $16 as the cost of the paint.

The Position of Numbers Relative to Each Other

Place Value of a Digit

Numbers count in groups of 10. That number is the same throughout the set of natural numbers and whole numbers. It is referred to as working within a base 10 numeration system. Only the numbers from zero to 9 are used to represent any number. The foundation for doing this involves **place value**. Numbers are written side by side. This is to show the amount in each place value.

For place value, let's look at how the number 10 is different from zero to 9. It has two digits instead of just one. The one is in the tens' place, and the zero is in the ones' place. Therefore, there is one group of tens and zero ones. 11 has one 10 and one 1. The introduction of numbers from 11 to 19 should be the next step. Each value within this range of numbers consists of one group of 10 and a specific number of leftover ones. Counting by tens can be practiced once the tens column is understood. This process consists of increasing the number in the tens place by one. For example, counting by 10 starting at 17 would result in the next four values being 27, 37, 47, and 57.

A place value chart can be used for understanding and learning about numbers that have more digits. Here is an example of a place value chart:

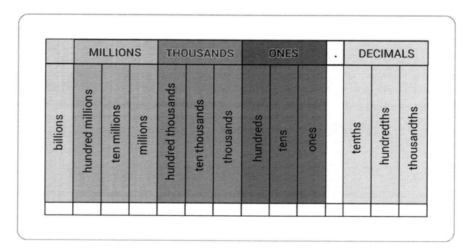

In the number 1,234, there are 4 ones and 3 tens. The 2 is in the hundreds' place, and the one is in the thousands' place. Note that each group of three digits is separated by a comma. The 2 has a value that is 10 times greater than the 3. Every place to the left has a value 10 times greater than the place to its right. Also, each group of three digits is also known as a *period*. 234 is in the ones' period.

The number 1,234 can be written out as *one-thousand, two hundred thirty-four*. The process of writing out numbers is known as the *decimal system*. It is also based on groups of 10. The place value chart is a helpful tool in using this system. In order to write out a number, it always starts with the digit(s) in the highest period. For example, in the number 23,815,467, the 23 is in highest place and is in the millions' period. The number is read *twenty-three million, eight hundred fifteen thousand, four hundred sixty-seven*. Each period is written separately through the use of commas. Also, no "ands" are used within the

number. Another way to think about the number 23,815,467 is through the use of an addition problem. For example:

$$23,815,467 = 20,000,000 + 3,000,000 + 800,000 + 10,000 + 5,000 + 400 + 60 + 7$$

This expression is known as *expanded form*. The actual number 23,815,467 is known as being in **standard form**.

In order to compare whole numbers with many digits, place value can be used. In each number to be compared, it is necessary to find the highest place value in which the numbers differ and to compare the value within that place value. For example, $4,523,345 < 4,532,456$ because of the values in the ten thousands place. A similar process can be used for decimals. However, number lines can also be used. Tick marks can be placed within two whole numbers on the number line that represent tenths, hundredths, etc. Each number being compared can then be plotted. The value farthest to the right on the number line is the largest.

Comparing, Classifying, and Ordering Real Numbers

Rational numbers are any number that can be written as a fraction or ratio. Within the set of rational numbers, several subsets exist that are referenced throughout the mathematics topics. **Counting numbers** are the first numbers learned as a child. Counting numbers consist of 1,2,3,4, and so on. **Whole numbers** include all counting numbers and zero (0,1,2,3,4,…). **Integers** include counting numbers, their opposites, and zero (…,-3,-2,-1,0,1,2,3,…). **Rational numbers** are inclusive of integers, fractions, and decimals that terminate, or end (1.7, 0.04213) or repeat ($0.136\overline{5}$).

A **number line** typically consists of integers (…3,2,1,0,-1,-2,-3…), and is used to visually represent the value of a rational number. Each rational number has a distinct position on the line determined by comparing its value with the displayed values on the line. For example, if plotting -1.5 on the number line below, it is necessary to recognize that the value of -1.5 is .5 less than -1 and .5 greater than -2. Therefore, -1.5 is plotted halfway between -1 and -2.

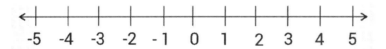

The number system that is used consists of only ten different digits or characters. However, this system is used to represent an infinite number of values. As mentioned, the **place value system** makes this infinite number of values possible. The position in which a digit is written corresponds to a given value. Starting from the decimal point (which is implied, if not physically present), each subsequent place value to the left represents a value greater than the one before it. Conversely, starting from the decimal point, each subsequent place value to the right represents a value less than the one before it.

In accordance with the **base-10 system**, the value of a digit increases by a factor of ten each place it moves to the left. For example, consider the number 7. Moving the digit one place to the left (70), increases its value by a factor of 10 ($7 \times 10 = 70$). Moving the digit two places to the left (700) increases its value by a factor of 10 twice ($7 \times 10 \times 10 = 700$). Moving the digit three places to the left (7,000) increases its value by a factor of 10 three times ($7 \times 10 \times 10 \times 10 = 7,000$), and so on.

Conversely, the value of a digit decreases by a factor of ten each place it moves to the right. (Note that multiplying by $\frac{1}{10}$ is equivalent to dividing by 10). For example, consider the number 40. Moving the digit one place to the right (4) decreases its value by a factor of 10 ($40 \div 10 = 4$). Moving the digit two places to the right (0.4), decreases its value by a factor of 10 twice ($40 \div 10 \div 10 = 0.4$) or ($40 \times \frac{1}{10} \times \frac{1}{10} = 0.4$). Moving the digit three places to the right (0.04) decreases its value by a factor of 10 three times ($40 \div 10 \div 10 \div 10 = 0.04$) or ($40 \times \frac{1}{10} \times \frac{1}{10} \times \frac{1}{10} = 0.04$), and so on.

Ordering Numbers

A common question type asks to order rational numbers from least to greatest or greatest to least. The numbers will come in a variety of formats, including decimals, percentages, roots, fractions, and whole numbers. These questions test for knowledge of different types of numbers and the ability to determine their respective values.

Before discussing ordering all numbers, let's start with decimals.

To compare decimals and order them by their value, utilize a method similar to that of ordering large numbers.

The main difference is where the comparison will start. Assuming that any numbers to left of the decimal point are equal, the next numbers to be compared are those immediately to the right of the decimal point. If those are equal, then move on to compare the values in the next decimal place to the right.

For example:

Which number is greater, 12.35 or 12.38?

Check that the values to the left of the decimal point are equal:

12 = 12

Next, compare the values of the decimal place to the right of the decimal:

12.3 = 12.3

Those are also equal in value.

Finally, compare the value of the numbers in the next decimal place to the right on both numbers:

12.3**5** and 12.3**8**

Here the 5 is less than the 8, so the final way to express this inequality is:

12.35 < 12.38

Comparing decimals is regularly exemplified with money because the "cents" portion of money ends in the hundredths place. When paying for gasoline or meals in restaurants, and even in bank accounts, if enough errors are made when calculating numbers to the hundredths place, they can add up to dollars and larger amounts of money over time.

Now that decimal ordering has been explained, let's expand and consider all real numbers. Whether the question asks to order the numbers from greatest to least or least to greatest, the crux of the question is the same—convert the numbers into a common format. Generally, it's easiest to write the numbers as whole numbers and decimals so they can be placed on a number line. Follow these examples to understand this strategy.

1) Order the following rational numbers from greatest to least:

$$\sqrt{36}, 0.65, 78\%, \frac{3}{4}, 7, 90\%, \frac{5}{2}$$

Of the seven numbers, the whole number (7) and decimal (0.65) are already in an accessible form, so concentrate on the other five.

First, the square root of 36 equals 6. (If the test asks for the root of a non-perfect root, determine which two whole numbers the root lies between.) Next, convert the percentages to decimals. A percentage means "per hundred," so this conversion requires moving the decimal point two places to the left, leaving 0.78 and 0.9.

Lastly, evaluate the fractions:

$$\frac{3}{4} = \frac{75}{100} = 0.75 \, ; \frac{5}{2} = 2\frac{1}{2} = 2.5$$

Now, the only step left is to list the numbers in the request order:

$$7, \sqrt{36}, \frac{5}{2}, 90\%, 78\%, \frac{3}{4}, 0.65$$

2) Order the following rational numbers from least to greatest:

$$2.5, \sqrt{9}, -10.5, 0.853, 175\%, \sqrt{4}, \frac{4}{5}$$

$$\sqrt{9} = 3$$

$$175\% = 1.75$$

$$\sqrt{4} = 2$$

$$\frac{4}{5} = 0.8$$

From least to greatest, the answer is:

$$-10.5, \frac{4}{5}, 0.853, 175\%, \sqrt{4}, 2.5, \sqrt{9}$$

Expressing Numeric Relationships

If a question asks to give words to a mathematical expression and says "equals," then an = sign must be included in the answer. Similarly, "less than or equal to" is expressed by the inequality symbol ≤, and "greater than or equal" to is expressed as ≥. Furthermore, "less than" is represented by <, and "greater than" is expressed by >.

Equations use the equals sign because the numeric expressions on either side of the symbol (=) are equivalent. In contrast, inequalities compare values or expressions that are unequal. Although not always true, linear equations that include a variable often have just one value for the variable that makes the statement true. Linear inequalities generally have an infinite number of values that make the statement true.

Inequalities are a concise mathematical way to express the relationship between unequal values. More specifically, they describe in what way the values are unequal. A value could be greater than (>); less than (<); greater than or equal to (≥); or less than or equal to (≤) another value. The statement "five times a number added to forty is more than sixty-five" can be expressed as $5x + 40 > 65$. Common words and phrases that express inequalities are:

Symbol	Phrase
<	is under, is below, smaller than, beneath
>	is above, is over, bigger than, exceeds
≤	no more than, at most, maximum
≥	no less than, at least, minimum

Conversions

Decimals and Percentages
Since a percentage is based on "per hundred," decimals and percentages can be converted by multiplying or dividing by 100. Practically speaking, this always amounts to moving the decimal point two places to the right or left, depending on the conversion. To convert a percentage to a decimal, move the decimal point two places to the left and remove the % sign. To convert a decimal to a percentage, move the decimal point two places to the right and add a "%" sign. Here are some examples:

65% = 0.65
0.33 = 33%
0.215 = 21.5%
99.99% = 0.9999
500% = 5.00
7.55 = 755%

Fractions and Percentages
Remember that a percentage is a number per one hundred. So a percentage can be converted to a fraction by making the number in the percentage the numerator and putting 100 as the denominator:

$$43\% = \frac{43}{100}$$

$$97\% = \frac{97}{100}$$

Note that the percent symbol (%) kind of looks like a 0, a 1, and another 0. So think of a percentage like 54% as 54 over 100.

To convert a fraction to a percent, follow the same logic. If the fraction happens to have 100 in the denominator, you're in luck. Just take the numerator and add a percent symbol:

$$\frac{28}{100} = 28\%$$

Otherwise, divide the numerator by the denominator to get a decimal:

$$\frac{9}{12} = 0.75$$

Then convert the decimal to a percentage:

$$0.75 = 75\%$$

Another option is to make the denominator equal to 100. Be sure to multiply the numerator and the denominator by the same number. For example:

$$\frac{3}{20} \times \frac{5}{5} = \frac{15}{100}$$

$$\frac{15}{100} = 15\%$$

Changing Fractions to Decimals

To change a fraction into a decimal, divide the denominator into the numerator until there are no remainders. There may be repeating decimals, so rounding is often acceptable. A straight line above the repeating portion denotes that the decimal repeats.

Example: Express 4/5 as a decimal.

Set up the division problem.

$$5\overline{)4}$$

5 does not go into 4, so place the decimal and add a zero.

$$5\overline{)4.0}$$

5 goes into 40 eight times. There is no remainder.

$$\begin{array}{r} 0.8 \\ 5\overline{)4.0} \\ -4.0 \\ \hline 0 \end{array}$$

The solution is 0.8.

Example: Express 33 1/3 as a decimal.

Since the whole portion of the number is known, set it aside to calculate the decimal from the fraction portion.

Set up the division problem.

$$3\overline{)1}$$

3 does not go into 1, so place the decimal and add zeros. 3 goes into 10 three times.

$$0.3$$
$$3\overline{)1.0}$$

This will repeat with a remainder of 1.

$$
\begin{array}{r}
0.333 \\
3\overline{)1.000} \\
-9 \\
\hline
10 \\
-9 \\
\hline
10
\end{array}
$$

So, we will place a line over the 3 to denote the repetition. The solution is written $0.\overline{3}$.

Changing Decimals to Fractions

To change decimals to fractions, place the decimal portion of the number, the numerator, over the respective place value, the denominator, then reduce, if possible.

Example: Express 0.25 as a fraction.

This is read as twenty-five hundredths, so put 25 over 100. Then reduce to find the solution.

$$\frac{25}{100} = \frac{1}{4}$$

Example: Express 0.455 as a fraction

This is read as four hundred fifty-five thousandths, so put 455 over 1000. Then reduce to find the solution.

$$\frac{455}{1000} = \frac{91}{200}$$

There are two types of problems that commonly involve percentages. The first is to calculate some percentage of a given quantity, where you convert the percentage to a decimal, and multiply the quantity by that decimal. Secondly, you are given a quantity and told it is a fixed percent of an unknown quantity. In this case, convert to a decimal, then divide the given quantity by that decimal.

Example: What is 30% of 760?

Convert the percent into a useable number. "Of" means to multiply.

$$30\% = 0.30$$

Set up the problem based on the givens, and solve.

$$0.30 \times 760 = 228$$

Example: 8.4 is 20% of what number?

Convert the percent into a useable number.

$$20\% = 0.20$$

The given number is a percent of the answer needed, so divide the given number by this decimal rather than multiplying it.

$$\frac{8.4}{0.20} = 42$$

Algebra

Mathematical and Real-World Linear Equations and Inequalities

Linear equations and linear inequalities are both comparisons of two algebraic expressions. However, unlike equations in which the expressions are equal, linear inequalities compare expressions that may be unequal. Linear equations typically have one value for the variable that makes the statement true. Linear inequalities generally have an infinite number of values that make the statement true.

When solving a linear equation, the desired result requires determining a numerical value for the unknown variable. If given a linear equation involving addition, subtraction, multiplication, or division, working backwards isolates the variable. Addition and subtraction are inverse operations, as are multiplication and division. Therefore, they can be used to cancel each other out.

For example, solve $4(t - 2) + 2t - 4 = 2(9 - 2t)$

Distributing: $4t - 8 + 2t - 4 = 18 - 4t$

Combining like terms: $6t - 12 = 18 - 4t$

Adding $4t$ to each side to move the variable: $10t - 12 = 18$

Adding 12 to each side to isolate the variable: $10t = 30$

Dividing each side by 10 to isolate the variable: $t = 3$

The answer can be checked by substituting the value for the variable into the original equation, ensuring that both sides calculate to be equal.

Linear inequalities express the relationship between unequal values. More specifically, they describe in what way the values are unequal. A value can be greater than (>), less than (<), greater than or equal to (≥), or less than or equal to (≤) another value. $5x + 40 > 65$ is read as *five times a number added to forty is greater than sixty-five.*

When solving a linear inequality, the solution is the set of all numbers that make the statement true. The inequality $x + 2 \geq 6$ has a solution set of 4 and every number greater than 4 (4.01; 5; 12; 107; etc.). Adding 2 to 4 or any number greater than 4 results in a value that is greater than or equal to 6. Therefore, $x \geq 4$ is the solution set.

To algebraically solve a linear inequality, follow the same steps as those for solving a linear equation. The inequality symbol stays the same for all operations *except* when multiplying or dividing by a negative number. If multiplying or dividing by a negative number while solving an inequality, the relationship reverses (the sign flips). In other words, > switches to < and vice versa. Multiplying or dividing by a positive number does not change the relationship, so the sign stays the same. An example is shown below.

Solve $-2x - 8 \leq 22$

Add 8 to both sides: $-2x \leq 30$

Divide both sides by -2: $x \geq -15$

Solutions of a linear equation or a linear inequality are the values of the variable that make a statement true. In the case of a linear equation, the solution set (list of all possible solutions) typically consists of a single numerical value. To find the solution, the equation is solved by isolating the variable. For example, solving the equation $3x - 7 = -13$ produces the solution $x = -2$. The only value for x which produces a true statement is -2. This can be checked by substituting -2 into the original equation to check that both sides are equal. In this case, $3(-2) - 7 = -13 \rightarrow -13 = -13$; therefore, -2 is a solution.

Although linear equations generally have one solution, this is not always the case. If there is no value for the variable that makes the statement true, there is no solution to the equation. Consider the equation $x + 3 = x - 1$. There is no value for x in which adding 3 to the value produces the same result as subtracting one from the value. Conversely, if any value for the variable makes a true statement, the equation has an infinite number of solutions. Consider the equation:

$$3x + 6 = 3(x + 2)$$

Any number substituted for x will result in a true statement (both sides of the equation are equal).

By manipulating equations like the two above, the variable of the equation will cancel out completely. If the remaining constants express a true statement (ex. $6 = 6$), then all real numbers are solutions to the equation. If the constants left express a false statement (ex. $3 = -1$), then no solution exists for the equation.

Solving a linear inequality requires all values that make the statement true to be determined. For example, solving $3x - 7 \geq -13$ produces the solution $x \geq -2$. This means that -2 and any number greater than -2 produces a true statement. Solution sets for linear inequalities will often be displayed using a number line. If a value is included in the set ($\geq$ or $\leq$), a shaded dot is placed on that value and an arrow extending in the direction of the solutions. For a variable > or $\geq$ a number, the arrow will point right on a number line, the direction where the numbers increase. If a variable is < or $\leq$ a number, the arrow will point left on a number line, which is the direction where the numbers decrease. If the value is not included in the set (> or <), an open (unshaded) circle on that value is used with an arrow in the appropriate direction.

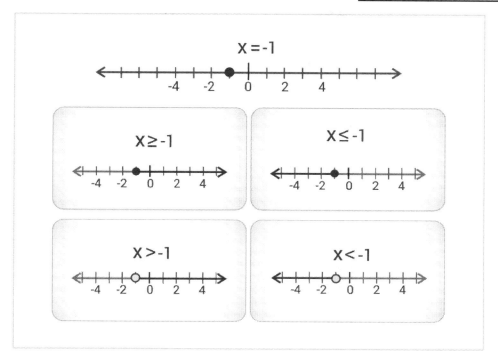

Similar to linear equations, a linear inequality may have a solution set consisting of all real numbers or can contain no solution. When solved algebraically, a linear inequality in which the variable cancels out and results in a true statement (ex. $7 \geq 2$) has a solution set of all real numbers. A linear inequality in which the variable cancels out and results in a false statement (ex. $7 \leq 2$) has no solution.

Translating Words into Math

To translate a word problem into an expression, test takers should look for key words indicating addition, subtraction, multiplication, or division:

- *Addition*: add, altogether, together, plus, increased by, more than, in all, sum, and total
- *Subtraction*: minus, less than, difference, decreased by, fewer than, remain, and take away
- *Multiplication*: times, twice, of, double, and triple
- *Division*: divided by, cut up, half, quotient of, split, and shared equally

If a question asks to give words to a mathematical expression and says "equals," then an = sign must be included in the answer. Similarly, "less than or equal to" is expressed by the inequality symbol ≤, and "greater than or equal" to is expressed as ≥. Furthermore, "less than" is represented by <, and "greater than" is expressed by >.

Linear Models and Relationships

Linear relationships describe the way two quantities change with respect to each other. The relationship is defined as linear because a line is produced if all the sets of corresponding values are graphed on a coordinate grid. When expressing the linear relationship as an equation, the equation is often written in the form $y = mx + b$ (slope-intercept form) where m and b are numerical values and x and y are variables (for example, $y = 5x + 10$). Given a linear equation and the value of either variable (x or y), the value of the other variable can be determined.

Suppose a teacher is grading a test containing 20 questions with 5 points given for each correct answer, adding a curve of 10 points to each test. This linear relationship can be expressed as the equation $y = 5x + 10$ where x represents the number of correct answers and y represents the test score. To determine the score of a test with a given number of correct answers, the number of correct answers is substituted into the equation for x and evaluated. For example, for 10 correct answers, 10 is substituted for x:

$$y = 5(10) + 10 \rightarrow y = 60$$

Therefore, 10 correct answers will result in a score of 60. The number of correct answers needed to obtain a certain score can also be determined. To determine the number of correct answers needed to score a 90, 90 is substituted for y in the equation (y represents the test score) and solved:

$$90 = 5x + 10 \rightarrow 80 = 5x \rightarrow 16 = x$$

Therefore, 16 correct answers are needed to score a 90.

Linear relationships may be represented by a table of 2 corresponding values. Certain tables may determine the relationship between the values and predict other corresponding sets. Consider the table below, which displays the money in a checking account that charges a monthly fee:

Month	0	1	2	3	4
Balance	$210	$195	$180	$165	$150

An examination of the values reveals that the account loses $15 every month (the month increases by one and the balance decreases by 15). This information can be used to predict future values. To determine what the value will be in month 6, the pattern can be continued, and it can be concluded that the balance will be $120. To determine which month the balance will be $0, $210 is divided by $15 (since the balance decreases $15 every month), resulting in month 14.

Similar to a table, a graph can display corresponding values of a linear relationship.

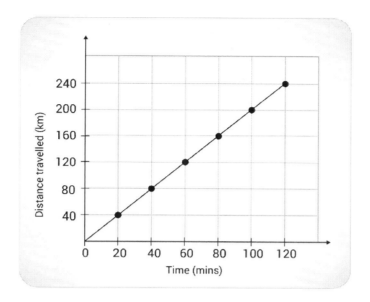

The graph above represents the relationship between distance traveled and time. To find the distance traveled in 80 minutes, the mark for 80 minutes is located at the bottom of the graph. By following this mark directly up on the graph, the corresponding point for 80 minutes is directly across from the 160-kilometer mark. This information indicates that the distance travelled in 80 minutes is 160 kilometers. To predict information not displayed on the graph, the way in which the variables change with respect to one another is determined. In this case, distance increases by 40 kilometers as time increases by 20 minutes. This information can be used to continue the data in the graph or convert the values to a table.

Identifying Variables for Linear Models

The first step to writing a linear model is to identify what the variables represent. A **variable** represents an unknown quantity, and in the case of a linear equation, a specific relationship exists between the two variables (usually x and y). Within a given scenario, the variables are the two quantities that are changing. The variable x is considered the independent variable and represents the inputs of a function. The variable y is considered the dependent variable and represents the outputs of a function. For example, if a scenario describes distance traveled and time traveled, distance would be represented by y and time represented by x. The distance traveled depends on the time spent traveling (time is independent). If a scenario describes the cost of a cab ride and the distance traveled, the cost would be represented by y and the distance represented by x. The cost of a cab ride depends on the distance traveled.

Identifying the Slope and Y-Intercept for Linear Models

The **slope of the graph of a line** represents the rate of change between the variables of an equation. In the context of a real-world scenario, the slope will tell the way in which the unknown quantities (variables) change with respect to each other. A scenario involving distance and time might state that someone is traveling at a rate of 45 miles per hour. The slope of the linear model would be 45. A scenario involving the cost of a cab ride and distance traveled might state that the person is charged $3 for each mile. The slope of the linear model would be 3.

The **y-intercept of a linear function** is the value of y when $x = 0$ (the point where the line intercepts the y-axis on the graph of the equation). It is sometimes helpful to think of this as a "starting point" for a linear function. Suppose for the scenario about the cab ride that the person is told that the cab company charges a flat fee of $5 plus $3 for each mile. Before traveling any distance ($x = 0$), the cost is $5. The y-intercept for the linear model would be 5.

Identifying Ordered Pairs for Linear Models

A linear equation with two variables can be written given a point (ordered pair) and the slope or given two points on a line. An ordered pair gives a set of corresponding values for the two variables (x and y). As an example, for a scenario involving distance and time, it is given that the person traveled 112.5 miles in 2 ½ hours. Knowing that x represents time and y represents distance, this information can be written as the ordered pair (2.5, 112.5).

Word Problems

Word problems can appear daunting, but prepared test takers shouldn't let the verbiage psyche them out. No matter the scenario or specifics, the key to answering them is to translate the words into a math problem. It is critical to keep in mind what the question is asking and what operations could lead to that answer. The following word problem resembles one of the question types most frequently encountered on the exam.

Walter's Coffee Shop sells a variety of drinks and breakfast treats.

Price List	
Hot Coffee	$2.00
Slow Drip Iced Coffee	$3.00
Latte	$4.00
Muffins	$2.00
Crepe	$4.00
Egg Sandwich	$5.00

Costs	
Hot Coffee	$0.25
Slow Drip Iced Coffee	$0.75
Latte	$1.00
Muffins	$1.00
Crepe	$2.00
Egg Sandwich	$3.00

Walter's utilities, rent, and labor costs him $500 per day. Today, Walter sold 200 hot coffees, 100 slow drip iced coffees, 50 lattes, 75 muffins, 45 crepes, and 60 egg sandwiches. What was Walter's total profit today?

To accurately answer this type of question, the first step is to determine the total cost of making his drinks and treats, then determine how much revenue he earned from selling those products. After arriving at these two totals, the profit is measured by deducting the total cost from the total revenue.

Walter's costs for today:

200 hot coffees	× $0.25	= $50
100 slow drip iced coffees	× $0.75	= $75
50 lattes	× $1.00	= $50
75 muffins	× $1.00	= $75
45 crepes	× $2.00	= $90
60 egg sandwiches	× $3.00	= $180
Utilities, Rent, and Labor		= $500
Total costs		= $1,020

Walter's revenue for today:

200 hot coffees	× $2.00	= $400
100 slow drip iced coffees	× $3.00	= $300
50 lattes	× $4.00	= $200
75 muffins	× $2.00	= $150
45 crepes	× $4.00	= $180
60 egg sandwiches	× $5.00	= $300
Total revenue		= $1,530

Walter's $Profit = Revenue - Costs = \$1,530 - \$1,020 = \$510$

This strategy can be applied to other question types. For example, calculating salary after deductions, balancing a checkbook, and calculating a dinner bill are common word problems similar to business planning. In all cases, the most important step is remembering to use the correct operations. When a balance is increased, addition is used. When a balance is decreased, the problem requires subtraction. Common sense and organization are one's greatest assets when answering word problems.

Solving for X in Proportions

Proportions are commonly used in word problems to find unknown values, such as x, that are some percent or fraction of a known number. Proportions are solved by cross-multiplying and then dividing to arrive at x. The following examples show how this is done:

1. $\dfrac{75\%}{90\%} = \dfrac{25\%}{x}$

To solve for x, the fractions must be cross-multiplied: ($75\%x = 90\% \times 25\%$). To make things easier, the percentages can be converted to decimals: ($0.9 \times 0.25 = 0.225 = 0.75x$). To get rid of the coefficient of x, each side must be divided by that same coefficient to get the answer $x = 0.3$. The question could ask for the answer as a percentage or fraction in lowest terms, which are 30% and $\frac{3}{10}$, respectively.

$$2. \frac{x}{12} = \frac{30}{96}$$

Cross-multiply: $96x = 30 \times 12$

Multiply: $96x = 360$

Divide: $x = 360 \div 96$

Answer: $x = 3.75$

$$3. \frac{0.5}{3} = \frac{x}{6}$$

Cross-multiply: $3x = 0.5 \times 6$

Multiply: $3x = 3$

Divide: $x = 3 \div 3$

Answer: $x = 1$

Observant test takers may have noticed there's a faster way to arrive at the answer. If there is an obvious operation being performed on the proportion, the same operation can be used on the other side of the proportion to solve for x. For example, in the first practice problem, 75% became 25% when divided by 3, and upon doing the same to 90%, the correct answer of 30% would have been found with much less legwork. However, these questions aren't always so intuitive, so it's a good idea for test takers to work through the steps, even if the answer seems apparent from the onset.

Unit Rate

Unit rate word problems ask test takers to calculate the rate or quantity of something in a different value. For example, a problem might say that a car drove a certain number of miles in a certain number of minutes and then ask how many miles per hour the car was traveling. These questions involve solving proportions. Consider the following examples:

1. Alexandra made $96 during the first 3 hours of her shift as a temporary worker at a law office. She will continue to earn money at this rate until she finishes in 5 more hours. How much does Alexandra make per hour? How much money will Alexandra have made at the end of the day?

This problem can be solved in two ways. The first is to set up a proportion, as the rate of pay is constant. The second is to determine her hourly rate, multiply the 5 hours by that rate, and then adding the $96.

To set up a proportion, the money already earned (numerator) is placed over the hours already worked (denominator) on one side of an equation. The other side has x over 8 hours (the total hours worked in the day).

It looks like this: $\frac{96}{3} = \frac{x}{8}$.

Now, cross-multiply yields $768 = 3x$.

To get x, the 768 is divided by 3, which leaves $x = 256$.

Alternatively, as x is the numerator of one of the proportions, multiplying by its denominator will reduce the solution by one step. Thus, Alexandra will make $256 at the end of the day. To calculate her hourly rate, the total is divided by 8, giving $32 per hour.

Alternatively, it is possible to figure out the hourly rate by dividing $96 by 3 hours to get $32 per hour. Now her total pay can be figured by multiplying $32 per hour by 8 hours, which comes out to $256.

2. Jonathan is reading a novel. So far, he has read 215 of the 335 total pages. It takes Jonathan 25 minutes to read 10 pages, and the rate is constant. How long does it take Jonathan to read one page? How much longer will it take him to finish the novel? Express the answer in time.

To calculate how long it takes Jonathan to read one page, 25 minutes is divided by 10 pages to determine the page per minute rate. Thus, it takes 2.5 minutes to read one page.

Jonathan must read 120 more pages to complete the novel. (This is calculated by subtracting the pages already read from the total.) Now, his rate per page is multiplied by the number of pages. Thus, $120 \times 2.5 = 300$. Expressed in time, 300 minutes is equal to 5 hours.

3. At a hotel, $\frac{4}{5}$ of the 120 rooms are booked for Saturday. On Sunday, $\frac{3}{4}$ of the rooms are booked. On which day are more of the rooms booked, and by how many more?

The first step is to calculate the number of rooms booked for each day. This is done by multiplying the fraction of the rooms booked by the total number of rooms.

$$\text{Saturday:} \frac{4}{5} \times 120 = \frac{4}{5} \times \frac{120}{1} = \frac{480}{5} = 96 \text{ rooms}$$

$$\text{Sunday:} \frac{3}{4} \times 120 = \frac{3}{4} \times \frac{120}{1} = \frac{360}{4} = 90 \text{ rooms}$$

Thus, more rooms were booked on Saturday by 6 rooms.

4. In a veterinary hospital, the veterinarian-to-pet ratio is 1:9. The ratio is always constant. If there are 45 pets in the hospital, how many veterinarians are currently in the veterinary hospital?

A proportion is set up to solve for the number of veterinarians: $\frac{1}{9} = \frac{x}{45}$

Cross-multiplying results in $9x = 45$, which works out to 5 veterinarians.

Alternatively, as there are always 9 times as many pets as veterinarians, it is possible to divide the number of pets (45) by 9. This also arrives at the correct answer of 5 veterinarians.

5. At a general practice law firm, 30% of the lawyers work solely on tort cases. If 9 lawyers work solely on tort cases, how many lawyers work at the firm?

The first step is to solve for the total number of lawyers working at the firm, which will be represented here with x. The problem states that 9 lawyers work solely on torts cases, and they make up 30% of the total lawyers at the firm. Thus, 30% multiplied by the total, x, will equal 9. Written as equation, this is: $30\% \times x = 9$.

It's easier to deal with the equation after converting the percentage to a decimal, leaving $0.3x = 9$. Thus, $x = \frac{9}{0.3} = 30$ lawyers working at the firm.

6. Xavier was hospitalized with pneumonia. He was originally given 35mg of antibiotics. Later, after his condition continued to worsen, Xavier's dosage was increased to 60mg. What was the percent increase of the antibiotics? Round the percentage to the nearest tenth.

An increase or decrease in percentage can be calculated by dividing the difference in amounts by the original amount and multiplying by 100. Written as an equation, the formula is:

$$\frac{new\ quantity - old\ quantity}{old\ quantity} \times 100$$

Here, the question states that the dosage was increased from 35mg to 60mg, so these values are plugged into the formula to find the percentage increase:

$$\frac{60 - 35}{35} \times 100 = \frac{25}{35} \times 100$$

$$0.7142 \times 100 = 71.4\%$$

Algebraic Properties

Solving Linear Equations

When asked to solve a linear equation, it requires determining a numerical value for the unknown variable. Given a linear equation involving addition, subtraction, multiplication, and division, isolation of the variable is done by working backward. Addition and subtraction are inverse operations, as are multiplication and division; therefore, they can be used to cancel each other out.

The first steps to solving linear equations are to distribute if necessary and combine any like terms that are on the same side of the equation. Sides of an equation are separated by an = sign. Next, the equation should be manipulated to get the variable on one side. Whatever is done to one side of an equation, must be done to the other side to remain equal. Then, the variable should be isolated by using inverse operations to undo the order of operations backward. Undo addition and subtraction, then undo multiplication and division. For example:

Solve $4(t - 2) + 2t - 4 = 2(9 - 2t)$

Distribute: $4t - 8 + 2t - 4 = 18 - 4t$

Combine like terms: $6t - 12 = 18 - 4t$

Add 4t to each side to move the variable: $10t - 12 = 18$

Add 12 to each side to isolate the variable: $10t = 30$

Divide each side by 10 to isolate the variable: $t = 3$

The answer can be checked by substituting the value for the variable into the original equation and ensuring both sides calculate to be equal.

Distributive Property

The **distributive property** states that multiplying a sum (or difference) by a number produces the same result as multiplying each value in the sum (or difference) by the number and adding (or subtracting) the products. Using mathematical symbols, the distributive property states $a(b + c) = ab + ac$. The expression $4(3 + 2)$ is simplified using the order of operations. Simplifying inside the parenthesis first produces 4×5, which equals 20. The expression $4(3 + 2)$ can also be simplified using the distributive property:

$$4(3 + 2)$$

$$4 \times 3 + 4 \times 2$$

$$12 + 8$$

$$20$$

Consider the following example: $4(3x - 2)$. The expression cannot be simplified inside the parenthesis because $3x$ and -2 are not like terms and therefore cannot be combined. However, the expression can be simplified by using the distributive property and multiplying each term inside of the parenthesis by the term outside of the parenthesis: $12x - 8$. The resulting equivalent expression contains no like terms, so it cannot be further simplified.

Consider the expression:

$$(3x + 2y + 1) - (5x - 3) + 2(3y + 4)$$

Again, there are no like terms, but the distributive property is used to simplify the expression. Note there is an implied one in front of the first set of parentheses and an implied -1 in front of the second set of parentheses. Distributing the 1, -1, and 2 produces:

$$1(3x) + 1(2y) + 1(1) - 1(5x) - 1(-3) + 2(3y) + 2(4)$$

$$3x + 2y + 1 - 5x + 3 + 6y + 8$$

This expression contains like terms that are combined to produce the simplified expression:

$$-2x + 8y + 12$$

Algebraic expressions are tested to be equivalent by choosing values for the variables and evaluating both expressions. For example, $4(3x - 2)$ and $12x - 8$ are tested by substituting 3 for the variable x and calculating to determine if equivalent values result.

FOIL Method

FOIL is a technique for generating polynomials through the multiplication of binomials. A **polynomial** is an expression of multiple variables (for example, x, y, z) in at least three terms involving only the four basic operations and exponents. FOIL is an acronym for First, Outer, Inner, and Last. "First" represents the multiplication of the terms appearing first in the binomials. "Outer" means multiplying the outermost terms. "Inner" means multiplying the terms inside. "Last" means multiplying the last terms of each binomial.

After completing FOIL and solving the operations, **like terms** are combined. To identify like terms, test takers should look for terms with the same variable and the same exponent. For example, in

$$4x^2 - x^2 + 15x + 2x^2 - 8$$

the $4x^2$, $-x^2$, and $2x^2$ are all like terms because they have the variable (x) and exponent (2).

Thus, after combining the like terms, the polynomial has been simplified to $5x^2 + 15x - 8$.

The purpose of FOIL is to simplify an equation involving multiple variables and operations. Although it sounds complicated, working through some examples will provide some clarity:

1. Simplify $(x + 10)(x + 4) =$

$$\underset{\text{First}}{(x \times x)} + \underset{\text{Outer}}{(x \times 4)} + \underset{\text{Inner}}{(10 \times x)} + \underset{\text{Last}}{(10 \times 4)}$$

After multiplying these binomials, it's time to solve the operations and combine like terms. Thus, the expression becomes: $x^2 + 4x + 10x + 40 = x^2 + 14x + 40$.

2. Simplify $2x(4x^3 - 7y^2 + 3x^2 + 4)$

Here, a monomial ($2x$) is multiplied into a polynomial ($4x^3 - 7y^2 + 3x^2 + 4$). Using the distributive property, the monomial gets multiplied by each term in the polynomial. This becomes $2x(4x^3) - 2x(7y^2) + 2x(3x^2) + 2x(4)$.

Now, each monomial is simplified, starting with the coefficients:

$$(2 \times 4)(x \times x^3) - (2 \times 7)(x \times y^2) + (2 \times 3)(x \times x^2) + (2 \times 4)(x)$$

When multiplying powers with the same base, their exponents are added. Remember, a variable with no listed exponent has an exponent of 1, and exponents of distinct variables cannot be combined. This produces the answer:

$$8x^{1+3} - 14xy^2 + 6x^{1+2} + 8x = 8x^4 - 14xy^2 + 6x^3 + 8x$$

3. Simplify $(8x^{10}y^2z^4) \div (4x^2y^4z^7)$

The first step is to divide the coefficients of the first two polynomials: $8 \div 4 = 2$. The second step is to divide exponents with the same variable, which requires subtracting the exponents. This results in:

$$2(x^{10-2}y^{2-4}z^{4-7}) = 2x^8y^{-2}z^{-3}$$

However, the most simplified answer should include only positive exponents. Thus, $y^{-2}z^{-3}$ needs to be converted into fractions, respectively $\frac{1}{y^2}$ and $\frac{1}{z^3}$. Since the $2x^8$ has a positive exponent, it is placed in the numerator, and $\frac{1}{y^2}$ and $\frac{1}{z^3}$ are combined into the denominator, leaving $\frac{2x^8}{y^2z^3}$ as the final answer.

Algebraic Expressions

Algebraic expressions look similar to equations, but they do not include the equal sign. Algebraic expressions are comprised of numbers, variables, and mathematical operations. Some examples of algebraic expressions are $8x + 7y - 12z$, $3a^2$, and $5x^3 - 4y^4$.

Algebraic expressions consist of variables, numbers, and operations. A **term** of an expression is any combination of numbers and/or variables, and terms are separated by addition and subtraction. For example, the expression $5x^2 - 3xy + 4 - 2$ consists of 4 terms: $5x^2$, -3xy, 4y, and -2. Note that each term includes its given sign (+ or −). The **variable** part of a term is a letter that represents an unknown quantity. The coefficient of a term is the number by which the variable is multiplied. For the term 4y, the variable is *y* and the coefficient is 4. Terms are identified by the power (or exponent) of its variable.

A number without a variable is referred to as a **constant**. If the variable is to the first power (x^1 or simply *x*), it is referred to as a **linear** term. A term with a variable to the second power (x^2) is **quadratic** and a term to the third power (x^3) is **cubic**. Consider the expression $x^3 + 3x - 1$. The constant is -1. The linear term is 3x. There is no quadratic term. The cubic term is x^3.

An algebraic expression can also be classified by how many terms exist in the expression. Any like terms should be combined before classifying. A **monomial** is an expression consisting of only one term. Examples of monomials are: 17, 2x, and $-5ab^2$. A **binomial** is an expression consisting of two terms separated by addition or subtraction. Examples include $2x - 4$ and $-3y^2 + 2y$. A **trinomial** consists of 3 terms. For example, $5x^2 - 2x + 1$ is a trinomial.

Algebraic expressions and equations can be used to represent real-life situations and model the behavior of different variables. For example, $2x + 5$ could represent the cost to play games at an arcade. In this case, 5 represents the price of admission to the arcade and 2 represents the cost of each game played. To calculate the total cost, use the number of games played for x, multiply it by 2, and add 5.

Adding and Subtracting Linear Algebraic Expressions

An algebraic expression is simplified by combining like terms. A term is a number, variable, or product of a number, and variables separated by addition and subtraction. For the algebraic expression $3x^2 - 4x + 5 - 5x^2 + x - 3$, the terms are $3x^2$, -4x, 5, $-5x^2$, x, and -3. **Like terms** have the same variables raised to the same powers (exponents). The like terms for the previous example are $3x^2$ and $-5x^2$, -4x and x, 5 and -3. To combine like terms, the coefficients (numerical factor of the term including sign) are added, and the variables and their powers are kept the same. Note that if a coefficient is not written, it is an implied coefficient of 1 ($x = 1x$). The previous example will simplify to $-2x^2 - 3x + 2$.

When adding or subtracting algebraic expressions, each expression is written in parenthesis. The negative sign is distributed when necessary, and like terms are combined. Consider the following:

$$\text{add } 2a + 5b - 2 \text{ to } a - 2b + 8c - 4$$

The sum is set as follows:

$$(a - 2b + 8c - 4) + (2a + 5b - 2)$$

In front of each set of parentheses is an implied positive one, which, when distributed, does not change any of the terms. Therefore, the parentheses are dropped and like terms are combined:

$$a - 2b + 8c - 4 + 2a + 5b - 2$$

$$3a + 3b + 8c - 6$$

Consider the following problem:

$$\text{Subtract } 2a + 5b - 2 \text{ from } a - 2b + 8c - 4$$

The difference is set as follows:

$$(a - 2b + 8c - 4) - (2a + 5b - 2)$$

The implied one in front of the first set of parentheses will not change those four terms. However, distributing the implied -1 in front of the second set of parentheses will change the sign of each of those three terms:

$$a - 2b + 8c - 4 - 2a - 5b + 2$$

Combining like terms yields the simplified expression:

$$-a - 7b + 8c - 2$$

Multiplying Algebraic Expressions

The **distributive property** states that multiplying a sum (or difference) by a number produces the same result as multiplying each value in the sum (or difference) by the number and adding (or subtracting) the products. Using mathematical symbols, the distributive property states $a(b + c) = ab + ac$. The expression $4(3 + 2)$ is simplified using the order of operations. Simplifying inside the parenthesis first produces 4×5, which equals 20. The expression $4(3 + 2)$ can also be simplified using the distributive property:

$$4(3 + 2)$$

$$4 \times 3 + 4 \times 2$$

$$12 + 8$$

$$20$$

Consider the following example: $4(3x - 2)$. The expression cannot be simplified inside the parenthesis because $3x$ and -2 are not like terms, and therefore cannot be combined. However, the expression can be simplified by using the distributive property and multiplying each term inside of the parenthesis by the term outside of the parenthesis: $12x - 8$. The resulting equivalent expression contains no like terms, so it cannot be further simplified.

Consider the expression:

$$(3x + 2y + 1) - (5x - 3) + 2(3y + 4)$$

Again, there are no like terms, but the distributive property is used to simplify the expression. Note there is an implied one in front of the first set of parentheses and an implied -1 in front of the second set of parentheses. Distributing the one, -1, and 2 produces:

$$1(3x) + 1(2y) + 1(1) - 1(5x) - 1(-3) + 2(3y) + 2(4)$$

$$3x + 2y + 1 - 5x + 3 + 6y + 8$$

This expression contains like terms that are combined to produce the simplified expression:

$$-2x + 8y + 12$$

Algebraic expressions are tested to be equivalent by choosing values for the variables and evaluating both expressions. For example, $4(3x - 2)$ and $12x - 8$ are tested by substituting 3 for the variable x and calculating to determine if equivalent values result.

Evaluating Algebraic Expressions

To evaluate the expression, the given values for the variables are substituted (or replaced) and the expression is simplified using the order of operations. Parenthesis should be used when substituting. Consider the following: Evaluate $a - 2b + ab$ for $a = 3$ and $b = -1$. To evaluate, any variable a is replaced with 3 and any variable b with -1, producing:

$$(3) - 2(-1) + (3)(-1)$$

Next, the order of operations is used to calculate the value of the expression, which is 2.

Here's another example:

$$\text{Evaluate } a - 2b + ab \text{ for } a = 3 \text{ and } b = -1$$

To evaluate an expression, the given values should be substituted for the variables and simplified using the order of operations. In this case:

$$(3) - 2(-1) + (3)(-1)$$

Parentheses are used when substituting.

Given an algebraic expression, students may be asked to simplify the expression. For example:

$$\text{Simplify } 5x^2 - 10x + 2 - 8x^2 + x - 1.$$

Simplifying algebraic expressions requires combining like terms. A term is a number, variable, or product of a number and variables separated by addition and subtraction. The terms in the above expression are: $5x^2, -10x, 2, -8x^2, x$, and -1. Like terms have the same variables raised to the same powers (exponents). To combine like terms, the coefficients (numerical factor of the term including sign) are added, while the variables and their powers are kept the same. The example above simplifies to:

$$-3x^2 - 9x + 1$$

Let's try two more.

Evaluate $\frac{1}{2}x^2 - 3, x = 4$.

The first step is to substitute in 4 for x in the expression:

$$\frac{1}{2}(4)^2 - 3$$

Then, the order of operations is used to simplify.

The exponent comes first, $\frac{1}{2}(16) - 3$, then the multiplication $8 - 3$, and then, after subtraction, the solution is 5.

Evaluate $4|5 - x| + 2y, x = 4, y = -3$.

The first step is to substitute 4 in for x and -3 in for y in the expression:

$$4|5 - 4| + 2(-3)$$

Then, the absolute value expression is simplified, which is:

$$|5 - 4| = |1| = 1$$

The expression is $4(1) + 2(-3)$ which can be simplified using the order of operations.

First is the multiplication, $4 + (-6)$; then addition yields an answer of -2.

Creating Algebraic Expressions

A linear expression is a statement about an unknown quantity expressed in mathematical symbols. The statement "five times a number added to forty" can be expressed as $5x + 40$. A linear equation is a statement in which two expressions (at least one containing a variable) are equal to each other. The statement "five times a number added to forty is equal to ten" can be expressed as $5x + 40 = 10$.

Real world scenarios can also be expressed mathematically. Suppose a job pays its employees $300 per week and $40 for each sale made. The weekly pay is represented by the expression $40x + 300$ where x is the number of sales made during the week.

Consider the following scenario: Bob had $20 and Tom had $4. After selling 4 ice cream cones to Bob, Tom has as much money as Bob. The cost of an ice cream cone is an unknown quantity and can be represented by a variable (x). The amount of money Bob has after his purchase is four times the cost of an ice cream cone subtracted from his original $20 \rightarrow 20 - 4x$. The amount of money Tom has after his sale is four times the cost of an ice cream cone added to his original $4 \rightarrow 4x + 4$. After the sale, the amount of money that Bob and Tom have are equal $\rightarrow 20 - 4x = 4x + 4$.

When expressing a verbal or written statement mathematically, it is key to understand words or phrases that can be represented with symbols. The following are examples:

Symbol	Phrase
$+$	added to, increased by, sum of, more than
$-$	decreased by, difference between, less than, take away
x	multiplied by, 3 (4, 5 . . .) times as large, product of
$\div$	divided by, quotient of, half (third, etc.) of
$=$	is, the same as, results in, as much as
$x, t, n, etc.$	a number, unknown quantity, value of

Using Structure to Isolate or Identify a Quantity of Interest

Formulas are mathematical expressions that define the value of one quantity given the value of one or more different quantities. A formula or equation expressed in terms of one variable can be manipulated

to express the relationship in terms of any other variable. The equation $y = 3x + 2$ is expressed in terms of the variable y. By manipulating the equation, it can be written as $x = \frac{y - 2}{3}$, which is expressed in terms of the variable x. To manipulate an equation or formula to solve for a variable of interest, how the equation would be solved if all other variables were numbers should be considered. The same steps for solving should be followed, leaving operations in terms of the variables, instead of calculating numerical values.

The formula $P = 2l + 2w$ expresses how to calculate the perimeter of a rectangle given its length and width. To write a formula to calculate the width of a rectangle given its length and perimeter, the previous formula relating the three variables should be used and the variable w should be solved. If P and l were numerical values, this would be a two-step linear equation solved by subtraction and division. To solve the equation $P = 2l + 2w$ for w, $2l$ should be subtracted from both sides: $P - 2l = 2w$. Then both sides should be divided by 2: $\frac{P - 2l}{2} = w$ or $\frac{P}{2} - l = w$.

The distance formula between two points on a coordinate plane can be found using the formula:

$$d = \sqrt{(x_2 - x_1)^2 + (y_2 - y_1)^2}$$

A problem might require determining the x-coordinate of one point (x_2), given its y-coordinate (y_2) and the distance (d) between that point and another given point (x_1, y_1). To do so, the above formula for x_1 should be solved just as a radical equation containing numerical values in place of the other variables. Both sides should be squared; the quantity should be subtracted $(y_2 - y_1)^2$; the square root of both sides should be taken; x_1 should be subtracted to produce:

$$\sqrt{d^2 - (y_2 - y_1)^2} + x_1 = x_2$$

Functions

A **function** is defined as a relationship between inputs and outputs where there is only one output value for a given input. As an example, the following function is in function notation: $f(x) = 3x - 4$. The $f(x)$ represents the output value for an input of x. If $x = 2$, the equation becomes:

$$f(2) = 3(2) - 4 = 6 - 4 = 2$$

The input of 2 yields an output of 2, forming the ordered pair $(2, 2)$. The following set of ordered pairs corresponds to the given function: $(2, 2), (0, -4), (-2, -10)$. The set of all possible inputs of a function is its **domain**, and all possible outputs is called the **range**. By definition, each member of the domain is paired with only one member of the range.

Functions can also be defined recursively. In this form, they are not defined explicitly in terms of variables. Instead, they are defined using previously-evaluated function outputs, starting with either $f(0)$ or $f(1)$. An example of a recursively-defined function is:

$$f(1) = 2, f(n) = 2f(n - 1) + 2n, n > 1$$

The domain of this function is the set of all integers.

Creating Functions Using Function Notation

A **function** is defined as a relationship between inputs and outputs where there is only one output value for a given input. As an example, the following function is in function notation:

$$f(x) = 3x - 4$$

The $f(x)$ represents the output value for an input of x. If $x = 2$, the equation becomes:

$$f(2) = 3(2) - 4 = 6 - 4 = 2$$

The input of 2 yields an output of 2, forming the ordered pair $(2, 2)$. The following set of ordered pairs corresponds to the given function:

$$(2, 2), (0, -4), (-2, -10)$$

The set of all possible inputs of a function is its **domain**, and all possible outputs is called the **range**. By definition, each member of the domain is paired with only one member of the range.

Functions can also be defined recursively. In this form, they are not defined explicitly in terms of variables. Instead, they are defined using previously-evaluated function outputs, starting with either $f(0)$ or $f(1)$. An example of a recursively-defined function is:

$$f(1) = 2, f(n) = 2f(n - 1) + 2n, n > 1$$

The domain of this function is the set of all integers.

A function $f(x)$ is a mathematical object which takes one number, x, as an input and gives a number in return. The input is called the **independent variable**. If the variable is set equal to the output, as in $y = f(x)$, then this is called the **dependent variable**. To indicate the dependent value a function, y, gives for a specific independent variable, x, the notation y = $f(x)$ is used.

The **domain** of a function is the set of values that the independent variable is allowed to take. Unless otherwise specified, the domain is any value for which the function is well defined. The **range** of the function is the set of possible outputs for the function.

In many cases, a function can be defined by giving an equation. For instance, $f(x) = x^2$ indicates that given a value for x, the output of f is found by squaring x.

Not all equations in x and y can be written in the form $y = f(x)$. An equation can be written in such a form if it satisfies the **vertical line test**: no vertical line meets the graph of the equation at more than a single point. In this case, y is said to be a *function of x*. If a vertical line meets the graph in two places, then this equation cannot be written in the form $y = f(x)$.

The graph of a function $f(x)$ is the graph of the equation $y = f(x)$. Thus, it is the set of all pairs (x, y) where $y = f(x)$. In other words, it is all pairs $(x, f(x))$. The x-intercepts are called the **zeros** of the function. The y-intercept is given by $f(0)$.

If, for a given function f, the only way to get $f(a) = f(b)$ is for $a = b$, then f is *one-to-one*. Often, even if a function is not one-to-one on its entire domain, it is one-to-one by considering a restricted portion of the domain.

A function $f(x) = k$ for some number k is called a **constant function**. The graph of a constant function is a horizontal line.

The function $f(x) = x$ is called the **identity function**. The graph of the identity function is the diagonal line pointing to the upper right at 45 degrees, $y = x$.

A function is called **monotone** if it is either always increasing or always decreasing. For example, the functions $f(x) = 3x$ and $f(x) = -x^5$ are monotone.

An **even function** looks the same when flipped over the y-axis: $f(x) = f(-x)$. The following image shows a graphic representation of an even function.

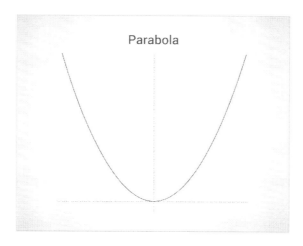
Parabola

An **odd function** looks the same when flipped over the y-axis and then flipped over the x-axis: $f(x) = -f(-x)$. The following image shows an example of an odd function.

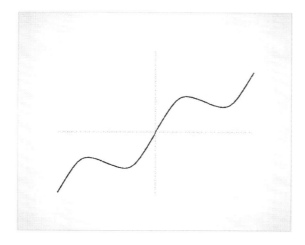

Domain and Range

The domain and range of a function can be found visually by its plot on the coordinate plane. In the function $f(x) = x^2 - 3$, for example, the domain is all real numbers because the parabola stretches as far left and as far right as it can go, with no restrictions. This means that any input value from the real number system will yield an answer in the real number system. For the range, the inequality $y \geq -3$ would be used to describe the possible output values because the parabola has a minimum at $y = -3$.

This means there will not be any real output values less than -3 because -3 is the lowest value it reaches on the y-axis.

These same answers for domain and range can be found by observing a table. The table below shows that from input values $x = -1$ to $x = 1$, the output results in a minimum of -3. On each side of $x = 0$, the numbers increase, showing that the range is all real numbers greater than or equal to -3.

x (domain/input)	y (range/output)
-2	1
-1	-2
0	-3
-1	-2
2	1

Function Behavior

Different types of functions behave in different ways. A function is defined to be increasing over a subset of its domain if for all $x_1 \geq x_2$ in that interval, $f(x_1) \geq f(x_2)$. Also, a function is decreasing over an interval if for all $x_1 \geq x_2$ in that interval, $f(x_1) \leq f(x_2)$. A point in which a function changes from increasing to decreasing can also be labeled as the **maximum value** of a function if it is the largest point the graph reaches on the y-axis. A point in which a function changes from decreasing to increasing can be labeled as the minimum value of a function if it is the smallest point the graph reaches on the y-axis. Maximum values are also known as **extreme values**. The graph of a continuous function does not have any breaks or jumps in the graph. This description is not true of all functions. A radical function, for example, $f(x) = \sqrt{x}$, has a restriction for the domain and range because there are no real negative inputs or outputs for this function. The domain can be stated as $x \geq 0$, and the range is $y \geq 0$.

A piecewise-defined function also has a different appearance on the graph. In the following function, there are three equations defined over different intervals. It is a function because there is only one y-value for each x-value, passing the Vertical Line Test. The domain is all real numbers less than or equal to 6. The range is all real numbers greater than zero. From left to right, the graph decreases to zero, then increases to almost 4, and then jumps to 6.

From input values greater than 2, the input decreases just below 8 to 4, and then stops.

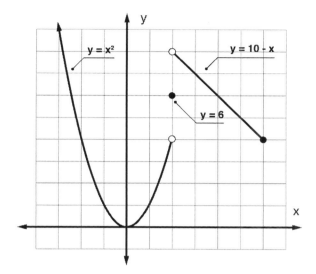

Logarithmic and exponential functions also have different behavior than other functions. These two types of functions are inverses of each other. The **inverse** of a function can be found by switching the place of x and y, and solving for y. When this is done for the exponential equation, $y = 2^x$, the function $y = \log_2 x$ is found. The general form of a **logarithmic function** is $y = \log_b x$, which says b raised to the y power equals x.

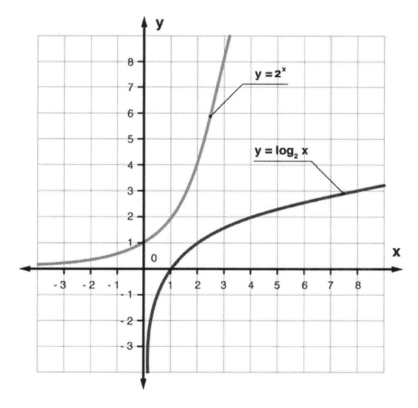

The thick black line on the graph above represents the logarithmic function:

$$y = \log_2 x$$

This curve passes through the point $(1, 0)$, just as all log functions do, because any value $b^0 = 1$. The graph of this logarithmic function starts very close to zero but does not touch the y-axis. The output value will never be zero by the definition of logarithms. The thinner gray line seen above represents the exponential function $y = 2^x$. The behavior of this function is opposite the logarithmic function because the graph of an inverse function is the graph of the original function flipped over the line $y = x$. The curve passes through the point $(0, 1)$ because any number raised to the zero power is one. This curve also gets very close to the x-axis but never touches it because an exponential expression never has an output of zero. The x-axis on this graph is called a horizontal asymptote. An **asymptote** is a line that represents a boundary for a function. It shows a value that the function will get close to, but never reach.

Functions can also be described as being even, odd, or neither. If $f(-x) = f(x)$, the function is even. For example, the function $f(x) = x^2 - 2$ is even. Plugging in $x = 2$ yields an output of $y = 2$. After changing the input to $x = -2$, the output is still $y = 2$. The output is the same for opposite inputs. Another way to observe an even function is by the symmetry of the graph. If the graph is symmetrical about the axis, then the function is even. If the graph is symmetric about the origin, then the function is odd. Algebraically, if $f(-x) = -f(x)$, the function is odd.

Also, a function can be described as **periodic** if it repeats itself in regular intervals. Common periodic functions are trigonometric functions. For example, $y = \sin x$ is a periodic function with period 2π because it repeats itself every 2π units along the x-axis.

Common Functions

Three common functions used to model different relationships between quantities are linear, quadratic, and exponential functions. **Linear functions** are the simplest of the three, and the independent variable x has an exponent of 1. Written in the most common form, $y = mx + b$, the coefficient of x tells how fast the function grows at a constant rate, and the b-value tells the starting point. A **quadratic** function has an exponent of 2 on the independent variable x. Standard form for this type of function is $y = ax^2 + bx + c$, and the graph is a parabola. These type functions grow at a changing rate. An **exponential** function has an independent variable in the exponent $y = ab^x$. The graph of these types of functions is described as **growth** or **decay**, based on whether the base, b, is greater than or less than 1. These functions are different from quadratic functions because the base stays constant. A common base is base e.

The following three functions model a linear, quadratic, and exponential function respectively: $y = 2x$, $y = x^2$, and $y = 2^x$. Their graphs are shown below. The first graph, modeling the linear function, shows that the growth is constant over each interval. With a horizontal change of 1, the vertical change is 2. It models a constant positive growth. The second graph shows the quadratic function, which is a curve that is symmetric across the y-axis. The growth is not constant, but the change is mirrored over the axis. The last graph models the exponential function, where the horizontal change of 1 yields a vertical change that increases more and more. The exponential graph gets very close to the x-axis, but never

touches it, meaning there is an asymptote there. The y-value can never be zero because the base of 2 can never be raised to an input value that yields an output of zero.

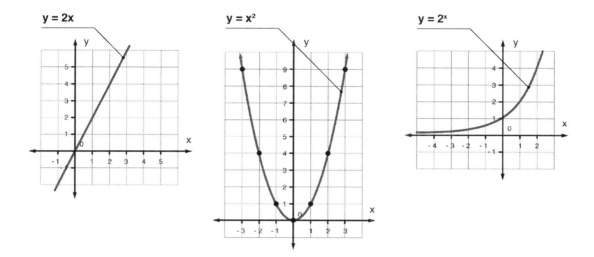

The three tables below show specific values for three types of functions. The third column in each table shows the change in the y-values for each interval. The first table shows a constant change of 2 for each equal interval, which matches the slope in the equation $y = 2x$. The second table shows an increasing change, but it also has a pattern. The increase is changing by 2 more each time, so the change is quadratic. The third table shows the change as factors of the base, 2. It shows a continuing pattern of factors of the base.

y = 2x		
x	y	Δy
1	2	
2	4	2
3	6	2
4	8	2
5	10	2

y = x²		
x	y	Δy
1	1	
2	4	3
3	9	5
4	16	7
5	25	9

y = 2ˣ		
x	y	Δy
1	2	
2	4	2
3	8	4
4	16	8
5	32	16

Given a table of values, the type of function can be determined by observing the change in y over equal intervals. For example, the tables below model two functions. The changes in interval for the x-values is 1 for both tables. For the first table, the y-values increase by 5 for each interval. Since the change is constant, the situation can be described as a linear function. The equation would be $y = 5x + 3$. For the second table, the change for y is 5, 20, 100, and 500, respectively. The increases are multiples of 5, meaning the situation can be modeled by an exponential function. The equation $y = 5^x + 3$ models this situation.

x	y
0	3
1	8
2	13
3	18
4	23

x	y
0	3
1	8
2	28
3	128
4	628

Quadratic equations can be used to model real-world area problems. For example, a farmer may have a rectangular field that he needs to sow with seed. The field has length $x + 8$ and width $2x$. The formula for area should be used: $A = lw$. Therefore:

$$A = (x + 8) \times 2x = 2x^2 + 16x$$

The possible values for the length and width can be shown in a table, with input x and output A. If the equation was graphed, the possible area values can be seen on the y-axis for given x-values.

Exponential growth and decay can be found in real-world situations. For example, if a piece of notebook paper is folded 25 times, the thickness of the paper can be found. To model this situation, a table can be used. The initial point is one-fold, which yields a thickness of 2 papers. For the second fold, the thickness is 4. Since the thickness doubles each time, the table below shows the thickness for the next few folds. Notice the thickness changes by the same factor each time. Since this change for a constant interval of folds is a factor of 2, the function is exponential. The equation for this is $y = 2^x$. For twenty-five folds, the thickness would be 33,554,432 papers.

x (folds)	y (paper thickness)
0	1
1	2
2	4
3	8
4	16
5	32

One exponential formula that is commonly used is the **interest formula**: $A = Pe^{rt}$. In this formula, interest is compounded continuously. A is the value of the investment after the time, t, in years. P is the initial amount of the investment, r is the interest rate, and e is the constant equal to approximately 2.718. Given an initial amount of $200 and a time of 3 years, if interest is compounded continuously at a rate of 6%, the total investment value can be found by plugging each value into the formula. The invested value at the end is $239.44. In more complex problems, the final investment may be given, and the rate may be the unknown. In this case, the formula becomes $239.44 = 200e^{r3}$. Solving for r requires isolating the exponential expression on one side by dividing by 200, yielding the equation

$1.20 = e^{r3}$. Taking the natural log of both sides results in $\ln(1.2) = r3$. Using a calculator to evaluate the logarithmic expression, $r = 0.06 = 6\%$.

When working with logarithms and exponential expressions, it is important to remember the relationship between the two. In general, the logarithmic form is $y = log_b x$ for an exponential form $b^y = x$. Logarithms and exponential functions are inverses of each other.

Interpreting and Building a Function Within a Context

Functions can be built out of the context of a situation. For example, the relationship between the money paid for a gym membership and the months that someone has been a member can be described through a function. If the one-time membership fee is $40 and the monthly fee is $30, then the function can be written $f(x) = 30x + 40$. The x-value represents the number of months the person has been part of the gym, while the output is the total money paid for the membership. The table below shows this relationship. It is a representation of the function because the initial cost is $40 and the cost increases each month by $30.

x (months)	y (money paid to gym)
0	40
1	70
2	100
3	130

Functions can also be built from existing functions. For example, a given function $f(x)$ can be transformed by adding a constant, multiplying by a constant, or changing the input value by a constant. The new function $g(x) = f(x) + k$ represents a vertical shift of the original function. In $f(x) = 3x - 2$, a vertical shift 4 units up would be:

$$g(x) = 3x - 2 + 4 = 3x + 2$$

Multiplying the function times a constant k represents a vertical stretch, based on whether the constant is greater than or less than 1. The function

$$g(x) = kf(x) = 4(3x - 2) = 12x - 8$$

represents a stretch. Changing the input x by a constant forms the function:

$$g(x) = f(x + k) = 3(x + 4) - 2 = 3x + 12 - 2 = 3x + 10$$

and this represents a horizontal shift to the left 4 units. If $(x - 4)$ was plugged into the function, it would represent a vertical shift.

A composition function can also be formed by plugging one function into another. In function notation, this is written:

$$(f \circ g)(x) = f(g(x))$$

For two functions $f(x) = x^2$ and $g(x) = x - 3$, the composition function becomes:

$$f(g(x)) = (x - 3)^2 = x^2 - 6x + 9$$

The composition of functions can also be used to verify if two functions are inverses of each other. Given the two functions $f(x) = 2x + 5$ and $g(x) = \frac{x-5}{2}$, the composition function can be found $(f \circ g)(x)$. Solving this equation yields:

$$f(g(x)) = 2\left(\frac{x-5}{2}\right) + 5 = x - 5 + 5 = x$$

It also is true that $g(f(x)) = x$. Since the composition of these two functions gives a simplified answer of x, this verifies that $f(x)$ and $g(x)$ are inverse functions. The domain of $f(g(x))$ is the set of all x-values in the domain of $g(x)$ such that $g(x)$ is in the domain of $f(x)$. Basically, both $f(g(x))$ and $g(x)$ have to be defined.

To build an inverse of a function, $f(x)$ needs to be replaced with y, and the x and y values need to be switched. Then, the equation can be solved for y. For example, given the equation $y = e^{2x}$, the inverse can be found by rewriting the equation $x = e^{2y}$. The natural logarithm of both sides is taken down, and the exponent is brought down to form the equation:

$$\ln(x) = \ln(e)\, 2y$$

ln (e)=1, which yields the equation $\ln(x) = 2y$. Dividing both sides by 2 yields the inverse equation

$$\frac{\ln(x)}{2} = y = f^{-1}(x)$$

The domain of an inverse function is the range of the original function, and the range of an inverse function is the domain of the original function. Therefore, an ordered pair (x, y) on either a graph or a table corresponding to $f(x)$ means that the ordered pair (y, x) exists on the graph of $f^{-1}(x)$. Basically, if $f(x) = y$, then $f^{-1}(y) = x$. For a function to have an inverse, it must be one-to-one. That means it must pass the **Horizontal Line Test**, and if any horizontal line passes through the graph of the function twice, a function is not one-to-one. The domain of a function that is not one-to-one can be restricted to an interval in which the function is one-to-one, to be able to define an inverse function.

Functions can also be formed from combinations of existing functions.

Given $f(x)$ and $g(x)$, the following can be built:

$$f + g$$

$$f - g$$

$$fg$$

$$\frac{f}{g}$$

The domains of $f + g, f - g,$ and fg are the intersection of the domains of f and g. The domain of $\frac{f}{g}$ is the same set, excluding those values that make $g(x) = 0$.

For example, if:

$$f(x) = 2x + 3$$

$$g(x) = x + 1$$

then

$$\frac{f}{g} = \frac{2x + 3}{x + 1}$$

Its domain is all real numbers except -1.

Algebraic Functions

A function is called **algebraic** if it is built up from polynomials by adding, subtracting, multiplying, dividing, and taking radicals. This means that, for example, the variable can never appear in an exponent. Thus, polynomials and rational functions are algebraic, but exponential functions are not algebraic. It turns out that logarithms and trigonometric functions are not algebraic either.

A function of the form:

$$f(x) = a_n x^n + a_{n-1} x^{n-1} + a_{n-2} x^{n-2} + \cdots + a_1 x + a_0$$

is called a **polynomial function**. The value of n is called the **degree** of the polynomial. In the case where $n = 1$, it is called a **linear function**. In the case where $n = 2$, it is called a **quadratic function**. In the case where $n = 3$, it is called a **cubic function**.

When n is even, the polynomial is called **even**, and not all real numbers will be in its range. When n is odd, the polynomial is called **odd**, and the range includes all real numbers.

The graph of a quadratic function $f(x) = ax^2 + bx + c$ will be a **parabola**. To see whether or not the parabola opens up or down, it's necessary to check the coefficient of x^2, which is the value a. If the coefficient is positive, then the parabola opens upward. If the coefficient is negative, then the parabola opens downward.

The quantity $D = b^2 - 4ac$ is called the **discriminant** of the parabola. If the discriminant is positive, then the parabola has two real zeros. If the discriminant is negative, then it has no real zeros. If the discriminant is zero, then the parabola's highest or lowest point is on the x-axis, and it has a single real zero.

The highest or lowest point of the parabola is called the **vertex**. The coordinates of the vertex are given by the point $(-\frac{b}{2a}, -\frac{D}{4a})$. The roots of a quadratic function can be found with the quadratic formula, which is:

$$x = \frac{-b \pm \sqrt{b^2 - 4ac}}{2a}$$

A **rational function** is a function $f(x) = \frac{p(x)}{q(x)}$, where p and q are both polynomials. The domain of f will be all real numbers except the (real) roots of q. At these roots, the graph of f will have a **vertical asymptote,** unless they are also roots of p. Here is an example to consider:

$$p(x) = p_n x^n + p_{n-1} x^{n-1} + p_{n-2} x^{n-2} + \cdots + p_1 x + p_0$$

$$q(x) = q_m x^m + q_{m-1} x^{m-1} + q_{m-2} x^{m-2} + \cdots + q_1 x + q_0$$

When the degree of p is less than the degree of q, there will be a **horizontal asymptote** of $y = 0$. If p and q have the same degree, there will be a horizontal asymptote of $y = \frac{p_n}{q_n}$. If the degree of p is exactly one greater than the degree of q, then f will have an oblique asymptote along the line:

$$y = \frac{p_n}{q_{n-1}} x + \frac{p_{n-1}}{q_{n-1}}$$

Exponential Functions

An **exponential function** is a function of the form $f(x) = b^x$, where b is a positive real number other than 1. In such a function, b is called the **base**.

The **domain** of an exponential function is all real numbers, and the **range** is all positive real numbers. There will always be a horizontal asymptote of $y = 0$ on one side. If b is greater than 1, then the graph will be increasing when moving to the right. If b is less than 1, then the graph will be decreasing when moving to the right. Exponential functions are one-to-one. The basic exponential function graph will go through the point (0, 1).

The following example demonstartes this more clearly:

Solve $5^{x+1} = 25$.

The first step is to get the x out of the exponent by rewriting the equation $5^{x+1} = 5^2$ so that both sides have a base of 5. Since the bases are the same, the exponents must be equal to each other. This leaves $x + 1 = 2$ or $x = 1$. To check the answer, the x-value of 1 can be substituted back into the original equation.

Logarithmic Functions

A **logarithmic function** is an inverse for an exponential function. The inverse of the base b exponential function is written as $\log_b(x)$, and is called the **base b logarithm**. The domain of a logarithm is all positive real numbers. It has the properties that $\log_b(b^x) = x$. For positive real values of x:

$$b^{\log_b(x)} = x$$

When there is no chance of confusion, the parentheses are sometimes skipped for logarithmic functions: $\log_b(x)$ may be written as $\log_b x$. For the special number e, the base e logarithm is called the **natural logarithm** and is written as $\ln x$. Logarithms are one-to-one.

When working with logarithmic functions, it is important to remember the following properties. Each one can be derived from the definition of the logarithm as the inverse to an exponential function:

- $\log_b 1 = 0$
- $\log_b b = 1$
- $\log_b b^p = p$
- $\log_b MN = \log_b M + \log_b N$
- $\log_b \frac{M}{N} = \log_b M - \log_b N$
- $\log_b M^p = p \log_b M$

When solving equations involving exponentials and logarithms, the following fact should be used:

If f is a one-to-one function, $a = b$ is equivalent to $f(a) = f(b)$.

Using this, together with the fact that logarithms and exponentials are inverses, allows for manipulations of the equations to isolate the variable as is demonstrated in the following example:

Solve $4 = \ln(x - 4)$.

Using the definition of a logarithm, the equation can be changed to $e^4 = e^{\ln(x-4)}$. The functions on the right side cancel with a result of $e^4 = x - 4$. This then gives $x = 4 + e^4$.

Trigonometric Functions

Trigonometric functions are built out of two basic functions, the **sine** and **cosine**, written as $\sin\theta$ and $\cos\theta$, respectively. Note that similar to logarithms, it is customary to drop the parentheses as long as the result is not confusing.

Sine and cosine are defined using the **unit circle**. If θ is the angle going counterclockwise around the origin from the x-axis, then the point on the unit circle in that direction will have the coordinates ($\cos\theta$, $\sin\theta$).

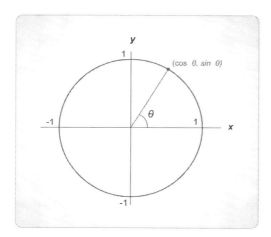

Since the angle returns to the start every 2π radians (or 360 degrees), the graph of these functions is **periodic**, with period 2π. This means that the graph repeats itself as one moves along the x-axis because $\sin\theta = \sin(\theta + 2\pi)$. Cosine works similarly.

From the unit circle definition, the sine function starts at 0 when $\theta = 0$. It grows to 1 as θ grows to $\pi/2$, and then back to 0 at $\theta = \pi$. Then it decreases to -1 as θ grows to $3\pi/2$, and goes back up to 0 at $\theta = 2\pi$.

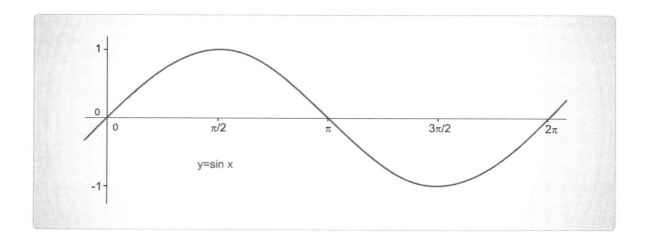

The graph of the cosine is similar. The cosine graph will start at 1, decreasing to 0 at $\pi/2$ and continuing to decrease to -1 at $\theta = \pi$. Then, it grows to 0 as θ grows to $3\pi/2$ and back up to 1 at $\theta = 2\pi$.

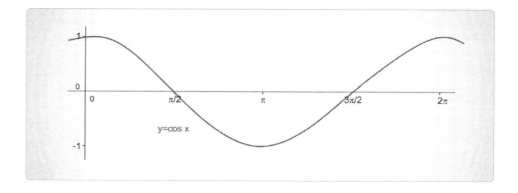

Another trigonometric function that is frequently used, is the **tangent** function. This is defined as the following equation: $\tan \theta = \frac{\sin \theta}{\cos \theta}$.

The tangent function is a period of π rather than 2π because the sine and cosine functions have the same absolute values after a change in the angle of π, but they flip their signs. Since the tangent is a ratio of the two functions, the changes in signs cancel.

The tangent function will be zero when sine is zero, and it will have a vertical asymptote whenever cosine is zero. The following graph shows the tangent function:

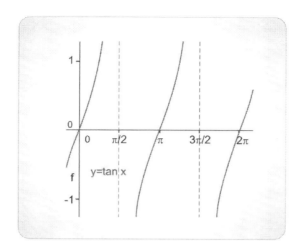

Three other trigonometric functions are sometimes useful. These are the **reciprocal** trigonometric functions, so named because they are just the reciprocals of sine, cosine, and tangent. They are the **cosecant**, defined as $\csc \theta = \frac{1}{\sin \theta}$, the **secant**, $\sec \theta = \frac{1}{\cos \theta}$, and the **cotangent**, $\cot \theta = \frac{1}{\tan \theta}$. Note that from the definition of tangent, $\cot \theta = \frac{\cos \theta}{\sin \theta}$.

In addition, there are three identities that relate the trigonometric functions to one another:

- $\cos \theta = \sin(\frac{\pi}{2} - \theta)$
- $\csc \theta = \sec\left(\frac{\pi}{2} - \theta\right)$
- $\cot \theta = \tan(\frac{\pi}{2} - \theta)$

Here is a list of commonly-needed values for trigonometric functions, given in radians, for the first quadrant:

Table for trigonometric functions

$\sin 0 = 0$	$\cos 0 = 1$	$\tan 0 = 0$
$\sin\dfrac{\pi}{6} = \dfrac{1}{2}$	$\cos\dfrac{\pi}{6} = \dfrac{\sqrt{3}}{2}$	$\tan\dfrac{\pi}{6} = \dfrac{\sqrt{3}}{3}$
$\sin\dfrac{\pi}{4} = \dfrac{\sqrt{2}}{2}$	$\cos\dfrac{\pi}{4} = \dfrac{\sqrt{2}}{2}$	$\tan\dfrac{\pi}{4} = 1$
$\sin\dfrac{\pi}{3} = \dfrac{\sqrt{3}}{2}$	$\cos\dfrac{\pi}{3} = \dfrac{1}{2}$	$\tan\dfrac{\pi}{3} = \sqrt{3}$
$\sin\dfrac{\pi}{2} = 1$	$\cos\dfrac{\pi}{2} = 0$	$\tan\dfrac{\pi}{2} = undefined$
$\csc 0 = undefined$	$\sec 0 = 1$	$\cot 0 = undefined$
$\csc\dfrac{\pi}{6} = 2$	$\sec\dfrac{\pi}{6} = \dfrac{2\sqrt{3}}{3}$	$\cot\dfrac{\pi}{6} = \sqrt{3}$
$\csc\dfrac{\pi}{4} = \sqrt{2}$	$\sec\dfrac{\pi}{4} = \sqrt{2}$	$\cot\dfrac{\pi}{4} = 1$
$\csc\dfrac{\pi}{3} = \dfrac{2\sqrt{3}}{3}$	$\sec\dfrac{\pi}{3} = 2$	$\cot\dfrac{\pi}{3} = \dfrac{\sqrt{3}}{3}$
$\csc\dfrac{\pi}{2} = 1$	$\sec\dfrac{\pi}{2} = undefined$	$\cot\dfrac{\pi}{2} = 0$

To find the trigonometric values in other quadrants, complementary angles can be used. The **complementary angle** is the smallest angle between the x-axis and the given angle.

Once the complementary angle is known, the following rule is used:

For an angle θ with complementary angle x, the absolute value of a trigonometric function evaluated at θ is the same as the absolute value when evaluated at x.

The correct sign for sine and cosine is determined by the x and y coordinates on the unit circle.

- Sine will be positive in quadrants I and II and negative in quadrants III and IV.
- Cosine will be positive in quadrants I and IV, and negative in II and III.
- Tangent will be positive in I and III, and negative in II and IV.

The signs of the reciprocal functions will be the same as the sign of the function of which they are the reciprocal. For example:

Find $\sin \frac{3\pi}{4}$.

The reference angle must be found first. This angle is in the II quadrant, and the angle between it and the x-axis is $\frac{\pi}{4}$. Now, $\sin \frac{\pi}{4} = \frac{\sqrt{2}}{2}$. Since this is in the II quadrant, sine takes on positive values (the y coordinate is positive in the II quadrant). Therefore, $\sin \frac{3\pi}{4} = \frac{\sqrt{2}}{2}$.

In addition to the six trigonometric functions defined above, there are inverses for these functions. However, since the trigonometric functions are not one-to-one, one can only construct inverses for them on a restricted domain.

Usually, the domain chosen will be $[0, \pi)$ for cosine and $(-\frac{\pi}{2}, \frac{\pi}{2}]$ for sine. The inverse for tangent can use either of these domains. The inverse functions for the trigonometric functions are also called **arc functions.** In addition to being written with a -1 as the exponent to denote that the function is an inverse, they will sometimes be written with an "a" or "arc" in front of the function name, so $\cos^{-1} \theta = a\cos \theta = \arccos \theta$.

When solving equations that involve trigonometric functions, there are often multiple solutions. For example, $2 \sin \theta = \sqrt{2}$ can be simplified to $\sin \theta = \frac{\sqrt{2}}{2}$. This has solutions $\theta = \frac{\pi}{4}, \frac{3\pi}{4}$, but in addition, because of the periodicity, any integer multiple of 2π can also be added to these solutions to find another solution.

The full set of solutions is $\theta = \frac{\pi}{4} + 2\pi k, \frac{3\pi}{4} + 2\pi k$ for all integer values of k. It is very important to remember to find all possible solutions when dealing with equations that involve trigonometric functions.

The name *trigonometric* comes from the fact that these functions play an important role in the geometry of triangles, particularly right triangles. Consider the right triangle shown in this figure:

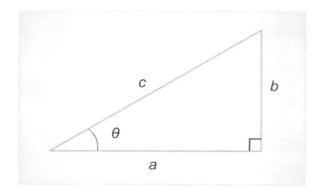

The following hold true:

- $c \sin \theta = b$.
- $c \cos \theta = a$.
- $\tan \theta = \frac{b}{a}$.

- $b \csc \theta = c$.
- $a \sec \theta = c$.
- $\cot \theta = \frac{a}{b}$.

It is important to remember that the angles of a triangle must add up to π radians (180 degrees).

Systems of Equations

A system of two linear equations in two variables is a set of equations that use the same variables, usually x and y. Here's a sample problem:

> An Internet provider charges an installation fee and a monthly charge. It advertises that two months of its offering costs $100 and six months costs $200. Find the monthly charge and the installation fee.

The two unknown quantities (variables) are the monthly charge and the installation fee. There are two different statements given relating the variables: two months added to the installation fee is $100; and six months added to the installation fee is $200. Using the variable x as the monthly charge and y as the installation fee, the statements can be written as the following: $2x + y = 100$; $6x + y = 200$. These two equations taken together form a system modeling the given scenario.

A system of linear inequalities consists of two linear inequalities making comparisons between two variables. Students may be given a scenario and asked to express it as a system of inequalities:

> A consumer study calls for at least 60 adult participants. It cannot use more than 25 men. Express these constraints as a system of inequalities.

This can be modeled by the system: $x + y \geq 60$; $x \leq 25$, where x represents the number of men and y represents the number of women. A solution to the system is an ordered pair that makes both inequalities true when substituting the values for x and y.

Algebraically Solving Linear Equations (or Inequalities) in One Variable

Linear equations in one variable and linear inequalities in one variable can be solved following similar processes. Although they typically have one solution, a linear equation can have no solution or can have a solution set of all real numbers. Solution sets for linear inequalities typically consist of an infinite number of values either greater or less than a given value (where the given value may or may not be included in the set). However, a linear inequality can have no solution or can have a solution set consisting of all real numbers.

Linear Equations in One Variable – Special Cases

Solving a linear equation produces a value for the variable that makes the algebraic statement true. If there is no value for the variable that would make the statement true, there is no solution to the equation. Here's a sample equation: $x + 3 = x - 1$. There is no value for x in which adding 3 to the value would produce the same result as subtracting 1 from that value. Conversely, if any value for the variable would make a true statement, the equation has an infinite number of solutions. Here's another sample equation: $3x + 6 = 3(x + 2)$. Any real number substituted for x would result in a true statement (both sides of the equation are equal).

By manipulating equations similar to the two above, the variable of the equation will cancel out completely. If the constants that are left express a true statement (ex., $6 = 6$), then all real numbers are

solutions to the equation. If the constants left express a false statement (ex., $3 = -1$), then there is no solution to the equation.

A question on this material may present a linear equation with an unknown value for either a constant or a coefficient of the variable and ask to determine the value that produces an equation with no solution or infinite solutions. For example:

$3x + 7 = 3x + 10 + n$; Find the value of n that would create an equation with an infinite number of solutions for the variable x.

To solve this problem, the equation should be manipulated so the variable x will cancel. To do this, $3x$ should be subtracted from both sides, which would leave $7 = 10 + n$. By subtracting 10 on both sides, it is determined that $n = -3$. Therefore, a value of -3 for n would result in an equation with a solution set of all real numbers.

If the same problem asked for the equation to have no solution, the value of n would be all real numbers except -3.

Linear Inequalities in One Variable – Special Cases

A linear inequality can have a solution set consisting of all real numbers or can contain no solution. When solved algebraically, a linear inequality in which the variable cancels out and results in a true statement (ex., $7 \geq 2$) has a solution set of all real numbers. A linear inequality in which the variable cancels out and results in a false statement (ex., $7 \leq 2$) has no solution.

Compound Inequalities

A compound inequality is a pair of inequalities joined by *and* or *or*. Given a compound inequality, to determine its solution set, both inequalities should be solved for the given variable. The solution set for a compound inequality containing *and* consists of all the values for the variable that make both inequalities true. If solving the compound inequality results in $x > -9$ and $x \leq 6$, the solution set would consist of all values between -2 and 3, including 3. This may also be written as follows: $-9 < x \leq 6$. Due to the graphs of their solution sets (shown below), compound inequalities such as these are referred to as conjunctions.

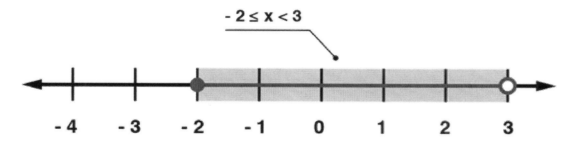

If there are no values that would make both inequalities of a compound inequality containing *and* true, then there is no solution. An example would be $x > 2$ and $x \leq 0$.

The solution set for a compound inequality containing *or* consists of all the values for the variable that make at least one of the inequalities true. The solution set for the compound inequality $x \leq -2$ or $x > 1$ consists of all values less than 2, 6, and all values greater than 6. Due to the graphs of their solution sets (shown below), compound inequalities such as these are referred to as disjunctions.

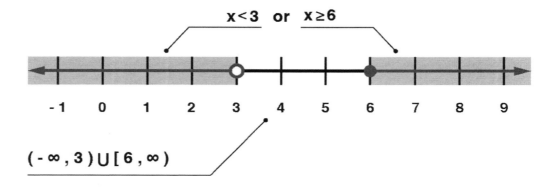

If the two inequalities for a compound inequality containing *or* "overlap," then the solution set contains all real numbers. An example would be $x > 2$ or $x < 7$. Any number would make at least one of these true.

Solving Systems of Two Linear Equations in Two Variables

A system of two linear equations in two variables is a set of equations that use the same variables (typically *x* and *y*). A solution to the system is an ordered pair that makes both equations true. One method for solving a system is by graphing. This method, however, is not always practical. Students may not have graph paper; or the solution may not consist of integers, making it difficult to identify the exact point of intersection on a graph. There are two methods for solving systems of equations algebraically: substitution and elimination. The method used will depend on the characteristics of the equations in the system.

Solving Systems of Equations with the Substitution Method

If one of the equations in a system has an isolated variable (*x*= or *y*=) or a variable that can be easily isolated, the substitution method can be used. Here's a sample system: $x + 3y = 7; 2x - 4y = 24$. The first equation can easily be solved for *x*. By subtracting $3y$ on both sides, the resulting equation is $x = 7 - 3y$. When one equation is solved for a variable, the expression that it is equal can be substituted into the other equation. For this example, $(7 - 3y)$ would be substituted for *x* into the second equation as follows: $2(7 - 3y) + 4y = 24$. Solving this equation results in $y = -5$. Once the value for one variable is known, this value should be substituted into either of the original equations to determine the value of the other variable. For the example, -5 would be substituted for *y* in either of the original equations. Substituting into the first equation results in $x + 3(-5) = 7$, and solving this equation yields $x = 22$. The solution to a system is an ordered pair, so the solution to the example is written as (22, 7). The solution can be checked by substituting it into both equations of the system to ensure it results in two true statements.

Solving Systems of Equations with the Elimination Method

The elimination method for solving a system of equations involves canceling out (or eliminating) one of the variables. This method is typically used when both equations of a system are written in standard form ($Ax + By = C$). An example is $2x + 3y = 12; 5x - y = 13$. To perform the elimination method, the equations in the system should be arranged vertically to be added together and then one or both of the equations should be multiplied so that one variable will be eliminated when the two are added. Opposites will cancel each other when added together. For example, $8x$ and $-8x$ will cancel each other when added. For the example above, writing the system vertically helps identify that the bottom equation should be multiplied by 3 to eliminate the variable *y*.

$$2x + 3y = 12 \quad \rightarrow \quad 2x + 3y = 12$$

$$3(5x - y = 13) \quad \rightarrow \quad 15x - 3y = 39$$

Adding the two equations together vertically results in $17x = 51$. Solving yields $x = 3$. Once the value for one variable is known, it can be substituted into either of the original equations to determine the value of the other variable. Once this is obtained, the solution can be written as an ordered pair (x, y) and checked in both equations of the system. In this example, the solution is (3, 2).

Systems of Equations with No Solution or an Infinite Number of Solutions

A system of equations can have one solution, no solution, or an infinite number of solutions. If, while solving a system algebraically, both variables cancel out, then the system has either no solution or has an infinite number of solutions. If the remaining constants result in a true statement (ex., $7 = 7$), then there is an infinite number of solutions. This would indicate coinciding lines. If the remaining constants result in a false statement, then there is no solution to the system. This would indicate parallel lines.

Graphing a System of Two Linear Equations in Two Variables

A solution for a system of equations is an ordered pair that makes both equations true. One method for solving a system of equations is to graph both lines on a coordinate plane. If the lines intersect, the point of intersection is the solution to the system. Every point on a line represents an ordered pair that makes its equation true. The ordered pair represented by this point of intersection lies on both lines and therefore makes both equations true. This ordered pair should be checked by substituting its values into both of the original equations of the system. Note that given a system of equations and an ordered pair, the ordered pair can be determined to be a solution or not by checking it in both equations.

If, when graphed, the lines representing the equations of a system do not intersect, then the two lines are parallel to each other or they are the same exact line. Parallel lines extend in the same direction without ever meeting. A system consisting of parallel lines has no solution. If the equations for a system represent the same exact line, then every point on the line is a solution to the system. In this case, there would be an infinite number of solutions. A system consisting of intersecting lines is referred to as independent; a system consisting of parallel lines is referred to as inconsistent; and a system consisting of coinciding lines is referred to as dependent.

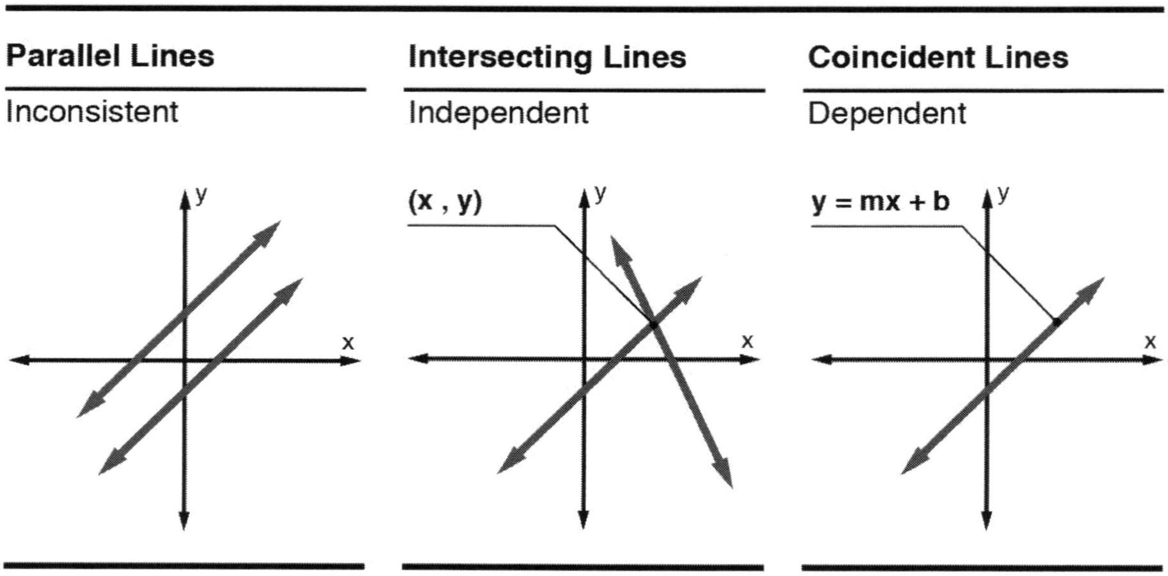

Parallel Lines	Intersecting Lines	Coincident Lines
Inconsistent	Independent	Dependent

(x , y)

y = mx + b

Graphing Solution Sets for Linear Inequalities in Two Variables

A graph of the solution set for a linear inequality shows the ordered pairs that make the statement true. The graph consists of a boundary line dividing the coordinate plane and shading on one side of the boundary. The boundary line should be graphed just as a linear equation would be graphed. If the inequality symbol is > or <, a dashed line can be used to indicate that the line is not part of the solution set. If the inequality symbol is ≥ or ≤, a solid line can be used to indicate that the boundary line is included in the solution set. An ordered pair (x, y) on either side of the line should be chosen to test in the inequality statement. If substituting the values for x and y results in a true statement ($15(3) + 25(2) > 90$), that ordered pair and all others on that side of the boundary line are part of the solution set. To indicate this, that region of the graph should be shaded. If substituting the ordered pair results in a false statement, the ordered pair and all others on that side are not part of the solution set.

Therefore, the other region of the graph contains the solutions and should be shaded.

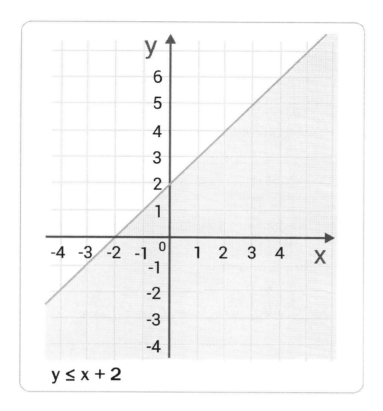

$$y \leq x + 2$$

A question may simply ask whether a given ordered pair is a solution to a given inequality. To determine this, the values should be substituted for the ordered pair into the inequality. If the result is a true statement, the ordered pair is a solution; if the result is a false statement, the ordered pair is not a solution.

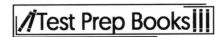

Geometry

Using Geometry to Calculate Quantities of Interest

Properties of Polygons and Circles

A **polygon** is a closed two-dimensional figure consisting of three or more sides. Polygons can be either convex or concave. A polygon that has interior angles all measuring less than 180° is **convex**. A **concave** polygon has one or more interior angles measuring greater than 180°. Examples are shown below.

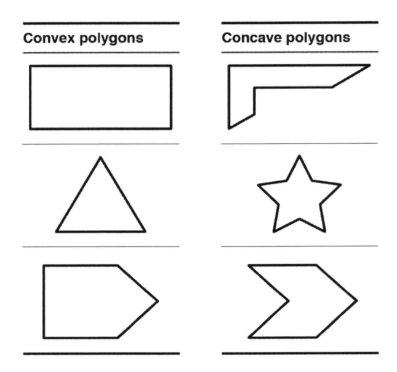

Polygons can be classified by the number of sides (also equal to the number of angles) they have. The following are the names of polygons with a given number of sides or angles:

# of sides	3	4	5	6	7	8	9	10
Name of polygon	Triangle	Quadrilateral	Pentagon	Hexagon	Septagon (or heptagon)	Octagon	Nonagon	Decagon

Equiangular polygons are polygons in which the measure of every interior angle is the same. The sides of equilateral polygons are always the same length. If a polygon is both equiangular and equilateral, the polygon is defined as a **regular polygon**. Examples are shown below.

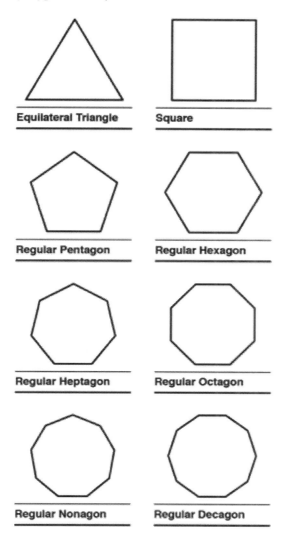

Triangles can be further classified by their sides and angles. A triangle with its largest angle measuring 90° is a **right triangle**.

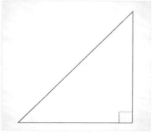

Right triangle

171

A triangle with the largest angle less than 90° is an **acute triangle**. A triangle with the largest angle greater than 90° is an **obtuse triangle**.

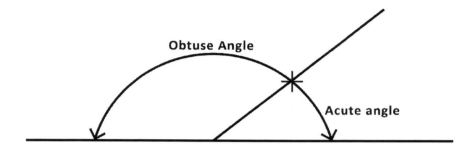

A triangle consisting of two equal sides and two equal angles is an **isosceles triangle**. A triangle with three equal sides and three equal angles is an **equilateral triangle**. A triangle with no equal sides or angles is a **scalene triangle**.

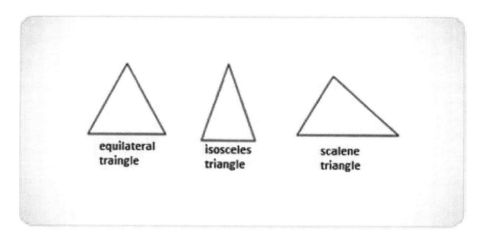

Quadrilaterals can be further classified according to their sides and angles. A quadrilateral with exactly one pair of parallel sides is called a **trapezoid**. A quadrilateral that shows both pairs of opposite sides parallel is a **parallelogram**. Parallelograms include rhombuses, rectangles, and squares. A **rhombus** has

four equal sides. A **rectangle** has four equal angles (90° each). A **square** has four 90° angles and four equal sides. Therefore, a square is both a rhombus and a rectangle.

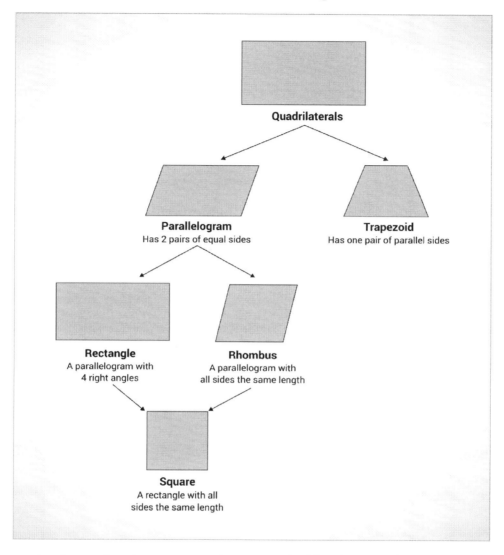

There are many key facts related to geometry that are applicable. The sum of the measures of the angles of a triangle are 180°, and for a quadrilateral, the sum is 360°. Rectangles and squares each have four right angles. A **right angle** has a measure of 90°.

Lines and Angles

In geometry, a **line** connects two points, has no thickness, and extends indefinitely in both directions beyond each point. If the length is finite, it's known as a **line segment** and has two **endpoints**. A **ray** is the straight portion of a line that has one endpoint and extends indefinitely in the other direction. An **angle** is formed when two rays begin at the same endpoint and extend indefinitely. The endpoint of an angle is called a **vertex**. **Adjacent angles** are two side-by-side angles formed from the same ray that have the same endpoint. Angles are measured in **degrees** or **radians**, which is a measure of **rotation**. A **full rotation** equals 360 degrees or 2π radians, which represents a circle. Half a rotation equals 180 degrees or π radians and represents a half-circle. Angle measurement is additive. When an angle is broken into two non-overlapping angles, the total measure of the larger angle equals the sum of the two

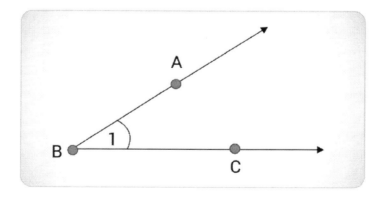

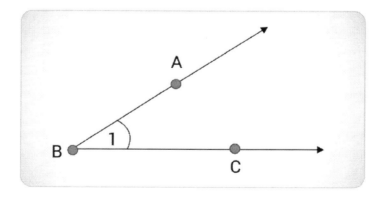

smaller angles. Lines are **coplanar** if they're located in the same plane. Two lines are **parallel** if they are coplanar, extend in the same direction, and never cross. If lines do cross, they're labeled as **intersecting lines** because they "intersect" at one point. If they intersect at more than one point, they're the same line. **Perpendicular lines** are coplanar lines that form a right angle at their point of intersection.

Two lines are parallel if they have the same slope and different intercept. Two lines are perpendicular if the product of their slope equals -1. Parallel lines never intersect unless they are the same line, and perpendicular lines intersect at a right angle. If two lines aren't parallel, they must intersect at one point. Determining equations of lines based on properties of parallel and perpendicular lines appears in word problems. To find an equation of a line, both the slope and a point the line goes through are necessary. Therefore, if an equation of a line is needed that's parallel to a given line and runs through a specified point, the slope of the given line and the point are plugged into the point-slope form of an equation of a line. Secondly, if an equation of a line is needed that's perpendicular to a given line running through a specified point, the negative reciprocal of the slope of the given line and the point are plugged into the point-slope form. Also, if the point of intersection of two lines is known, that point will be used to solve the set of equations. Therefore, to solve a system of equations, the point of intersection must be found. If a set of two equations with two unknown variables has no solution, the lines are parallel.

An **angle** consists of two rays that have a common endpoint. This common endpoint is called the **vertex** of the angle. The two rays can be called sides of the angle. The angle below has a vertex at point B and the sides consist of ray BA and ray BC. An angle can be named in three ways:

1. Using the vertex and a point from each side, with the vertex letter in the middle.
2. Using only the vertex. This can only be used if it is the only angle with that vertex.
3. Using a number that is written inside the angle.

The angle below can be written $\angle ABC$ (read angle ABC), $\angle CBA$, $\angle B$, or $\angle 1$.

An angle divides a plane, or flat surface, into three parts: the angle itself, the interior (inside) of the angle, and the exterior (outside) of the angle. The figure below shows point *M* on the interior of the angle and point *N* on the exterior of the angle.

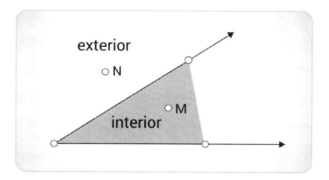

Angles can be measured in units called degrees, with the symbol °. The degree measure of an angle is between 0° and 180° and can be obtained by using a protractor.

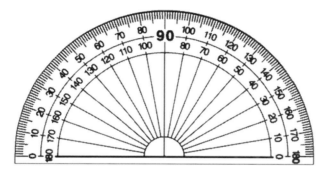

A straight angle (or simply a line) measures exactly 180°. A right angle's sides meet at the vertex to create a square corner. A right angle measures exactly 90° and is typically indicated by a box drawn in the interior of the angle. An acute angle has an interior that is narrower than a right angle. The measure of an acute angle is any value less than 90° and greater than 0°. For example, 89.9°, 47°, 12°, and 1°. An obtuse angle has an interior that is wider than a right angle. The measure of an obtuse angle is any value greater than 90° but less than 180°. For example, 90.1°, 110°, 150°, and 179.9°.

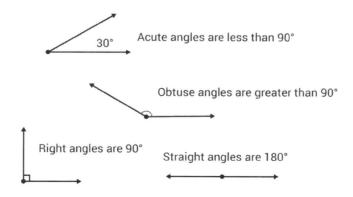

Perimeter and Area

Perimeter is the measurement of a distance around something or the sum of all sides of a polygon. Think of perimeter as the length of the boundary, like a fence. In contrast, **area** is the space occupied by a defined enclosure, like a field enclosed by a fence.

When thinking about perimeter, think about walking around the outside of something. When thinking about area, think about the amount of space or **surface area** something takes up.

Square

The perimeter of a square is measured by adding together all of the sides. Since a square has four equal sides, its perimeter can be calculated by multiplying the length of one side by 4. Thus, the formula is $P = 4 \times s$, where s equals one side. For example, the following square has side lengths of 5 meters:

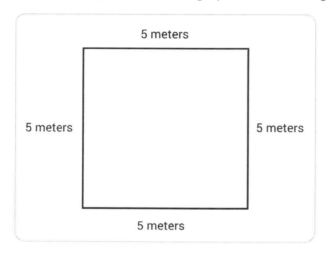

The perimeter is 20 meters because 4 times 5 is 20.

The area of a square is the length of a side squared. For example, if a side of a square is 7 centimeters, then the area is 49 square centimeters. The formula for this example is:

$$A = s^2 = 7^2 = 49 \text{ square centimeters}$$

An example is if the rectangle has a length of 6 inches and a width of 7 inches, then the area is 42 square inches:

$$A = lw = 6(7) = 42 \text{ square inches}$$

Rectangle

Like a square, a rectangle's perimeter is measured by adding together all of the sides. But as the sides are unequal, the formula is different. A rectangle has equal values for its lengths (long sides) and equal values for its widths (short sides), so the perimeter formula for a rectangle is:

$$P = l + l + w + w = 2l + 2w$$

l equals length
w equals width

The area is found by multiplying the length by the width, so the formula is $A = l \times w$.

For example, if the length of a rectangle is 10 inches and the width 8 inches, then the perimeter is 36 inches because:

$$P = 2l + 2w = 2(10) + 2(8) = 20 + 16 = 36 \text{ inches}$$

Triangle

A triangle's perimeter is measured by adding together the three sides, so the formula is $P = a + b + c$, where a, b, and c are the values of the three sides. The area is the product of one-half the base and height so the formula is:

$$A = \frac{1}{2} \times b \times h$$

It can be simplified to:

$$A = \frac{bh}{2}$$

The base is the bottom of the triangle, and the height is the distance from the base to the peak. If a problem asks to calculate the area of a triangle, it will provide the base and height.

For example, if the base of the triangle is 2 feet and the height 4 feet, then the area is 4 square feet. The following equation shows the formula used to calculate the area of the triangle:

$$A = \frac{1}{2}bh = \frac{1}{2}(2)(4) = 4 \text{ square feet}$$

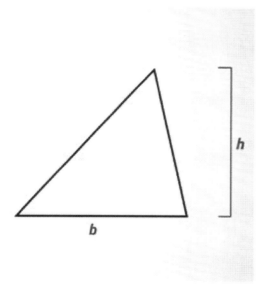

Circle

A circle's perimeter—also known as its circumference—is measured by multiplying the diameter by π.

Diameter is the straight line measured from a point on one side of the circle to a point directly across on the opposite side of the circle.

π is referred to as pi and is equal to 3.14 (with rounding).

So the formula is $\pi \times d$.

This is sometimes expressed by the formula $C = 2 \times \pi \times r$, where r is the radius of the circle. These formulas are equivalent, as the radius equals half of the diameter.

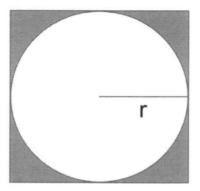

The area of a circle is calculated through the formula $A = \pi \times r^2$. The test will indicate either to leave the answer with π attached or to calculate to the nearest decimal place, which means multiplying by 3.14 for π.

The **arc of a circle** is the distance between two points on the circle. The length of the arc of a circle in terms of **degrees** is easily determined if the value of the central angle is known. The length of the arc is simply the value of the central angle. In this example, the length of the arc of the circle in degrees is 75°.

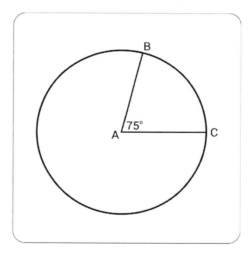

To determine the length of the arc of a circle in distance, the values for both the central angle and the radius must be known. This formula is:

$$\frac{central\ angle}{360°} = \frac{arc\ length}{2\pi r}$$

The equation is simplified by cross-multiplying to solve for the arc length.

In the following example, to solve for arc length, substitute the values of the central angle (75°) and the radius (10 inches) into the equation above.

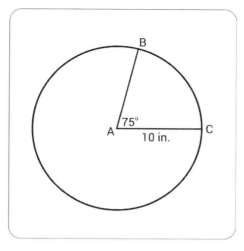

$$\frac{75°}{360°} = \frac{\text{arc length}}{2(3.14)(10\text{in.})}$$

To solve the equation, first cross-multiply: 4710 = 360(arc length). Next, divide each side of the equation by 360. The result of the formula is that the arc length is 13.1 (rounded).

Irregular Shapes

The perimeter of an irregular polygon is found by adding the lengths of all of the sides. In cases where all of the sides are given, this will be very straightforward, as it will simply involve finding the sum of the provided lengths. Other times, a side length may be missing and must be determined before the perimeter can be calculated. Consider the example below:

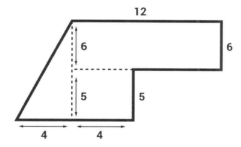

All of the side lengths are provided except for the angled side on the left. Test takers should notice that this is the hypotenuse of a right triangle. The other two sides of the triangle are provided (the base is 4 and the height is 6 + 5 = 11). The Pythagorean Theorem can be used to find the length of the hypotenuse, remembering that $a^2 + b^2 = c^2$.

Substituting the side values provided yields $(4)^2 + (11)^2 = c^2$.

Therefore, $c = \sqrt{16 + 121} = 11.7$

Finally, the perimeter can be found by adding this new side length with the other provided lengths to get the total length around the figure:

$$4+4+5+8+6+12+11.7=50.7$$

Although units are not provided in this figure, remember that reporting units with a measurement is important.

The area of an irregular polygon is found by decomposing, or breaking apart, the figure into smaller shapes. When the area of the smaller shapes is determined, these areas are added together to produce the total area of the area of the original figure. Consider the same example provided before:

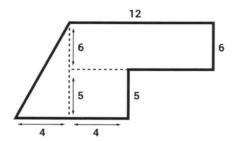

The irregular polygon is decomposed into two rectangles and a triangle. The area of the large rectangles $(A = l \times w \rightarrow A = 12 \times 6)$ is 72 square units. The area of the small rectangle is 20 square units:

$$(A = 4 \times 5)$$

The area of the triangle $(A = \frac{1}{2} \times b \times h \rightarrow A = \frac{1}{2} \times 4 \times 11)$ is 22 square units. The sum of the areas of these figures produces the total area of the original polygon:

$$A = 72 + 20 + 22 \rightarrow A = 114 \text{ square units}$$

Volume and Surface Area of Three-Dimensional Shapes
Geometry in three dimensions is similar to geometry in two dimensions. The main new feature is that three points now define a unique plane that passes through each of them. Three-dimensional objects can be made by putting together two-dimensional figures in different surfaces. Below, some of the possible three-dimensional figures will be provided, along with formulas for their volumes and surface areas.

Volume is the measurement of how much space an object occupies, like how much space is in the cube. Volume questions will ask how much of something is needed to completely fill the object. The most common surface area and volume questions deal with spheres, cubes, and rectangular prisms.

Surface area of a three-dimensional figure refers to the number of square units needed to cover the entire surface of the figure. This concept is similar to using wrapping paper to completely cover the outside of a box. For example, if a triangular pyramid has a surface area of 17 square inches (written $17in^2$), it will take 17 squares, each with sides one inch in length, to cover the entire surface of the pyramid. Surface area is also measured in square units.

A **rectangular prism** is a box whose sides are all rectangles meeting at 90° angles. Such a box has three dimensions: length, width, and height. If the length is x, the width is y, and the height is z, then the volume is given by $V = xyz$.

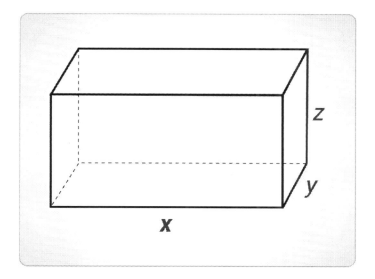

The surface area will be given by computing the surface area of each rectangle and adding them together. There is a total of six rectangles. Two of them have sides of length x and y, two have sides of length y and z, and two have sides of length x and z. Therefore, the total surface area will be given by:

$$SA = 2xy + 2yz + 2xz$$

A **cube** is a special type of rectangular solid in which its length, width, and height are the same. If this length is s, then the formula for the volume of a cube is $V = s \times s \times s$. The surface area of a cube is $SA = 6s^2$.

A **rectangular pyramid** is a figure with a rectangular base and four triangular sides that meet at a single vertex. If the rectangle has sides of length x and y, then the volume will be given by $V = \frac{1}{3}xyh$.

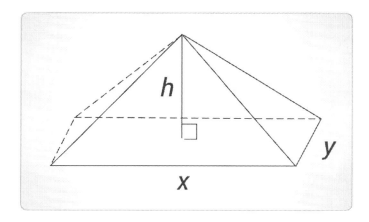

Many three-dimensional figures (solid figures) can be represented by nets consisting of rectangles and triangles. The surface area of such solids can be determined by adding the areas of each of its faces and bases. Finding the surface area using this method requires calculating the areas of rectangles and triangles. To find the area (A) of a rectangle, the length (l) is multiplied by the width (w) → $A = l \times w$. The area of a rectangle with a length of 8cm and a width of 4cm is calculated: $A = (8cm) \times (4cm) \rightarrow A = 32cm^2$.

To calculate the area (A) of a triangle, the product of $\frac{1}{2}$, the base (b), and the height (h) is found:

$$A = \frac{1}{2} \times b \times h$$

Note that the height of a triangle is measured from the base to the vertex opposite of it forming a right angle with the base. The area of a triangle with a base of 11cm and a height of 6cm is calculated:

$$A = \frac{1}{2} \times (11cm) \times (6cm)$$

$$A = 33cm^2$$

Consider the following triangular prism, which is represented by a net consisting of two triangles and three rectangles.

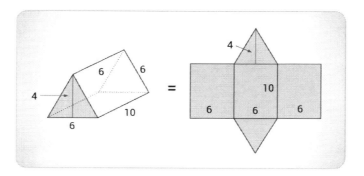

The surface area of the prism can be determined by adding the areas of each of its faces and bases. The surface area (SA) = area of triangle + area of triangle + area of rectangle + area of rectangle + area of rectangle.

$$SA = \left(\frac{1}{2} \times b \times h\right) + \left(\frac{1}{2} \times b \times h\right) + (l \times w) + (l \times w) + (l \times w)$$

$$SA = \left(\frac{1}{2} \times 6 \times 4\right) + \left(\frac{1}{2} \times 6 \times 4\right) + (6 \times 10) + (6 \times 10) + (6 \times 10)$$

$$SA = (12) + (12) + (60) + (60) + (60)$$

$$SA = 204 \; square \; units$$

A **sphere** is a set of points all of which are equidistant from some central point. It is like a circle, but in three dimensions. The volume of a sphere of radius r is given by:

$$V = \frac{4}{3}\pi r^3$$

The surface area is given by $A = 4\pi r^2$.

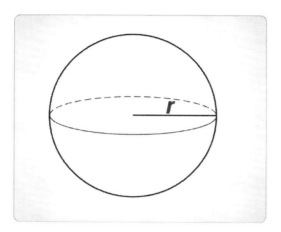

The volume of a **cylinder** is then found by adding a third dimension onto the circle. Volume of a cylinder is calculated by multiplying the area of the base (which is a circle) by the height of the cylinder. Doing so results in the equation $V = \pi r^2 h$. The volume of a **cone** is $\frac{1}{3}$ of the volume of a cylinder. Therefore, the formula for the volume of a cone is:

$$\frac{1}{3}\pi r^2 h$$

Proportional Reasoning

Transformations

Given a figure drawn on a plane, many changes can be made to that figure, including rotation, translation, and reflection. **Rotations** turn the figure about a point, **translations** slide the figure, and **reflections** flip the figure over a specified line. When performing these transformations, the original figure is called the **pre-image**, and the figure after transformation is called the **image**.

More specifically, **translation** means that all points in the figure are moved in the same direction by the same distance. In other words, the figure is slid in some fixed direction. Of course, while the entire figure is slid by the same distance, this does not change any of the measurements of the figures involved. The result will have the same distances and angles as the original figure.

In terms of Cartesian coordinates, a translation means a shift of each of the original points (x, y) by a fixed amount in the x and y directions, to become $(x + a, y + b)$.

Another procedure that can be performed is called **reflection**. To do this, a line in the plane is specified, called the **line of reflection**. Then, take each point and flip it over the line so that it is the same distance from the line but on the opposite side of it. This does not change any of the distances or angles involved, but it does reverse the order in which everything appears.

To reflect something over the x-axis, the points (x, y) are sent to $(x, -y)$. To reflect something over the y-axis, the points (x, y) are sent to the points $(-x, y)$. Flipping over other lines is not something easy to express in Cartesian coordinates. However, by drawing the figure and the line of reflection, the distance to the line and the original points can be used to find the reflected figure.

Example: Reflect this triangle with vertices (-1, 0), (2, 1), and (2, 0) over the y-axis. The pre-image is shown below.

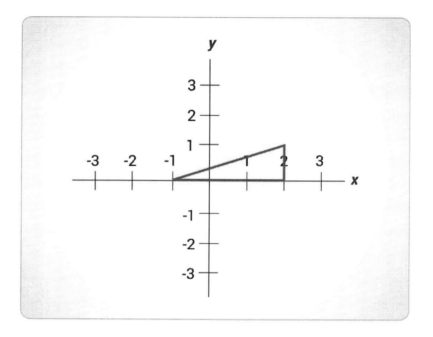

To do this, flip the x values of the points involved to the negatives of themselves, while keeping the y values the same. The image is shown here.

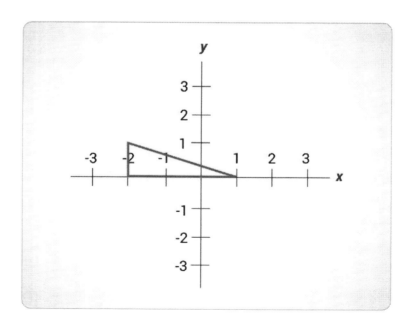

The new vertices will be (1, 0), (-2, 1), and (-2, 0).

Another procedure that does not change the distances and angles in a figure is **rotation**. In this procedure, pick a center point, then rotate every vertex along a circle around that point by the same angle. This procedure is also not easy to express in Cartesian coordinates, and this is not a requirement on this test. However, as with reflections, it's helpful to draw the figures and see what the result of the rotation would look like. This transformation can be performed using a compass and protractor.

Each one of these transformations can be performed on the coordinate plane without changes to the original dimensions or angles.

If two figures in the plane involve the same distances and angles, they are called **congruent figures**. In other words, two figures are congruent when they go from one form to another through reflection, rotation, and translation, or a combination of these.

Remember that rotation and translation will give back a new figure that is identical to the original figure, but reflection will give back a mirror image of it.

To recognize that a figure has undergone a rotation, check to see that the figure has not been changed into a mirror image, but that its orientation has changed (that is, whether the parts of the figure now form different angles with the *x* and *y* axes).

To recognize that a figure has undergone a translation, check to see that the figure has not been changed into a mirror image, and that the orientation remains the same.

To recognize that a figure has undergone a reflection, check to see that the new figure is a mirror image of the old figure.

Keep in mind that sometimes a combination of translations, reflections, and rotations may be performed on a figure.

Dilation

A **dilation** is a transformation that preserves angles, but not distances. This can be thought of as stretching or shrinking a figure. If a dilation makes figures larger, it is called an **enlargement**. If a dilation makes figures smaller, it is called a **reduction**. The easiest example is to dilate around the origin. In this case, multiply the *x* and *y* coordinates by a **scale factor**, k, sending points (x, y) to (kx, ky).

As an example, draw a dilation of the following triangle, whose vertices will be the points (-1, 0), (1, 0), and (1, 1).

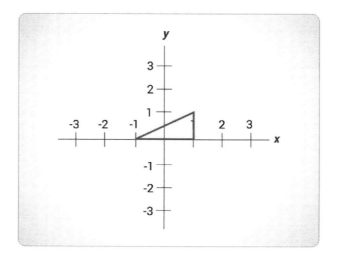

For this problem, dilate by a scale factor of 2, so the new vertices will be (-2, 0), (2, 0), and (2, 2).

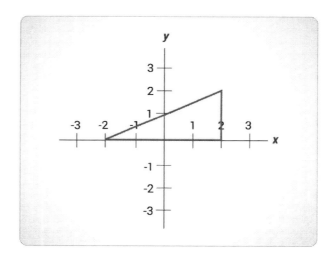

Note that after a dilation, the distances between the vertices of the figure will have changed, but the angles remain the same. The two figures that are obtained by dilation, along with possibly translation, rotation, and reflection, are all *similar* to one another. Another way to think of this is that similar figures have the same number of vertices and edges, and their angles are all the same. Similar figures have the same basic shape but are different in size.

Symmetry

Using the types of transformations above, if an object can undergo these changes and not appear to have changed, then the figure is symmetrical. If an object can be split in half by a line and flipped over

that line to lie directly on top of itself, it is said to have **line symmetry**. An example of both types of figures is seen below.

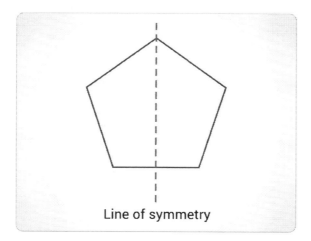

Line of symmetry

If an object can be rotated about its center to any degree smaller than 360, and it lies directly on top of itself, the object is said to have **rotational symmetry**. An example of this type of symmetry is shown below. The pentagon has an order of 5.

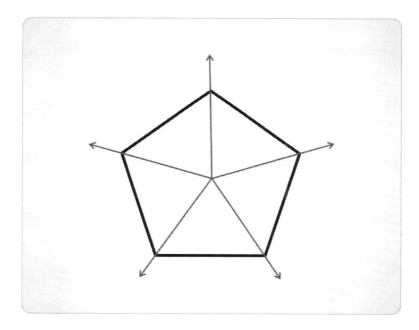

The rotational symmetry lines in the figure above can be used to find the angles formed at the center of the pentagon. Knowing that all of the angles together form a full circle, at 360 degrees, the figure can be split into 5 angles equally. By dividing the 360° by 5, each angle is 72°.

Given the length of one side of the figure, the perimeter of the pentagon can also be found using rotational symmetry. If one side length was 3 cm, that side length can be rotated onto each other side length four times. This would give a total of 5 side lengths equal to 3 cm. To find the perimeter, or distance around the figure, multiply 3 by 5. The perimeter of the figure would be 15 cm.

If a line cannot be drawn anywhere on the object to flip the figure onto itself or rotated less than or equal to 180 degrees to lay on top of itself, the object is asymmetrical. Examples of these types of figures are shown below.

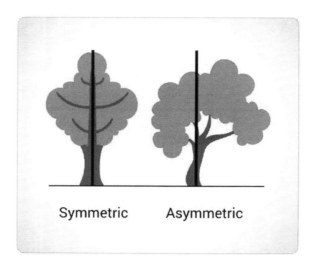

Similarity, Congruence, and Triangles

Triangles are similar if they have the same shape, the same angle measurements, and their sides are proportional to one another. Triangles are congruent if the angles of the triangles are equal in measurement and the sides of the triangles are equal in measurement.

There are five ways to show that a triangle is congruent.

- SSS (Side-Side-Side Postulate): When all three corresponding sides are equal in length, then the two triangles are congruent.

- SAS (Side-Angle-Side Postulate): If a pair of corresponding sides and the angle in between those two sides are equal, then the two triangles are congruent.

- ASA (Angle-Side-Angle Postulate): If a pair of corresponding angles are equal and the side within those angles are equal, then the two triangles are equal.

- AAS (Angle-Angle-Side Postulate): When a pair of corresponding angles for two triangles and a non-included side are equal, then the two triangles are congruent.

- HL (Hypotenuse-Leg Theorem): If two right triangles have the same hypotenuse length, and one of the other sides are also the same length, then the two triangles are congruent.

If two triangles are discovered to be similar or congruent, this information can assist in determining unknown parts of triangles, such as missing angles and sides.

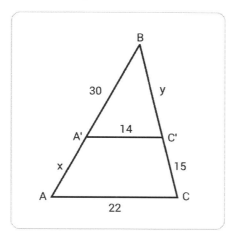

In the triangle shown above, *AC* and *A'C'* are parallel lines. Therefore, *BA* is a transversal that intersects the two parallel lines. The corresponding angles *BA'C'* and *BAC* are congruent. In a similar way, *BC* is also a transversal. Therefore, angle *BC'A'* and *BCA* are congruent. If two triangles have two congruent angles, the triangles are similar. If the triangles are similar, their corresponding sides are proportional.

Therefore, the following equation is established:

$$\frac{30 + x}{30} = \frac{22}{14} = \frac{y + 15}{y}$$

$$\frac{30 + x}{30} = \frac{22}{14}$$

$$x = 17.1$$

$$\frac{22}{14} = \frac{y + 15}{y}$$

$$y = 26.25$$

The example below involves the question of congruent triangles. The first step is to examine whether the triangles are congruent. If the triangles are congruent, then the measure of a missing angle can be found.

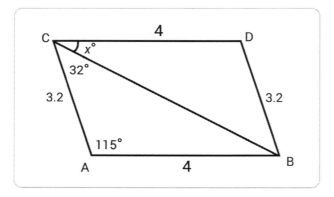

The above diagram provides values for angle measurements and side lengths in triangles *CAB* and *CDB*. Note that side *CA* is 3.2 and side *DB* is 3.2. Side *CD* is 4 and side *AB* is 4. Furthermore, line *CB* is congruent to itself by the reflexive property. Therefore, the two triangles are congruent by SSS (Side-Side-Side). Because the two triangles are congruent, all of the corresponding parts of the triangles are also congruent. Therefore, angle *x* is congruent to the inside of the angle for which a measurement is not provided in Triangle *CAB*. Thus, $115° + 32° = 147°$. A triangle measures 180°, therefore $180° - 147° = 33°$. *Angle x* = 33°, because the two triangles are reversed.

Complementary Angle Theorem

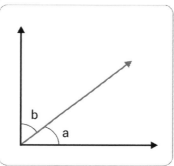

Two angles are complementary if the sum of the two angles equals 90°.

In the above diagram $Angle\ a + Angle\ b = 90°$. Therefore, the two angles are complementary. Certain trigonometric rules are also associated with complementary angles.

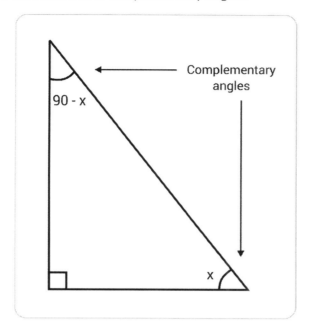

In the diagram above of a right triangle, if Angle *A* and Angle *C* are determined to be complementary angles, then certain relationships can be stated between the trigonometry of those angles.

$$sin(90° - x) = cos\ x$$

$$cos(90° - x) = sin\ x$$

For example, the sine of 80 degrees equals the cosine of $(90° - 80°)$, which is the cos $(10°)$.

This is true because the sine of an angle in a right triangle is equal to the cosine of its complement. Sine is known as the conjunction of cosine, and cosine is known as the conjunction of sine.

Examples:

1. $cos5° = sin\ x°$?
2. $sin(90° - x) = ?\ sin(90° - x) =?$

For problem number 1, the student should remember that $sin(90° - x) = cos\ x$. Cos 5° would be the same as $sin(90 - 5)°$. Therefore, $cos5° = sin85°$.

For problem number 2, the student would use the same fact that $sin(90° - x)° = cos\ x$.

An *acute angle* is an angle that is less than 90°. If Angle *A* and Angle *B* are acute angles of a right triangle, then $sinA = cosB$. Therefore, the sine of any acute angle in a right triangle is equal to the cosine of its complement, and the cosine of any acute angle is equal to the sine of its complement.

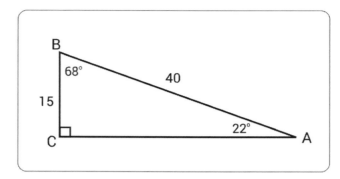

The example above is a right triangle. If only the value of angle *BAC* (which is 22°) was provided, the student would be able to figure out the value for angle *CBA* (68°) by knowing that a triangle is made up of 180°(180° − 90° − 22° = 68°). From the information given about acute angles on the previous page, the following statement is true:

Sine (angle *BAC*) = $\frac{15}{40}$, which is equivalent to the Cos (angle *CBA*) = $\frac{15}{40}$

Right Triangles: Pythagorean Theorem and Trigonometric Ratio

The value of a missing side of a right triangle may be determined two ways. The first way is to apply the Pythagorean Theorem, and the second way is to apply Trigonometric Ratios. The Pythagorean Theorem states that for every right triangle, the square of the length of the hypotenuse is equal to the sum of the squares of the lengths of the remaining two sides. The hypotenuse is the longest side of a right triangle and is also the side opposite the right angle.

According to the diagram, $a^2 + b^2 = c^2$, where *c* represents the hypotenuse, and *a* and *b* represent the lengths of the remaining two sides of the right triangle.

The Pythagorean Theorem may be applied a multitude of ways. For example, a person wishes to build a garden in the shape of a rectangle, having the dimensions of 5 feet by 8 feet. The garden's design includes a diagonal board to separate various types of plants. The Pythagorean Theorem can be used to determine the length of the diagonal board.

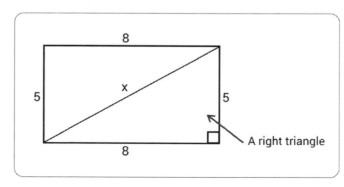

A right triangle

$$a^2 + b^2 = c^2$$

$$5^2 + 8^2 = c^2$$

$$25 + 64 = c^2$$

$$c = \sqrt{89}$$

$$c = 9.43$$

To solve for unknown sides of a right triangle using trigonometric ratios, the sine, cosine, and tangent are required.

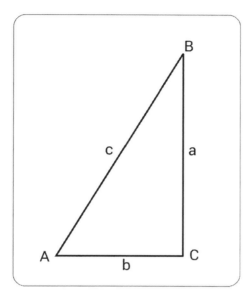

In the image above, angles are denoted by capital letters, and sides are denoted by lowercase letters. When examining angle A, b is the adjacent side, a is the opposite side, and c is the hypotenuse side. The various ratios of the lengths of the sides of the right triangle are used to find the sine, cosine, and tangent of angle A.

Thus, $\sin(A) = \dfrac{opposite}{hypotenuse}$, $\cos(A) = \dfrac{adjacent}{hypotenuse}$, and $\tan(A) = \dfrac{opposite}{adjacent}$. After substituting variables for the sides of the right triangle, $\sin(A) = \dfrac{a}{c}$, $\cos(A) = \dfrac{b}{c}$, and $\tan(A) = \dfrac{a}{b}$.

As a real-world example, the height of a tree can be discovered by using the information above. Surveying equipment can determine the tree's angle of inclination is 55.3 degrees, and the distance from the tree is 10 feet.

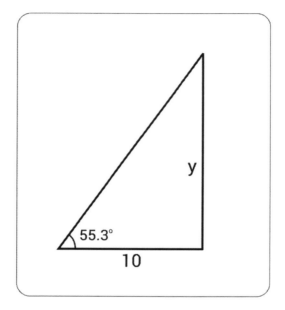

To find the height of the tree, substitute the known values into the trigonometric ratio of tangent:

$$\tan(55.3) = \frac{y}{10}$$

$$10 \times \tan(55.3) = y$$

$$10 \times 1.44418 = y$$

$$y = 14.4418$$

Plane Geometry

Algebraic equations can be used to describe geometric figures in the plane. The method for doing so is to use the **Cartesian coordinate plane**. The idea behind these Cartesian coordinates (named for mathematician and philosopher Descartes) is that from a specific point on the plane, known as the **origin**, one can specify any other point by saying *how far to the right or left* and *how far up or down*.

The plane is covered with a grid. The two directions, right to left and bottom to top, are called **axes** (singular **axis**). When working with x and y variables, the x variable corresponds to the right and left axis, and the y variable corresponds to the up and down axis.

Any point on the grid is found by specifying how far to travel from the center along the x-axis and how far to travel along the y-axis. The ordered pair can be written as (x, y). A positive x value means go to

the right on the *x*-axis, while a negative *x* value means to go to the left. A positive *y* value means to go up, while a negative value means to go down. Several points are shown as examples in the figure.

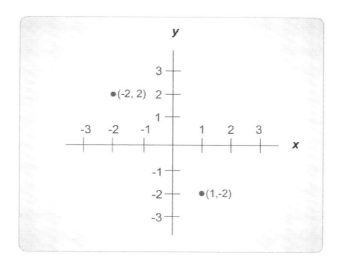

Cartesian Coordinate Plane

The coordinate plane can be divided into four **quadrants**. The upper-right part of the plane is called the first quadrant, where both *x* and *y* are positive. The second quadrant is the upper-left, where x is negative but y is positive. The third quadrant is the lower left, where both x and y are negative. Finally, the fourth quadrant is in the lower right, where x is positive but y is negative. These quadrants are often written with Roman numerals:

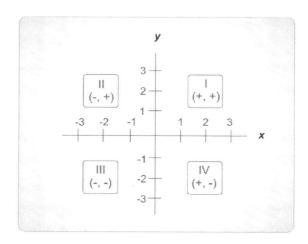

In addition to graphing individual points as shown above, the graph lines and curves in the plane can be graphed corresponding to equations. In general, if there is some equation involving *x* and *y*, then the graph of that equation consists of all the points (x, y) in the Cartesian coordinate plane, which satisfy this equation.

Given the equation $y = x + 2$, the point $(0, 2)$ is in the graph, since $2 = 0 + 2$ is a true equation. However, the point $(1, 4)$ will *not* be in the graph, because $4 = 1 + 2$ is false.

Proving Theorems with Coordinates

Many important formulas and equations exist in geometry that use coordinates. The distance between two points (x_1, y_1) and (x_2, y_2) is:

$$d = \sqrt{(x_2 - x_1)^2 + (y_2 - y_1)^2}$$

The slope of the line containing the same two points is $m = \frac{y_2 - y_1}{x_2 - x_1}$. Also, the midpoint of the line segment with endpoints (x_1, y_1) and (x_2, y_2) is:

$$M = \left(\frac{x_1 + x_2}{2}, \frac{y_1 + y_2}{2} \right)$$

The equations of a circle, parabola, ellipse, and hyperbola can also be used to prove theorems algebraically. Knowing when to use which formula or equation is extremely important, and knowing which formula applies to which property of a given geometric shape is an integral part of the process. In some cases, there are a number of ways to prove a theorem; however, only one way is required.

Formulas for Ratios

If a line segment with endpoints (x_1, y_1) and (x_2, y_2) is partitioned into two equal parts, the formula for midpoint is used. Recall this formula is $M = \left(\frac{x_1 + x_2}{2}, \frac{y_1 + y_2}{2} \right)$ and the ratio of line segments is 1:1. However, if the ratio needs to be anything other than 1:1, a different formula must be used. Consider a ratio that is $a : b$. This means the desired point that partitions the line segment is $\frac{a}{a+b}$ of the way from (x_1, y_1) to (x_2, y_2). The actual formula for the coordinate is:

$$\left(\frac{bx_1 + ax_2}{a + b}, \frac{by_1 + ay_2}{a + b} \right)$$

Computing Side Length, Perimeter, and Area

The side lengths of each shape can be found by plugging the endpoints into the distance formula:

$$d = \sqrt{(x_2 - x_1)^2 + (y_2 - y_1)^2}$$

between two ordered pairs (x_1, y_1) and (x_2, y_2). The distance formula is derived from the Pythagorean theorem. Once the side lengths are found, they can be added together to obtain the perimeter of the given polygon. Simplifications can be made for specific shapes such as squares and equilateral triangles. For example, one side length can be multiplied by 4 to obtain the perimeter of a square. Also, one side length can be multiplied by 3 to obtain the perimeter of an equilateral triangle. A similar technique can be used to calculate areas. For polygons, both side length and height can be found by using the same distance formula. Areas of triangles and quadrilaterals are straightforward through the use of $A = \frac{1}{2}bh$ or $A = bh$, depending on the shape. To find the area of other polygons, their shapes can be partitioned into rectangles and triangles. The areas of these simpler shapes can be calculated and then added together to find the total area of the polygon.

Circles on the Coordinate Plane

If a circle is placed on the coordinate plane with the center of the circle at the origin (0,0), then point (x, y) is a point on the circle. Furthermore, the line extending from the center to point (x, y) is the radius, or r. By applying the Pythagorean Theorem ($a^2 + b^2 = c^2$), it can be stated that $x^2 + y^2 = r^2$. However, the center of the circle does not always need to be on the origin of the coordinate plane.

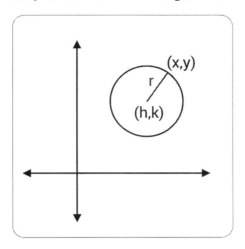

In the diagram above, the center of the circle is noted by (h, k). By applying the distance formula, the equation becomes: $= \sqrt{(x - h)^2 + (y - k)^2}$. When squaring both sides of the equation, the result is the standard form of a circle with the center (h, k) and radius r. Namely, $r^2 = (x - h)^2 + (y - k)^2$, where r = radius and center = (h, k). The following examples may be solved by using this information:

Example: Graph the equation $-x^2 + y^2 = 25$

To graph this equation, first note that the center of the circle is (0, 0). The radius is the positive square root of 25 or 5.

Example: Find the equation for the circle below.

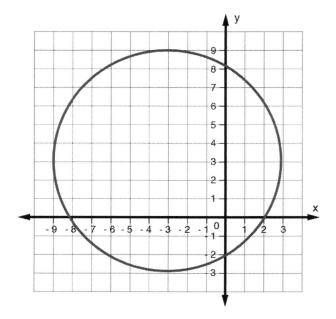

To find the equation for the circle, note that its center is not zero. Therefore, to find the circle's center, draw vertical and horizontal diameters to examine where they intersect. The center is located at point: (-3, 3). Next, count the number of spaces from the center to the outside of the circle. This number is 6. Therefore, 6 is the radius. Finally, plug in the numbers that are known into the standard equation for a circle:

$$36 = \left(x - (-3)\right)^2 + (y - 3)^2$$

$$or$$

$$36 = (x + 3)^2 + (y - 3)^2$$

It is possible to determine whether a point lies on a circle or not within the coordinate plane. For example, a circle has a center of (2, -5), and a radius of 6 centimeters. The first step is to apply the equation of a circle, which is $r^2 = (x - h)^2 + (y - k)^2$, where r = radius and the center = (h, k). Next, substitute the numbers for the center point and the number for the radius. This action simplifies the equation to $36 = (x - 2)^2 + (y + 5)^2$. Note that the radius of 6 was squared to get 36.

To prove that the point (2, -1) lies on the circle, apply the equation of the circle that was just used and input the values of (2, -1) for x and y in the equation.

$$36 = (x - 2)^2 + (y + 5)^2$$

$$36 = (2 - 2)^2 + (1 + 5)^2$$

$$36 = (0)^2 + (6)^2$$

$$36 = 36$$

Because the left side of the equation equals the right side of the equation, point (2, 1) lies on the given circle.

Statistics and Probability

Calculating Probabilities, Including Related Sample Spaces

Probability, represented by variable p, always has a value from 0 to 1. The total probability for all the possible outcomes (sample space) should equal 1.

Counting Techniques

There are many counting techniques that can help solve problems involving counting possibilities. For example, the **Addition Principle** states that if there are m choices from Group 1 and n choices from Group 2, then $n + m$ is the total number of choices possible from Groups 1 and 2. For this to be true, the groups can't have any choices in common. The **Multiplication Principle** states that if Process 1 can be completed n ways and Process 2 can be completed m ways, the total number of ways to complete both Process 1 and Process 2 is $n \times m$. For this rule to be used, both processes must be independent of each other. Counting techniques also involve permutations. A **permutation** is an arrangement of elements in a set for which order must be considered. For example, if three letters from the alphabet are chosen, ABC and BAC are two different permutations. The multiplication rule can be used to determine the total number of possibilities. If each letter can't be selected twice, the total number of possibilities is:

$$26 \times 25 \times 24 = 15,600$$

A formula can also be used to calculate this total. In general, the notation $P(n, r)$ represents the number of ways to arrange r objects from a set of n and, the formula is:

$$P(n, r) = \frac{n!}{(n - r)!}$$

In the previous example:

$$P(26, 3) = \frac{26!}{23!} = 15,600$$

Contrasting permutations, a **combination** is an arrangement of elements in which order doesn't matter. In this case, ABC and BAC are the same combination. In the previous scenario, there are six permutations that represent each single combination. Therefore, the total number of possible combinations is:

$$15,600 \div 6 = 2,600$$

In general, $C(n, r)$ represents the total number of combinations of n items selected r at a time where order doesn't matter, and the formula is:

$$C(n, r) = \frac{n!}{(n - r)! \ r!}$$

Therefore, the following relationship exists between permutations and combinations:

$$C(n,r) = \frac{P(n,r)}{n!} = \frac{P(n,r)}{P(r,r)}$$

Fundamental Counting Principle

The **fundamental counting principle** states that if there are m possible ways for an event to occur, and n possible ways for a second event to occur, there are $m \cdot n$ possible ways for both events to occur. For example, there are two events that can occur after flipping a coin and six events that can occur after rolling a die, so there are $2 \times 6 = 12$ total possible event scenarios if both are done simultaneously. This principle can be used to find probabilities involving finite sample spaces and independent trials because it calculates the total number of possible outcomes. For this principle to work, the events must be independent of each other.

Independence and Conditional Probability

Sample Subsets

A sample can be broken up into subsets that are smaller parts of the whole. For example, consider a sample population of females. The sample can be divided into smaller subsets based on the characteristics of each female. There can be a group of females with brown hair and a group of females that wear glasses. There also can be a group of females that have brown hair *and* wear glasses. This "and" relates to the **intersection** of the two separate groups of brunettes and those with glasses. Every female in that intersection group has both characteristics. Similarly, there also can be a group of females that either have brown hair *or* wear glasses. The "or" relates to the union of the two separate groups of brunettes and glasses. Every female in this group has at least one of the characteristics. Finally, the group of females who do not wear glasses can be discussed. This "not" relates to the **complement** of the glass-wearing group. No one in the complement has glasses. **Venn diagrams** are useful in highlighting these ideas. When discussing statistical experiments, this idea can also relate to events instead of characteristics.

Verifying Independent Events

Two events aren't always independent. For example, females with glasses and brown hair aren't independent characteristics. There definitely can be overlap because females with brown hair can wear glasses. Also, two events that exist at the same time don't have to have a relationship. For example, even if all females in a given sample are wearing glasses, the characteristics aren't related. In this case, the probability of a brunette wearing glasses is equal to the probability of a female being a brunette multiplied by the probability of a female wearing glasses. This mathematical test of $P(A \cap B) = P(A)P(B)$ verifies that two events are independent.

Conditional Probability

Conditional probability is the probability that event A will happen given that event B has already occurred. An example of this is calculating the probability that a person will eat dessert once they have eaten dinner. This is different than calculating the probability of a person just eating dessert. The formula for the conditional probability of event A occurring given B is $P(A|B) = \frac{P(A \text{ and } B)}{P(B)}$, and it's defined to be the probability of both A and B occurring divided by the probability of event B occurring. If A and B are independent, then the probability of both A and B occurring is equal to $P(A)P(B)$, so $P(A|B)$ reduces to just $P(A)$. This means that A and B have no relationship, and the probability of A occurring is the same as the conditional probability of A occurring given B. Similarly, $P(B|A) = \frac{P(B \text{ and } A)}{P(A)} = P(B)$ if A and B are independent.

Independent Versus Related Events

To summarize, conditional probability is the probability that an event occurs given that another event has happened. If the two events are related, the probability that the second event will occur changes if the other event has happened. However, if the two events aren't related and are therefore independent, the first event to occur won't impact the probability of the second event occurring.

Measuring Probabilities with Two-Way Frequency Tables

When measuring event probabilities, two-way frequency tables can be used to report the raw data and then used to calculate probabilities. If the frequency tables are translated into relative frequency tables, the probabilities presented in the table can be plugged directly into the formulas for conditional probabilities. By plugging in the correct frequencies, the data from the table can be used to determine if events are independent or dependent.

Differing Probabilities

The probability that event A occurs differs from the probability that event A occurs given B. When working within a given model, it's important to note the difference. $P(A|B)$ is determined using the formula $P(A|B) = \frac{P(A \text{ and } B)}{P(B)}$ and represents the total number of A's outcomes left that could occur after B occurs. $P(A)$ can be calculated without any regard for B. For example, the probability of a student finding a parking spot on a busy campus is different once class is in session.

The Addition Rule

The probability of event A or B occurring isn't equal to the sum of each individual probability. The probability that both events can occur at the same time must be subtracted from this total. This idea is shown in the **addition rule**:

$$P(A \text{ or } B) = P(A) + P(B) - P(A \text{ and } B)$$

The addition rule is another way to determine the probability of compound events that aren't mutually exclusive. If the events are mutually exclusive, the probability of both A and B occurring at the same time is 0.

Computing Probabilities

Simple and Compound Events

A **simple event** consists of only one outcome. The most popular simple event is flipping a coin, which results in either heads or tails. A **compound event** results in more than one outcome and consists of more than one simple event. An example of a compound event is flipping a coin while tossing a die. The result is either heads or tails on the coin and a number from one to six on the die. The probability of a simple event is calculated by dividing the number of possible outcomes by the total number of outcomes. Therefore, the probability of obtaining heads on a coin is $\frac{1}{2}$, and the probability of rolling a 6 on a die is $\frac{1}{6}$. The probability of compound events is calculated using the basic idea of the probability of simple events. If the two events are independent, the probability of one outcome is equal to the product of the probabilities of each simple event. For example, the probability of obtaining heads on a coin and rolling a 6 is equal to:

$$\frac{1}{2} \times \frac{1}{6} = \frac{1}{12}$$

The probability of either A or B occurring is equal to the sum of the probabilities minus the probability that both A and B will occur. Therefore, the probability of obtaining either heads on a coin or rolling a 6 on a die is:

$$\frac{1}{2} + \frac{1}{6} - \frac{1}{12} = \frac{7}{12}$$

The two events aren't mutually exclusive because they can happen at the same time. If two events are mutually exclusive, and the probability of both events occurring at the same time is zero, the probability of event A or B occurring equals the sum of both probabilities. An example of calculating the probability of two mutually exclusive events is determining the probability of pulling a king or a queen from a deck of cards. The two events cannot occur at the same time.

Uniform and Non-Uniform Probability Models

A **uniform probability model** is one where each outcome has an equal chance of occurring, such as the probabilities of rolling each side of a die. A **non-uniform probability model** is one where each outcome has an unequal chance of occurring. In a uniform probability model, the conditional probability formulas for $P(B|A)$ and $P(A|B)$ can be multiplied by their respective denominators to obtain two formulas for $P(A \text{ and } B)$. Therefore, the multiplication rule is derived as:

$$P(A \text{ and } B) = P(A)P(B|A) = P(B)P(A|B)$$

In a model, if the probability of either individual event is known and the corresponding conditional probability is known, the multiplication rule allows the probability of the joint occurrence of A and B to be calculated.

Basic Statistics

The field of statistics describes relationships between quantities that are related, but not necessarily in a deterministic manner. For example, a graduating student's salary will often be higher when the student graduates with a higher GPA, but this is not always the case. Likewise, people who smoke tobacco are more likely to develop lung cancer, but, in fact, it is possible for non-smokers to develop the disease as well. **Statistics** describes these kinds of situations, where the likelihood of some outcome depends on the starting data.

Descriptive statistics involves analyzing a collection of data to describe its broad properties such average (or mean), what percent of the data falls within a given range, and other such properties. An example of this would be taking all of the test scores from a given class and calculating the average test score. **Inferential statistics** attempts to use data about a subset of some population to make inferences about the rest of the population. An example of this would be taking a collection of students who received tutoring and comparing their results to a collection of students who did not receive tutoring, then using that comparison to try to predict whether the tutoring program in question is beneficial.

To be sure that inferences have a high probability of being true for the whole population, the subset that is analyzed needs to resemble a miniature version of the population as closely as possible. For this reason, statisticians like to choose random samples from the population to study, rather than picking a specific group of people based on some similarity. For example, studying the incomes of people who live in Portland does not tell anything useful about the incomes of people who live in Tallahassee.

A statistical question is answered by collecting data with variability. Data consists of facts and/or statistics (numbers), and variability refers to a tendency to shift or change. Data is a broad term, inclusive of things like height, favorite color, name, salary, temperature, gas mileage, and language. Questions requiring data as an answer are not necessarily statistical questions. If there is no variability in the data, then the question is not statistical in nature. Consider the following examples: what is Mary's favorite color? How much money does your mother make? What was the highest temperature last week? How many miles did your car get on its last tank of gas? How much taller than Bob is Ed?

None of the above are statistical questions because each case lacks variability in the data needed to answer the question. The questions on favorite color, salary, and gas mileage each require a single piece of data, whether a fact or statistic. Therefore, variability is absent. Although the temperature question requires multiple pieces of data (the high temperature for each day), a single, distinct number is the answer. The height question requires two pieces of data, Bob's height and Ed's height, but no difference in variability exists between those two values. Therefore, this is not a statistical question. Statistical questions typically require calculations with data.

Consider the following statistical questions:

How many miles per gallon of gas does the 2016 Honda Civic get? To answer this question, data must be collected. This data should include miles driven and gallons used. Different cars, different drivers, and different driving conditions will produce different results. Therefore, variability exists in the data. To answer the question, the mean (average) value could be determined.

Are American men taller than German men? To answer this question, data must be collected. This data should include the heights of American men and the heights of German men. All American men are not the same height and all German men are not the same height. Some American men are taller than some German men and some German men are taller than some American men. Therefore, variability exists in the data. To answer the question, the median values for each group could be determined and compared.

The following are more examples of statistical questions: What proportion of 4th graders have a favorite color of blue? How much money do teachers make? Is it colder in Boston or Chicago?

An **experiment** is the method in which a hypothesis is tested using a trial-and-error process. A cause and the effect of that cause are measured, and the hypothesis is accepted or rejected. Experiments are usually completed in a controlled environment where the results of a control population are compared to the results of a test population. The groups are selected using a randomization process in which each group has a representative mix of the population being tested. Finally, an **observational study** is similar to an experiment. However, this design is used when there cannot be a designed control and test population because of circumstances (e.g., lack of funding or unrealistic expectations). Instead, existing control and test populations must be used, so this method has a lack of randomization.

Statistics involves making decisions and predictions about larger data sets based on smaller data sets. Basically, the information from one part or subset can help predict what happens in the entire data set or population at large. The entire process involves guessing, and the predictions and decisions may not be 100 percent correct all of the time; however, there is some truth to these predictions, and the decisions do have mathematical support. The smaller data set is called a **sample** and the larger data set (in which the decision is being made) is called a **population**. A **random sample** is used as the sample, which is an unbiased collection of data points that represents the population as well as it can. There are many methods of forming a random sample, and all adhere to the fact that every potential data point has a predetermined probability of being chosen.

Mean, Median, and Mode

The center of a set of data (statistical values) can be represented by its mean, median, or mode. These are sometimes referred to as measures of central tendency.

Mean

The first property that can be defined for this set of data is the **mean**. This is the same as the average. To find the mean, add up all the data points, then divide by the total number of data points. For example, suppose that in a class of 10 students, the scores on a test were 50, 60, 65, 65, 75, 80, 85, 85, 90, 100. Therefore, the average test score will be:

$$\frac{50 + 60 + 65 + 65 + 75 + 80 + 85 + 85 + 90 + 100}{10} = 75.5$$

The mean is a useful number if the distribution of data is normal (more on this later), which roughly means that the frequency of different outcomes has a single peak and is roughly equally distributed on both sides of that peak. However, it is less useful in some cases where the data might be split or where there are some outliers. **Outliers** are data points that are far from the rest of the data. For example, suppose there are 10 executives and 90 employees at a company. The executives make $1000 per hour, and the employees make $10 per hour.

Therefore, the average pay rate will be:

$$\frac{\$1000 \times 10 + \$10 \times 90}{100} = \$109 \; per \; hour$$

In this case, this average is not very descriptive since it's not close to the actual pay of the executives *or* the employees.

Median

Another useful measurement is the **median**. In a data set, the median is the point in the middle. The middle refers to the point where half the data comes before it and half comes after, when the data is recorded in numerical order. For instance, these are the speeds of the fastball of a pitcher during the last inning that he pitched (in order from least to greatest):

90, 92, 93, 93, 95, 96, 97, 97, 97

There are nine total numbers, so the middle or *median* number is the 5th one, which is 95.

In cases where the number of data points is an even number, then the average of the two middle points is taken. In the previous example of test scores, the two middle points are 75 and 80. Since there is no single point, the average of these two scores needs to be found. The average is:

$$\frac{75 + 80}{2} = 77.5$$

The median is generally a good value to use if there are a few outliers in the data. It prevents those outliers from affecting the "middle" value as much as when using the mean.

Since an outlier is a data point that is far from most of the other data points in a data set, this means an outlier also is any point that is far from the median of the data set. The outliers can have a substantial effect on the mean of a data set, but they usually do not change the median or mode, or do not change them by a large quantity. For example, consider the data set (3, 5, 6, 6, 6, 8). This has a median of 6 and a mode of 6, with a mean of $\frac{34}{6} \approx 5.67$. Now, suppose a new data point of 1000 is added so that the data set is now (3, 5, 6, 6, 6, 8, 1000). The median and mode, which are both still 6, remain unchanged. However, the average is now $\frac{1034}{7}$, which is approximately 147.7. In this case, the median and mode will be better descriptions for most of the data points.

The reason for outliers in a given data set is a complicated problem. It is sometimes the result of an error by the experimenter, but often they are perfectly valid data points that must be taken into consideration.

Mode

One additional measure to define for *X* is the **mode**. This is the data point that appears most frequently. If two or more data points all tie for the most frequent appearance, then each of them is considered a mode. In the case of the test scores, where the numbers were 50, 60, 65, 65, 75, 80, 85, 85, 90, 100, there are two modes: 65 and 85.

Quartiles and Percentiles

The **first quartile** of a set of data *X* refers to the largest value from the first ¼ of the data points. In practice, there are sometimes slightly different definitions that can be used, such as the median of the first half of the data points (excluding the median itself if there are an odd number of data points). The term also has a slightly different use: when it is said that a data point lies *in the first quartile*, it means it is less than or equal to the median of the first half of the data points. Conversely, if it lies *at* the first quartile, then it is equal to the first quartile.

When it is said that a data point lies in the **second quartile**, it means it is between the first quartile and the median.

The **third quartile** refers to data that lies between ½ and ¾ of the way through the data set. Again, there are various methods for defining this precisely, but the simplest way is to include all of the data that lie between the median and the median of the top half of the data.

Data that lies in the **fourth quartile** refers to all of the data above the third quartile.

Percentiles may be defined in a similar manner to quartiles. Generally, this is defined in the following manner:

If a data point lies **in the n-th percentile**, this means it lies in the range of the first *n*% of the data.

If a data point lies **at the n-th percentile**, then it means that *n*% of the data lies below this data point.

Standard Deviation

Given a data set *X* consisting of data points $(x_1, x_2, x_3, \dots x_n)$, the **variance** of *X* is defined to be:

$$\frac{\sum_{i=1}^{n}(x_i - \bar{X})^2}{n}$$

This means that the variance of X is the average of the squares of the differences between each data point and the mean of X.

Given a data set X consisting of data points $(x_1, x_2, x_3, \dots x_n)$, the **standard deviation** of X is defined to be:

$$s_x = \sqrt{\frac{\sum_{i=1}^{n}(x_i - \bar{X})^2}{n}}$$

In other words, the standard deviation is the square root of the variance.

Both the variance and the standard deviation are measures of how much the data tend to be spread out. When the standard deviation is low, the data points are mostly clustered around the mean. When the standard deviation is high, this generally indicates that the data are quite spread out, or else that there are a few substantial outliers.

As a simple example, compute the standard deviation for the data set (1, 3, 3, 5). First, compute the mean, which will be:

$$\frac{1 + 3 + 3 + 5}{4} = \frac{12}{4} = 3$$

Now, find the variance of X with the formula:

$$\sum_{i=1}^{4}(x_i - \bar{X})^2 = (1 - 3)^2 + (3 - 3)^2 + (3 - 3)^2 + (5 - 3)^2$$

$$-2^2 + 0^2 + 0^2 + 2^2 = 8$$

Therefore, the variance is $\frac{8}{4} = 2$. Taking the square root, the standard deviation will be $\sqrt{2}$.

Note that the standard deviation only depends upon the mean, not upon the median or mode(s). Generally, if there are multiple modes that are far apart from one another, the standard deviation will be high. A high standard deviation does not always mean there are multiple modes, however.

Describing a Set of Data

A set of data can be described in terms of its center, spread, shape and any unusual features. The center of a data set can be measured by its mean, median, or mode. The spread of a data set refers to how far the data points are from the center (mean or median). A data set with data points clustered around the center will have a small spread. A data set covering a wide range will have a large spread.

When a data set is displayed as a graph like the one below, the shape indicates if a sample is normally distributed, symmetrical, or has measures of skewness. When graphed, a data set with a normal distribution will resemble a bell curve.

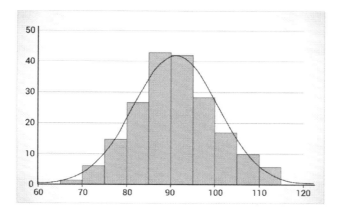

If the data set is symmetrical, each half of the graph when divided at the center is a mirror image of the other. If the graph has fewer data points to the right, the data is skewed right. If it has fewer data points to the left, the data is skewed left.

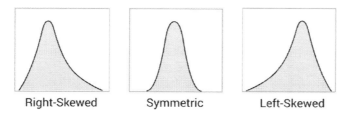

A description of a data set should include any unusual features such as gaps or outliers. A gap is a span within the range of the data set containing no data points. An outlier is a data point with a value either extremely large or extremely small when compared to the other values in the set.

The graphs above can be referred to as **unimodal** since they all have a single peak. In contrast, a bimodal graph has two peaks.

Practice Questions

1. Which of the following numbers has the greatest value?
   a. 1.4378
   b. 1.07548
   c. 1.43592
   d. 0.89409

2. The value of 6 x 12 is the same as:
   a. 2 x 4 x 4 x 2
   b. 7 x 4 x 3
   c. 6 x 6 x 3
   d. 3 x 3 x 4 x 2

3. This chart indicates how many sales of CDs, vinyl records, and MP3 downloads occurred over the last year. Approximately what percentage of the total sales was from CDs?

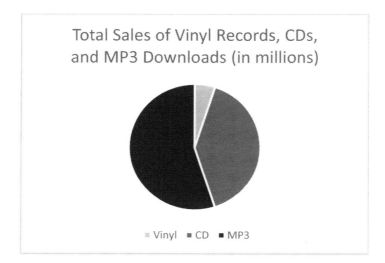

   a. 55%
   b. 25%
   c. 40%
   d. 5%

4. Alan currently weighs 200 pounds, but he wants to lose weight to get down to 175 pounds. What is this difference in kilograms? (1 pound is approximately equal to 0.45 kilograms.)
   a. 9 kg
   b. 11.25 kg
   c. 78.75 kg
   d. 90 kg

5. Johnny earns $2334.50 from his job each month. He pays $1437 for monthly expenses. Johnny is planning a vacation in 3 months' time that he estimates will cost $1750 total. How much will Johnny have left over from three months' of saving once he pays for his vacation?
 a. $948.50
 b. $584.50
 c. $852.50
 d. $942.50

6. Solve the following:

$$(\sqrt{36} \times \sqrt{16}) - 3^2$$

 a. 30
 b. 21
 c. 15
 d. 13

7. In Jim's school, there are 3 girls for every 2 boys. There are 650 students in total. Using this information, how many students are girls?
 a. 260
 b. 130
 c. 65
 d. 390

8. Kimberley earns $10 an hour babysitting, and after 10 p.m., she earns $12 an hour, with the amount paid being rounded to the nearest hour accordingly. On her last job, she worked from 5:30 p.m. to 11 p.m. In total, how much did Kimberley earn on her last job?
 a. $45
 b. $57
 c. $62
 d. $42

9. Arrange the following numbers from least to greatest value:

$0.85, \frac{4}{5}, \frac{2}{3}, \frac{91}{100}$

 a. $0.85, \frac{4}{5}, \frac{2}{3}, \frac{91}{100}$

 b. $\frac{4}{5}, 0.85, \frac{91}{100}, \frac{2}{3}$

 c. $\frac{2}{3}, \frac{4}{5}, 0.85, \frac{91}{100}$

 d. $0.85, \frac{91}{100}, \frac{4}{5}, \frac{2}{3}$

10. Keith's bakery had 252 customers go through its doors last week. This week, that number increased to 378. Express this increase as a percentage.
 a. 26%
 b. 50%
 c. 35%
 d. 12%

11. Simplify the following expression:

$$4\frac{2}{3} - 3\frac{4}{9}$$

    a. $1\frac{1}{3}$

    b. $1\frac{2}{9}$

    c. 1

    d. $1\frac{2}{3}$

12. Jessica buys 10 cans of paint. Red paint costs $1 per can and blue paint costs $2 per can. In total, she spends $16. How many red cans did she buy?

    a. 2

    b. 3

    c. 4

    d. 5

13. Six people apply to work for Janice's company, but she only needs four workers. How many different groups of four employees can Janice choose?

    a. 6

    b. 10

    c. 15

    d. 36

14. Which of the following is equivalent to the value of the digit 3 in the number 792.134?

    a. 3×10

    b. 3×100

    c. $\frac{3}{10}$

    d. $\frac{3}{100}$

15. In the following expression, which operation should be completed first? $5 \times 6 + (5 + 4) \div 2 - 1$.

    a. Multiplication

    b. Addition

    c. Division

    d. Parentheses

16. How will the number 847.89632 be written if rounded to the nearest hundredth?

    a. 847.90

    b. 900

    c. 847.89

    d. 847.896

17. The perimeter of a 6-sided polygon is 56 cm. The length of three of the sides are 9 cm each. The length of two other sides are 8 cm each. What is the length of the missing side?

    a. 11 cm

    b. 12 cm

    c. 13 cm

    d. 10 cm

18. Which of the following is a mixed number?

   a. $16\frac{1}{2}$
   b. 16
   c. $\frac{16}{3}$
   d. $\frac{1}{4}$

19. Change 9.3 to a fraction.

   a. $9\frac{3}{7}$
   b. $\frac{903}{1000}$
   c. $\frac{9.03}{100}$
   d. $9\frac{3}{10}$

20. What is the value of b in this equation?

$5b - 4 = 2b + 17$

   a. 13
   b. 24
   c. 7
   d. 21

21. Express the solution to the following problem in decimal form:

$$\frac{3}{5} \times \frac{7}{10} \div \frac{1}{2}$$

   a. 0.042
   b. 84%
   c. 0.84
   d. 0.42

22. Katie works at a clothing company and sold 192 shirts over the weekend. One third of the shirts that were sold were patterned, and the rest were solid. Which mathematical expression would calculate the number of solid shirts Katie sold over the weekend?

   a. $192 \times \frac{1}{3}$
   b. $192 \div \frac{1}{3}$
   c. $192 \times (1 - \frac{1}{3})$
   d. $192 \div 3$

23. Which four-sided shape is always a rectangle?
   a. Rhombus
   b. Square
   c. Parallelogram
   d. Quadrilateral

24. A rectangle was formed out of pipe cleaner. Its length was $\frac{1}{2}$ ft, and its width was $\frac{11}{2}$ inches. What is its area in square inches?

    a. $\frac{11}{4}$ inch^2

    b. $\frac{11}{2}$ inch^2

    c. 22 inches^2

    d. 33 inches^2

25. How will $\frac{4}{5}$ be written as a percent?

    a. 40 percent

    b. 125 percent

    c. 90 percent

    d. 80 percent

26. If Danny takes 48 minutes to walk 3 miles, how long should it take him to walk 5 miles maintaining the same speed?

    a. 32 min

    b. 64 min

    c. 80 min

    d. 96 min

27. A solution needs 5 ml of saline for every 8 ml of medicine given. How much saline is needed for 45 ml of medicine?

    a. $\frac{225}{8}$ ml

    b. 72 ml

    c. 28 ml

    d. $\frac{45}{8}$ ml

28. What unit of volume is used to describe the following 3-dimensional shape?

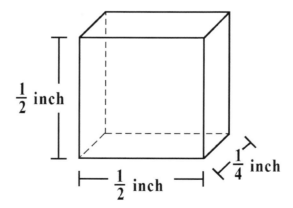

    a. Square inches

    b. Inches

    c. Cubic inches

    d. Squares

29. Which common denominator would be used in order to evaluate $\frac{2}{3} + \frac{4}{5}$?
    a. 15
    b. 3
    c. 5
    d. 10

30. The diameter of a circle measures 5.75 centimeters. What tool could be used to draw such a circle?
    a. Ruler
    b. Meter stick
    c. Compass
    d. Yard stick

31. A piggy bank contains 12 dollars' worth of nickels. A nickel weighs 5 grams, and the empty piggy bank weighs 1050 grams. What is the total weight of the full piggy bank?
    a. 1,110 grams
    b. 1,200 grams
    c. 2,150 grams
    d. 2,250 grams

32. Last year, the New York City area received approximately $27\frac{3}{4}$ inches of snow. The Denver area received approximately 3 times as much snow as New York City. How much snow fell in Denver?
    a. $71\frac{3}{4}$ inches
    b. $27\frac{1}{4}$ inches
    c. $89\frac{1}{4}$ inches
    d. $83\frac{1}{4}$ inches

33. Which of the following would be an instance in which ordinal numbers are used?
    a. Katie scored a 9 out of 10 on her quiz.
    b. Matthew finished second in the spelling bee.
    c. Jacob missed one day of school last month.
    d. Kim was 5 minutes late to school this morning.

34. How will the following number be written in standard form: $(1 \times 10^4) + (3 \times 10^3) + (7 \times 10^1) + (8 \times 10^0)$
    a. 137
    b. 13,780
    c. 1,378
    d. 13,078

35. What is the area of the regular hexagon shown below?

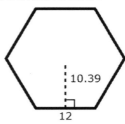

a. 72
b. 124.68
c. 374.04
d. 748.08

36. The area of a given rectangle is 24 square centimeters. If the measure of each side is multiplied by 3, what is the area of the new figure?
a. 48 cm²
b. 72 cm²
c. 216 cm²
d. 13,824 cm²

37. Which of the following is the definition of a prime number?
a. A number that factors only into itself and 1
b. A number greater than one that factors only into itself and 1
c. A number less than 10
d. A number divisible by 10

38. Add and express in reduced form $\frac{5}{12} + \frac{4}{9}$
a. $\frac{9}{17}$
b. $\frac{1}{3}$
c. $\frac{31}{36}$
d. $\frac{3}{5}$

39. Which of the following is the correct order of operations that could be used on a difficult math problem that contained grouping symbols?
a. Parentheses, Exponents, Multiplication, Division, Addition, Subtraction
b. Exponents, Parentheses, Multiplication, Division, Addition, Subtraction
c. Parentheses, Exponents, Addition, Multiplication, Division, Subtraction
d. Parentheses, Exponents, Division, Addition, Subtraction, Multiplication

40. Convert $\frac{5}{8}$ to a decimal.
a. 0.62
b. 1.05
c. 0.63
d. 1.60

41. Subtract $9,576 - 891$.
    a. 10,467
    b. 9,685
    c. 8,325
    d. 8,685

42. If a teacher was showing a class how to round 245.2678 to the nearest thousandth, which place value would be used to decide whether to round up or round down?
    a. Ten-thousandth
    b. Thousandth
    c. Hundredth
    d. Thousand

43. Gary is driving home to see his parents for Christmas. He travels at a constant speed of 60 miles per hour for a total of 350 miles. How many minutes will it take him to travel home if he takes a break for 10 minutes every 100 miles?

44. Kelly is selling cookies to raise money for the chorus. She has 500 cookies to sell. She sells 45% of the cookies to the sixth graders. At the second lunch, she sells 40% of what's left to the seventh graders. If she sells 60% of the remaining cookies to the eighth graders, how many cookies does Kelly have left at the end of all lunches?

45. What is the value of the following expression?

$$\sqrt{8^2 + 6^2}$$

46. Sam is twice as old as his sister, Lisa. Their oldest brother, Ray, will be 25 in three years. If Lisa is 13 years younger than Ray, how old is Sam?

47. What is the perimeter of the following figure rounded to the nearest tenth?

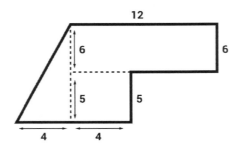

48. Solve the following:

$$\left(\sqrt{36} \times \sqrt{16}\right) - 3^2$$

49. What is the overall median of Dwayne's current scores: 78, 92, 83, 97?

50. The total perimeter of a rectangle is 36 cm. If the length is 12 cm, what is the width?

Answer Explanations

1. A: Compare each numeral after the decimal point to figure out which overall number is greatest. In answers *A* (1.43785) and *C* (1.43592), both have the same tenths (4) and hundredths (3). However, the thousandths is greater in answer *A* (7), so *A* has the greatest value overall.

2. D: By grouping the four numbers in the answer into factors of the two numbers of the question (6 and 12), it can be determined that:

$$(3 \times 2) \times (4 \times 3) = 6 \times 12$$

Alternatively, each of the answer choices could be prime factored or multiplied out and compared to the original value. 6×12 has a value of 72 and a prime factorization of $2^3 \times 3^2$.

The answer choices respectively have values of 64, 84, 108, 72, and 144 and prime factorizations of

$$2^6, 2^2 \times 3 \times 7, 2^2 \times 3^3$$

and $2^3 \times 3^2$, so answer *D* is the correct choice.

3. C: The sum total percentage of a pie chart must equal 100%. Since the CD sales take up less than half of the chart and more than a quarter (25%), it can be determined to be 40% overall.

This can also be measured with a protractor. The angle of a circle is 360°.

Since 25% of 360° would be 90° and 50% would be 180°, the angle percentage of CD sales falls in between; therefore, it would be Choice *C*.

4. B: Using the conversion rate, multiply the projected weight loss of 25 lb by 0.45 $\frac{kg}{lb}$ to get the amount in kilograms (11.25 kg).

5. D: First, subtract $1437 from $2334.50 to find Johnny's monthly savings; this equals $897.50.

Then, multiply this amount by 3 to find out how much he will have (in three months) before he pays for his vacation: this equals $2692.50.

Finally, subtract the cost of the vacation ($1750) from this amount to find how much Johnny will have left: $942.50.

6. C: Follow the *order of operations* in order to solve this problem. Solve the parentheses first, and then follow the remainder as usual.

$$(6 \times 4) - 9$$

This equals $24 - 9$ or 15, answer *C*.

7. D: Three girls for every two boys can be expressed as a ratio: 3:2. This can be visualized as splitting the school into 5 groups: 3 girl groups and 2 boy groups. The number of students which are in each group can be found by dividing the total number of students by 5:

$$\frac{650 \text{ students}}{5 \text{ groups}} = \frac{130 \text{ students}}{\text{group}}$$

To find the total number of girls, multiply the number of students per group (130) by the number of girl groups in the school (3). This equals 390, Choice *D*.

8. C: Kimberley worked 4.5 hours at the rate of $10/h and 1 hour at the rate of $12/h. The problem states that her pay is rounded to the nearest hour, so the 4.5 hours would round up to 5 hours at the rate of $10/h:

$$(5h)\left(\frac{\$10}{h}\right) + (1h)\left(\frac{\$12}{h}\right) = \$50 + \$12 = \$62$$

9. C: The first step is to depict each number using decimals.

$$\frac{91}{100} = 0.91$$

Dividing the numerator by denominator of $\frac{4}{5}$ to convert it to a decimal yields 0.80, while $\frac{2}{3}$ becomes 0.66 recurring. Rearrange each expression in ascending order, as found in answer *C*.

10. B: First, calculate the difference between the larger value and the smaller value.

$$378 - 252 = 126$$

11. B: Simplify each mixed number of the problem into a fraction by multiplying the denominator by the whole number and adding the numerator:

$$\frac{14}{3} - \frac{31}{9}$$

Since the first denominator is a multiple of the second, simplify it further by multiplying both the numerator and denominator of the first expression by 3 so that the denominators of the fractions are equal.

$$\frac{42}{9} - \frac{31}{9} = \frac{11}{9}$$

Simplifying this further, divide the numerator 11 by the denominator 9; this leaves 1 with a remainder of 2. To write this as a mixed number, place the remainder over the denominator, resulting in $1\frac{2}{9}$.

12. C: We are trying to find x, the number of red cans. The equation can be set up like this:

$$x + 2(10 - x) = 16$$

The left x is actually multiplied by $1, the price per red can. Since we know Jessica bought 10 total cans, $10 - x$ is the number blue cans that she bought. We multiply the number of blue cans by $2, the price per blue can.

That should all equal $16, the total amount of money that Jessica spent. Working that out gives us:

$$x + 20 - 2x = 16$$

$$20 - x = 16$$

$$x = 4$$

13. C: Janice will be choosing 4 employees out of a set of 6 applicants, so this will be given by the choice function. The following equation shows the choice function worked out:

$$\binom{6}{4} = \frac{6!}{4!\,(6-4)!} = \frac{6!}{4!\,(2)!}$$

$$\frac{6 \cdot 5 \cdot 4 \cdot 3 \cdot 2 \cdot 1}{4 \cdot 3 \cdot 2 \cdot 1 \cdot 2 \cdot 1} = \frac{6 \cdot 5}{2} = 15$$

14. D: $\frac{3}{100}$. Each digit to the left of the decimal point represents a higher multiple of 10 and each digit to the right of the decimal point represents a quotient of a higher multiple of 10 for the divisor.

The first digit to the right of the decimal point is equal to the value $\div$ 10. The second digit to the right of the decimal point is equal to the value $\div$ (10×10), or the value $\div$ 100.

15. D: Using the order of operations, multiplication and division are computed first from left to right. Multiplication is on the left; therefore, multiplication should be performed first.

16. A: 847.90. The hundredths place value is located two digits to the right of the decimal point (the digit 9 in the original number). The digit to the right of the place value is examined to decide whether to round up or keep the digit.

In this case, the digit 6 is 5 or greater so the hundredth place is rounded up. When rounding up, if the digit to be increased is a 9, the digit to its left is increased by one and the digit in the desired place value is made a zero. Therefore, the number is rounded to 847.90.

17. C: Perimeter is found by calculating the sum of all sides of the polygon. $9 + 9 + 9 + 8 + 8 + s = 56$, where s is the missing side length. Therefore, 43 plus the missing side length is equal to 56. The missing side length is 13 cm.

18. A: $16\frac{1}{2}$. A mixed number contains both a whole number and either a fraction or a decimal. Therefore, the mixed number is $16\frac{1}{2}$.

19. D: $9\frac{3}{10}$

To convert a decimal to a fraction, remember that any number to the left of the decimal point will be a whole number. Then, sense 0.3 goes to the tenths place, it can be placed over 10.

20. C: To solve for the value of b, both sides of the equation need to be equalized.

Start by cancelling out the lower value of -4 by adding 4 to both sides:

$$5b - 4 = 2b + 17$$
$$5b - 4 + 4 = 2b + 17 + 4$$
$$5b = 2b + 21$$

The variable b is the same on each side, so subtract the lower 2b from each side:

$$5b = 2b + 21$$
$$5b - 2b = 2b + 21 - 2b$$
$$3b = 21$$

Then divide both sides by 3 to get the value of b:

$$3b = 21$$

$$\frac{3b}{3} = \frac{21}{3}$$

$$b = 7$$

21. C: The first step in solving this problem is expressing the result in fraction form. Separate this problem first by solving the division operation of the last two fractions. When dividing one fraction by another, invert or flip the second fraction and then multiply the numerator and denominator.

$$\frac{7}{10} \times \frac{2}{1} = \frac{14}{10}$$

Next, multiply the first fraction with this value:

$$\frac{3}{5} \times \frac{14}{10} = \frac{42}{50}$$

In this instance, you can find the decimal form by converting the fraction into $\frac{x}{100}$, where x is the number from which the final decimal is found. Multiply both the numerator and denominator by 2 to get the fraction as an expression of $\frac{x}{100}$.

$$\frac{42}{50} \times \frac{2}{2} = \frac{84}{100}$$

In decimal form, this would be expressed as 0.84.

22. C: $\frac{1}{3}$ of the shirts sold were patterned. Therefore, $1 - \frac{1}{3} = \frac{2}{3}$ of the shirts sold were solid. Anytime "of" a quantity appears in a word problem, multiplication needs to be used. Therefore:

$$192 \times \frac{2}{3} = 192 \times \frac{2}{3}$$

$$\frac{384}{3} = 128 \text{ solid shirts were sold}$$

The entire expression is:

$$192 \times \left(1 - \frac{1}{3}\right)$$

23. B: A rectangle is a specific type of parallelogram. It has 4 right angles. A square is a rhombus that has 4 right angles. Therefore, a square is always a rectangle because it has two sets of parallel lines and 4 right angles.

24. D: Area = length x width. The answer must be in square inches, so all values must be converted to inches. $\frac{1}{2}$ ft is equal to 6 inches. Therefore, the area of the rectangle is equal to:

$$6 \times \frac{11}{2} = \frac{66}{2} = 33 \text{ square inches}$$

25. D: 80 percent. To convert a fraction to a percent, the fraction is first converted to a decimal. To do so, the numerator is divided by the denominator:

$$4 \div 5 = 0.8$$

To convert a decimal to a percent, the number is multiplied by 100:

$$0.8 \times 10 = 80\%$$

26. C: 80 min. To solve the problem, a proportion is written consisting of ratios comparing distance and time. One way to set up the proportion is:

$$\frac{3}{48} = \frac{5}{x} \left(\frac{distance}{time} = \frac{distance}{time}\right) \text{ where } x \text{ represents the unknown value of time}$$

To solve a proportion, the ratios are cross-multiplied:

$$(3)(x) = (5)(48) \rightarrow 3x = 240$$

The equation is solved by isolating the variable, or dividing by 3 on both sides, to produce $x = 80$.

27. A: Every 8 ml of medicine requires 5 ml. The 45 ml first needs to be split into portions of 8 ml. This results in $\frac{45}{8}$ portions. Each portion requires 5 ml. Therefore:

$$\frac{45}{8} \times 5 = 45 \times \frac{5}{8}$$

$$\frac{225}{8} \text{ ml is necessary}$$

28. C: Volume of this 3-dimensional figure is calculated using length x width x height. Each measure of length is in inches. Therefore, the answer would be labeled in cubic inches.

29. A: A common denominator must be found. The least common denominator is 15 because it has both 5 and 3 as factors. The fractions must be rewritten using 15 as the denominator.

30. C: A compass is a tool that can be used to draw a circle. The compass would be drawn by using the length of the radius, which is half of the diameter.

31. D: A dollar contains 20 nickels. Therefore, if there are 12 dollars' worth of nickels, there are:

$$12 \times 20 = 240 \text{ nickels}$$

Each nickel weighs 5 grams. Therefore, the weight of the nickels is:

$$240 \times 5 = 1,200 \text{ grams}$$

Adding in the weight of the empty piggy bank, the filled bank weighs 2,250 grams.

32. D: 3 must be multiplied times $27\frac{3}{4}$. In order to easily do this, the mixed number should be converted into an improper fraction.

$$27\frac{3}{4} = 27 \times 4 + \frac{3}{4} = \frac{111}{4}$$

Therefore, Denver had approximately:

$$3 \times \frac{111}{4} = \frac{333}{4} \text{ inches of snow}$$

The improper fraction can be converted back into a mixed number through division.

$$\frac{333}{4} = 83\frac{1}{4} \text{ inches}$$

33. B: Ordinal numbers represent a ranking. Placing second in a competition is a ranking among the other participants of the spelling bee.

34. D: 13,078. The power of 10 by which a digit is multiplied corresponds with the number of zeros following the digit when expressing its value in standard form. Therefore:

$$(1 \times 10^4) + (3 \times 10^3) + (7 \times 10^1) + (8 \times 10^0)$$

$$10,000 + 3,000 + 70 + 8 = 13,078$$

35. C: 374.04. The formula for finding the area of a regular polygon is $A = \frac{1}{2} \times a \times P$ where a is the length of the apothem (from the center to any side at a right angle), and P is the perimeter of the figure.

The apothem a is given as 10.39, and the perimeter can be found by multiplying the length of one side by the number of sides (since the polygon is regular):

$$P = 12 \times 6 \rightarrow P = 72$$

To find the area, substitute the values for a and P into the formula:

$$A = \frac{1}{2} \times a \times P \rightarrow A = \frac{1}{2} \times (10.39) \times (72)$$

$$A = 374.04$$

36. C: 216cm. Because area is a two-dimensional measurement, the dimensions are multiplied by a scale that is squared to determine the scale of the corresponding areas. The dimensions of the rectangle are multiplied by a scale of 3. Therefore, the area is multiplied by a scale of 3^2 (which is equal to 9):

$$24cm \times 9 = 216cm$$

37. B: A number is prime because its only factors are itself and 1. Positive numbers (greater than one) can be prime numbers.

38. C: $\frac{31}{36}$

Set up the problem and find a common denominator for both fractions.

$$\frac{5}{12} + \frac{4}{9}$$

Multiply each fraction across by 1 to convert to a common denominator.

$$\frac{5}{12} \times \frac{3}{3} + \frac{4}{9} \times \frac{4}{4}$$

Once over the same denominator, add across the top. The total is over the common denominator.

$$\frac{15 + 16}{36} = \frac{31}{36}$$

39. A: Order of operations follows PEMDAS—Parentheses, Exponents, Multiplication and Division from left to right, and Addition and Subtraction from left to right.

40. C: 0.63

Divide 5 by 8, which results in 0.63.

41. D: 8,685

Set up the problem, with the larger number on top. Begin subtracting with the far-right column (ones). Borrow 10 from the column to the left, when necessary.

42. A: The place value to the right of the thousandth place, which would be the ten-thousandth place, is what gets used. The value in the thousandth place is 7. The number in the place value to its right is greater than 4, so the 7 gets bumped up to 8. Everything to its right turns to a zero, to get 245.2680. The zero is dropped because it is part of the decimal.

43.

		3	8	0

380 miles. To find the total driving time, the total distance of 350 miles can be divided by the constant speed of 60 miles per hour. This yields a time of 5.8333 hours, which is then rounded. Once the driving time is computed, the break times need to be found. If Gary takes a break for 10 minutes every 100 miles, he will take 3 breaks on his trip. This will yield a total of 30 minutes of break time. Since the answer is needed in minutes, 5.8333 can be converted to minutes by multiplying by 60, giving a driving time of 350 minutes. Adding the break time of 30 minutes to the driving time of 350 minutes gives a total travel time of 380 minutes.

44.

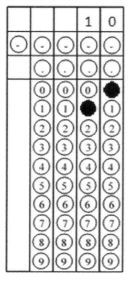

66 Cookies. If the sixth graders bought 45% of the cookies, the number they bought is found by multiplying 0.45 by 500. They bought 225 cookies. The number of cookies left is:

$$500 - 225 = 275$$

During the second lunch, the seventh graders bought 40% of the cookies, which is found by multiplying 0.40 by the remaining 275 cookies. The seventh graders bought 110 cookies. This leaves 165 cookies to sell to the eighth graders. If they bought 60% of the remaining cookies, then they bought 99 cookies. Subtracting 99 from 165 cookies leaves Kelly with 66 cookies remaining after the three lunches.

45.

10. 8 squared is 64, and 6 squared is 36. These should be added together to get:

$$64 + 36 = 100$$

Then, the last step is to find the square root of 100 which is 10.

46.

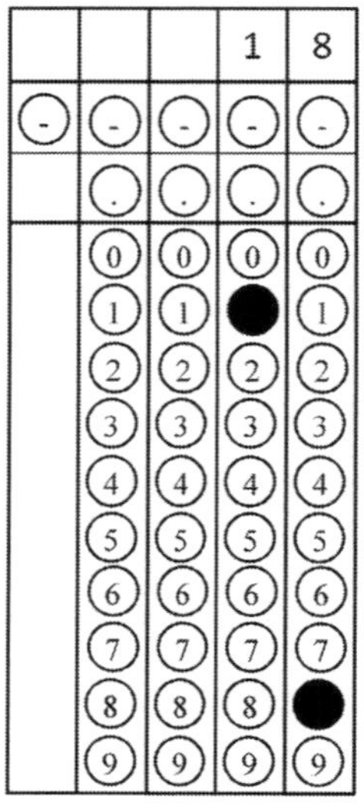

			1	8
⊖	⊖	⊖	⊖	⊖
	⊙	⊙	⊙	⊙
	⓪	⓪	⓪	⓪
	①	①	●	①
	②	②	②	②
	③	③	③	③
	④	④	④	④
	⑤	⑤	⑤	⑤
	⑥	⑥	⑥	⑥
	⑦	⑦	⑦	⑦
	⑧	⑧	⑧	●
	⑨	⑨	⑨	⑨

18; If Ray will be 25 in three years, then he is currently 22. The problem states that Lisa is 13 years younger than Ray, so she must be 9. Sam's age is twice that, which means that the correct answer is 18.

47.

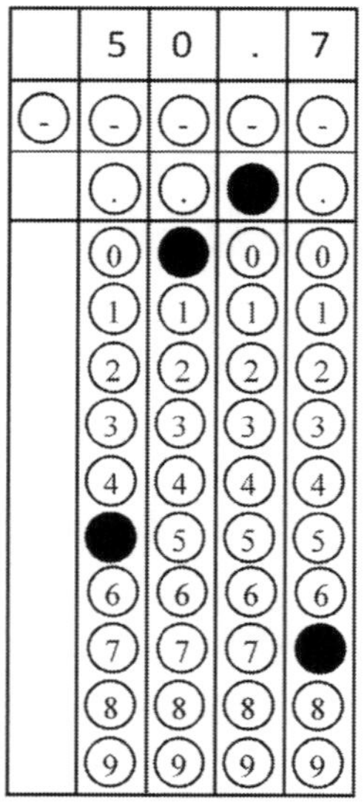

	5	0	.	7
⊖	⊖	⊖	⊖	⊖
	⊙	⊙	●	⊙
	⓪	●	⓪	⓪
	①	①	①	①
	②	②	②	②
	③	③	③	③
	④	④	④	④
	●	⑤	⑤	⑤
	⑥	⑥	⑥	⑥
	⑦	⑦	⑦	●
	⑧	⑧	⑧	⑧
	⑨	⑨	⑨	⑨

50.7; The values for the missing sides must first be found before the perimeter can be calculated. The missing side that is the hypotenuse of the right triangle can be calculated using the Pythagorean Theorem as follows:

$$11^2 + 4^2 = x^2$$

$$121 + 16 = x^2$$

$$137 = x^2$$

$$x = 11.7$$

The other missing side is equal to the value of the length of the larger rectangle less than the value of the side of the square $12 - 4 = 8$. Then, all the sides can be added together to find the perimeter:

$$12 + 6 + 8 + 5 + 4 + 4 + 11.7 = 50.7$$

48.

			1	5
⊝	⊝	⊝	⊝	⊝
	⊙	⊙	⊙	⊙
⓪	⓪	⓪	⓪	⓪
①	①	●		①
②	②	②	②	②
③	③	③	③	③
④	④	④	④	④
⑤	⑤	⑤	●	⑤
⑥	⑥	⑥	⑥	⑥
⑦	⑦	⑦	⑦	⑦
⑧	⑧	⑧	⑧	⑧
⑨	⑨	⑨	⑨	⑨

15; Follow the *order of operations* in order to solve this problem. Solve the parentheses first, and then follow the remainder as usual.

$$(6 \times 4) - 9$$

This equals $24 - 9$ or 15.

49.

87.5; For an even number of total values, the *median* is calculated by finding the *mean* or average of the two middle values once all values have been arranged in ascending order from least to greatest. In this case, $(92 + 83) \div 2$ would equal the median 87.5.

50.

6; The formula for the perimeter of a rectangle is $P = 2L + 2W$, where P is the perimeter, L is the length, and W is the width. The first step is to substitute all of the data into the formula:

$$36 = 2(12) + 2W$$

Simplify by multiplying 2×12:

$$36 = 24 + 2W$$

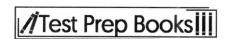

Simplifying this further by subtracting 24 on each side, which gives:

$$36 - 24 = 24 - 24 + 2W$$

$$12 = 2W$$

Divide by 2:

$$6 = W$$

The width is 6 cm. Remember to test this answer by substituting this value into the original formula:

$$36 = 2(12) + 2(6)$$

Social Studies

U.S. History

North American Geography, Peoples, and Cultures Prior to European Colonization

During the last Ice Age, large glaciers covered much of North America, trapping seawater and lowering the sea level. This exposed a land bridge between northwest Asia and North America that has been named **Beringia**. Nomadic peoples from Siberia used the Beringia land bridge to cross into North America. When the Ice Age ended a few thousand years later, the glaciers melted, the sea level rose, and Beringia disappeared beneath the ocean. The nomads who had crossed into North America before the Ice Age ended were searching for prey such as mastodons and wooly mammoths. These nomadic peoples lived in small groups and slowly spread out across North and South America during the next several millennia.

Several thousand years ago, the **Neolithic Revolution** occurred. This event occurred at different times in different places around the world, but it marked an important turning point in human history. The Neolithic Revolution included the development of agriculture and the domestication of animals. In North America, the primary crops were maize (corn), tomatoes, pumpkins, chilies, potatoes, and beans. There were only a few animals for Native Americans to domesticate, including the dog (in North America) and the llama (in Central America). Dogs were used to guard and hunt, and llamas produced wool and also transported goods.

The development of agriculture and domesticated animals led to a major change in North American societies: instead of small nomadic bands of hunter-gatherers, many native peoples became semi-sedentary. They began to form settlements, create social hierarchies, and develop new religious beliefs. However, many Native Americans remained semi-nomadic and lived in villages for only part of the year. Most Native American villages engaged in long distance trade with other groups. In general, most Native American peoples practiced animist beliefs—in other words, they thought that everything in the natural world had some spiritual power, including plants, animals, and rocks. In most Native American societies, women were responsible for farming while men hunted and fished.

Although most Native Americans continued to live in small villages, some larger settlements eventually developed. In Central America in 1325 A.D., the Aztec built the city of Tenochtitlan, located on the site of present day Mexico City, which was thought to contain more than 200,000 inhabitants. This meant it could have been one of the largest cities in the world before Spanish conquistadors destroyed it in 1521. In North America, Native Americans built the city of Cahokia (near the modern city of St. Louis), which may have had up to 40,000 residents before it was abandoned sometime around 1300. Cahokia also included a large circle of timber posts that may have had been used for astronomical predictions, similar to Stonehenge in Britain.

North and South America were rich in natural resources and also featured a variety of climates and geographies. This led to the development of various Native American societies before European colonists arrived.

The Iroquois and Algonquians were the major tribes in the eastern part of North America. They spoke different languages and frequently fought each other. The vast forests, especially in the Northeast, Midwest, and Southeast, provided wood for building wigwams and longhouses, as well as tools and

weapons. Native Americans also used major waterways such as the Ohio and Mississippi Rivers to engage in trade, participate in small- and large-scale fishing activities, and launch raids against each other.

Several different tribes also lived between the Mississippi River and the Rocky Mountains—an area later known as the Great Plains. The Sioux, Comanche, Blackfoot, and Cheyenne were among these more nomadic peoples who hunted buffalo. They lived in teepees that were easy to take apart and rebuild in a new location as the tribe moved across the vast prairies. Native Americans sometimes set fire to the grasslands in order to control vegetation and improve grazing conditions.

The Pueblo people—including the Zuni, Hope, and Acoma tribes—lived in the deserts of the southwest. They used clay bricks to build large apartment complexes called pueblos, digging into the side of cliffs to create shelter. Since the environment was so dry, they had to build complex irrigation systems in order to grow crops.

The Tlingit, Chinook, and Salish people lived on the Pacific coast where they survived by hunting and fishing. Salmon were a staple of their diet and culture, playing an important part in their eating habits as well as their religious beliefs. The tribes developed techniques to create waterproof baskets and woven raincoats due to the wet climate of the Pacific Northwest.

The Aleuts and Inuit lived in the arctic and subarctic regions of North America where they hunted seals and whales. Because of the harsh conditions, they built semi-subterranean shelters that were insulated against the cold.

Founding and Development of European Colonies in North America

European monarchs began dispatching naval expeditions in the 15th century in an effort to bypass Muslim powers in the Middle East, such as the Ottoman Empire, which controlled the overland trade routes that led to Asia. European nation-states were competing against each other for wealth and power and wanted to have direct contact with Asia. The spice trade was especially lucrative, but Europeans also sought to spread Christianity and earn a reputation for making exciting discoveries. Several technological improvements helped European explorers establish trading outposts in Africa, Asia, and the Americas. Tools such as the astrolabe and quadrant allowed sailors to navigate more accurately, and the Portuguese built a new ship called a caravel that was faster and more maneuverable.

Europeans had discovered a route around the southern tip of Africa that led to Asia. However, the journey was long and dangerous. Christopher Columbus, a Genoese sailor, lobbied several European monarchs and sought financial support for an expedition across the Atlantic Ocean. He mistakenly believed he could reach Asia much more quickly via a western route across the Atlantic and was unaware that his path was blocked. Eventually, King Ferdinand and Queen Isabella of Castile accepted his proposal and provided funding for three ships as well as sailors and provisions. They agreed to appoint Columbus governor of any territory he discovered and granted him ten percent of the revenue from the new lands. After five weeks at sea, Columbus landed on San Salvador and explored the Caribbean islands for about three months before returning home.

Interaction of European, Native American, and African Peoples During the Colonial Period

News of Columbus's success sparked a number of other expeditions and the British, French, Dutch, Spanish, and Portuguese all eventually laid claim to lands in the New World. Columbus himself made three more voyages to the Americas. The French and Dutch focused mostly on the lucrative fur trade in North America. The Spanish and Portuguese sought gold in Central and South America but also tried to convert Native Americans to Christianity. British settlers also sought economic opportunity and created the first British colony at Jamestown, Virginia, in 1607. However, the Puritans who landed at Plymouth Rock in 1620 left for the New World in order to establish their ideal religious community.

Connecticut, New Hampshire, Massachusetts, and Rhode Island were considered the "New England colonies." The settlements in New England were based around an economy focused on fishing and lumber. These colonies maintained puritanical and Congregationalist religious beliefs. While English Puritans mostly settled in New England, a wide variety of colonists settled in the mid-Atlantic region. English, Scottish, Dutch, and Swedish settlers came to Delaware, New York, New Jersey, and Pennsylvania. As a result, the mid-Atlantic colonies were more religiously diverse and tolerant than the settlements in New England. Agriculture was the foundation of the economy in mid-Atlantic colonies. This meant that settlements were more dispersed. Government and administration were based on counties instead of towns.

The southern colonies, including Virginia, Maryland, the Carolinas, and Georgia, were also organized by county. The southern economy focused on labor-intensive crops such as tobacco and rice, and as a result, landowners relied on indentured servants and African slaves. Slaves were present in most colonies, but were more common in the south.

Political power was also distributed differently among the colonies. Some colonies, such as New York and Virginia, were royal colonies ruled directly by the king. Pennsylvania was a proprietary colony—the king allowed William Penn to appoint officials and govern the colony as he saw fit. Corporate colonies, such as Rhode Island and Connecticut, were administered by a group of investors. But, by the early 1700s, the king had revoked the charters of most proprietary and corporate colonies and assumed direct control himself.

Native Americans continued to play an important role in the early history of Britain's North American colonies. Squanto was an Algonquian Indian who helped English settlers in Massachusetts survive by teaching them how to plant native crops. Some Native American tribes were friendly towards the colonists and traded with them.

However, Native Americans and Europeans often came into conflict, frequently over land disputes. The Native Americans and Europeans had very different concepts of land use and ownership. Native Americans did not understand the concept of landownership or sale. When they entered into agreements with the colonists, Native Americans thought they were allowing the settlers to farm the land temporarily, rather than retain it in perpetuity. On the other hand, colonists were frustrated when Native Americans continued to hunt and fish on lands they had "sold." These, and other disagreements, eventually led to bloody conflicts that gradually weakened Native American tribes.

Native Americans were also vulnerable to diseases to which the Europeans had developed immunity. These diseases included bubonic plague, cholera, chicken pox, pneumonic plague, influenza, measles, scarlet fever, typhus, smallpox, and tuberculosis. These diseases killed millions of Native Americans and

were sometimes used as a biological weapon. Historians estimate that as much as 80 percent of the Native American population died through disease and warfare.

Origins of the American Revolution and Founding of the United States

The French colonies in Canada also threatened the British settlements. France and Britain had been enemies for centuries. Religious differences reinforced their hostility; the British were Protestant and the French were mostly Catholic. Far fewer colonists settled in "New France," but they often clashed with the British, especially over the lucrative fur trade. Both the British and French sought to dominate the trade in beaver pelts, which were used to make hats in Europe. The British and French fought a series of colonial wars between 1689 and 1748 that failed to resolve the struggle for dominance in North America.

Eventually, the contest culminated in the **French and Indian War** (which was part of the **Seven Years' War**), which ended in 1763. The French initially enjoyed the upper hand because they were able to persuade more Native American tribes to support them. The Native Americans felt the French were less likely to encroach on their territory than the land-hungry British. The Native Americans launched devastating raids along the British colonial frontier. However, the British eventually emerged victorious after they blockaded the French colonies in Canada. This prevented the French from bringing in reinforcements or from resupplying their Native American allies with gunpowder and ammunition. Native American raids subsided and eventually the French surrendered almost all of their colonial possessions in North America. Some historians consider this war the first global conflict because battles were also fought in Europe, Asia, and Africa.

The French defeat radically altered the balance of power in North America. Previously, Native Americans had been able to play the French and British against each other, but now they were without many of their French allies. In addition, the French and Indian War also set the stage for the American Revolution. Although victorious, the British monarchy spent an enormous amount of money and the war doubled the national debt. In order to pay off the debts, King George III began imposing taxes upon the North American colonies, which eventually led to revolution.

Since 1651, the British crown had tried to control trade within its empire, which eventually led to tension and discontent in the North American colonies. That year, the monarchy introduced the Navigation Acts, which prevented the North American colonies from trading directly with other European powers—all goods had to be shipped to Britain first. This was an attempt to keep wealth within the British Empire and to prevent other empires from profiting from their colonies. This was an example of **mercantilism**—an economic policy that formed the foundation of Britain's empire. Mercantilism called for government regulation in the form of tariffs, a tax on imports from other countries. This raised prices on foreign goods and encouraged British imperial subjects to purchase goods made in Britain or the colonies. This reduced imports and maximized exports, thus enriching the British Empire.

The **Molasses Act** in 1731 was another outgrowth of mercantilism. This law imposed a higher tax on the molasses that colonists purchased from the Dutch, French, or Spanish colonies. The tax was unpopular with the colonists and British imperial officials eventually decided not to enforce the tax. The Molasses Act had threatened to disrupt the pattern of triangular trade that had emerged in the Atlantic world. First, ships from Britain's North American colonies carried rum to Africa where it was traded for slaves and gold. Then, the ships took the slaves to French and Spanish colonies in the Caribbean and exchanged them for sugar or molasses. In the last part of the triangular trade system, merchants sailed back to North America where the sugar and molasses was used to make rum, and the cycle could start over again.

In addition to economic connections, many other bonds also bridged the Atlantic Ocean. Most colonists shared a common language, common religion, and common culture. However, as the colonies grew in population, they began to develop local institutions and a separate sense of identity. For example, it became common for ministers to receive their education at seminaries in North America rather than Britain. Newspapers also began to focus on printing more local news as well. Perhaps most importantly, the colonies began to exercise more control over their own political affairs. The British government retained control over international issues, such as war and trade, but the colonists controlled their own domestic affairs. Colonies began to form their own political assemblies and elect landowners who represented local districts. In addition, communications between the colonies and Britain were very slow because it took months for a ship to cross the Atlantic and return with a response.

A number of political acts by the British monarchy also led to more discontent among the colonies. After the French and Indian War ended in 1763, the king declared that the colonists could not settle west of the Appalachian Mountains. This was known as the **Proclamation of 1763**. Many colonists were frustrated because they had expected this territory would be open for expansion after the French had been defeated.

Additionally, taxes were imposed in an effort to help reduce the debt Britain amassed during the French and Indian War. In 1764, Parliament passed the **Sugar Act**, which reduced the tax on molasses but also provided for greater enforcement powers. Some colonists protested by organizing boycotts on British goods. One year later, in 1765, Parliament passed the **Quartering Act**, which required colonists to provide housing and food to British troops. This law was also very unpopular and led to protests in the North American colonies.

The **Stamp Act** of 1765 required the colonists to pay a tax on legal documents, newspapers, magazines and other printed materials. Colonial assemblies protested the tax and petitioned the British government in order to have it repealed. Merchants also organized boycotts and established correspondence committees in order to share information. Eventually, Parliament repealed the Stamp Act but simultaneously reaffirmed the Crown's right to tax the colonies.

In 1767, Parliament introduced the **Townshend Acts**, which imposed a tax on goods the colonies imported from Britain, such as tea, lead, paint, glass, and paper. The colonies protested again and British imperial officials were assaulted in some cases. The British government sent additional troops to North America to restore order. The arrival of troops in Boston only led to more tension that eventually culminated in the **Boston Massacre** in 1770, where five colonists were killed and eight were wounded. Except for the duty on tea, most of Townshend Act taxes were repealed after the Boston Massacre.

Parliament passed the **Tea Act** in 1773 and, although it actually reduced the tax on tea, it was another unpopular piece of legislation. The Tea Act allowed the British East India Company to sell its products directly, effectively cutting out colonial merchants and stirring more Anglo-American anger and resentment. This resulted in the **Boston Tea Party** in 1773, an incident in which colonial tea merchants disguised themselves as Indians before storming several British ships that were anchored in Boston harbor. Once aboard, the disguised colonists dumped more than 300 chests of tea into the water.

Because the British government was unable to identify the perpetrators, Parliament passed a series of laws that punished the entire colony of Massachusetts. These acts were known as the **Coercive** or **Intolerable Acts**. The first law closed the port of Boston until the tea had been paid for (an estimated

$1.7 million in today's currency). The second act curtailed the authority of Massachusetts' colonial government. Instead of being elected by colonists, most government officials were now appointed by the king. In addition, the act restricted town meetings, the basic form of government in Massachusetts, and limited most villages to one meeting per year. This act angered colonists throughout the thirteen colonies because they feared their rights could be stripped away as well. A third act allowed for British soldiers to be tried in Britain if they were accused of a crime. The fourth act once again required colonists to provide food and shelter to British soldiers.

Colonists responded by forming the **First Continental Congress** in 1774, and all the colonies except for Georgia sent delegates. The delegates sought a compromise with the British government instead of launching an armed revolt. The First Continental Congress sent a petition to King George III affirming their loyalty but demanding the repeal of the Intolerable Acts. The delegates organized a boycott of imports from and exports to Britain until their demands were met.

The colonists began to form militias and gather weapons and ammunition. The first battle of the revolution began at Lexington and Concord in April 1775 when British troops tried to seize a supply of gunpowder and were confronted by about eighty Minutemen. A brief skirmish left eight colonists dead and ten wounded. Colonial reinforcements poured in and harassed the British force as they retreated to Boston. Although the battle did not result in many casualties, it marked the beginning of war.

A month later, the **Second Continental Congress** convened in Philadelphia. The delegates formed an army and appointed George Washington as commander in chief. Delegates were still reluctant to repudiate their allegiance to King George III and did not do so until they issued the **Declaration of Independence** on July 4, 1776. The Declaration drew on the ideas of the Enlightenment and declared that the colonists had the right to life, liberty, and the pursuit of happiness. The Declaration stated that the colonists had to break away from Britain because King George III had violated their rights.

After the **Battle of Lexington and Concord**, British troops retreated to Boston and the colonial militias laid siege to the city. Colonists built fortifications on Bunker Hill outside the city and British troops attacked the position in June 1775. The colonists inflicted heavy casualties on the British and killed a number of officers. However, the defenders ran out of ammunition and British troops captured Bunker Hill on the third assault. Although it was a defeat for the colonists, the **Battle of Bunker Hill** demonstrated that they could stand and fight against the disciplined and professional British army.

The British army initially had the upper hand and defeated colonial forces in a number of engagements. The Americans did not achieve a victory until the **Battle of Trenton** in December 1776. Washington famously crossed the Delaware River on Christmas Day and launched a surprise attack against Hessian mercenaries. They captured more than 1,000 soldiers and suffered very minimal casualties. The victory at Trenton bolstered American morale and showed that they could defeat professional European soldiers.

The **Battle of Saratoga** in New York in the fall of 1777 was an important turning point in the **American War for Independence**. American troops surrounded and captured more than 6,000 British soldiers. This victory convinced the French king to support the revolutionaries by sending troops, money, weapons, and ships to the American continent. French officers who fought alongside the Patriots brought back many ideas with them that eventually sparked a revolution in France in 1789.

French support was very important in the last major battle of the revolution at Yorktown, Virginia, in 1781. American troops laid siege to General Cornwallis's British forces at Yorktown. The French fleet defeated a British naval squadron sent to relieve Cornwallis. French and American troops began attacking the British fortifications in Yorktown; a sustained artillery bombardment by American guns eventually forced Cornwallis to surrender. This ended the Revolutionary War, and in 1783 the British signed the Treaty of Paris. Britain recognized the United States as an independent country and set the Mississippi River as the nation's western border. However, British troops continued to occupy several forts in the Great Lakes region.

In addition, tens of thousands of colonists who remained loyal to the British Empire fled the United States after the war. They were known as loyalists and many thousands had joined militias and fought against the patriots. Some loyalists fled to Canada or Britain but many remained in the United States. Many Native American tribes had sided with the British as well in an attempt to curb western expansion. No Native American leaders signed the Treaty of Paris and they refused to give up their territories, which led to further conflict as the new American nation began to expand westward.

Adoption of the Constitution and Bill of Rights

America's first system of government was actually laid out in the **Articles of Confederation**, and not the Constitution. The Articles of Confederation were ratified during the Revolutionary War and went into effect in 1781. The Articles of Confederation created a relatively weak central government and allowed individual states to retain most of the power. Under this system, the national government did not have a president or judiciary. Each state had only one vote in the Confederation Congress and most major decisions required unanimous approval by all thirteen states. Despite this requirement, the Confederation Congress did pass some important legislation, including the **Northwest Ordinance**, which organized the land west of Appalachian Mountains. The territories eventually became the states of Ohio, Indiana, Michigan, Illinois, Wisconsin, and Minnesota. However, Congress did not have the power to tax and could only request money from the states without any way to enforce its demands. A Revolutionary War veteran named Daniel Shays led an armed insurrection in western Massachusetts in 1787. Although Shay's Rebellion was defeated, it drew attention to the weaknesses of the Articles of Confederation.

The **Constitutional Convention** met in Philadelphia a few months after the rebellion in order to create a stronger federal government. However, delegates disagreed over how to structure the new system. The Virginia Plan was one proposal that included a bicameral legislature where states were awarded representation based on their population size. This would benefit more populous states at the expense of smaller states. The other main proposal was the **New Jersey Plan**, which retained many elements of the Articles of Confederation, including a unicameral legislature with one vote per state. This plan would put states on an equal footing regardless of population.

Eventually, delegates agreed to support the **Connecticut Compromise** (also known as the **Great Compromise**), which incorporated elements from both the Virginia and New Jersey Plans. Under the new Constitution, Congress would be a bicameral body. In the House of Representatives, states would be allocated seats based on population, but in the Senate each state would have two votes. The Constitution also included a president and judiciary that would each serve to check the power of other branches of government. In addition, Congress had the power to tax and had more enforcement powers.

Slavery was another contentious issue during the Constitutional Convention. Slavery was more common in the Southern states and less common in the North. The Southern states wanted slaves to be counted when calculating representation in Congress but not when it came to assessing taxes. Northern states wanted the opposite and eventually the two sides agreed to the **Three-Fifths Compromise** where slaves

were counted as three-fifths of a person for the purposes of both taxation and representation. The Constitution also included a provision that allowed slave owners to recover slaves who had escaped and permitted the international slave trade to continue until 1808.

Once the Constitution had been drafted, nine of the thirteen states had to ratify it. Vigorous debate erupted over whether or not the Constitution should be approved. Two different political factions emerged. The Federalists supported the Constitution because they felt a stronger central government was necessary in order to promote economic growth and improve national security. Several leading federalists, including Alexander Hamilton, John Jay, and James Madison, published a series of articles urging voters to support the Constitution. However, the Anti-Federalists, including Thomas Jefferson and Patrick Henry, felt that the Constitution took too much power away from the states and gave it to the national government. They also thought there weren't enough protections for individual rights and lobbied for the addition of a Bill of Rights that guaranteed basic liberties. Ultimately, the Constitution was ratified in 1788 and the Bill of Rights was approved a year later.

The Electoral College unanimously elected George Washington as the nation's first president in 1789. Despite this appearance of unity, deep political divisions led to the formation of the nation's first party system. Washington supported the Federalist ideology and appointed several Federalists to his cabinet, including Alexander Hamilton as secretary of the treasury. The Anti-Federalist faction evolved into the Democratic-Republican Party and favored stronger state governments instead of a powerful federal government. As settlers moved into the new Northwest Territories, Washington helped pacify Indians who opposed further expansion. He also successfully put down a rebellion in western Pennsylvania by farmers opposed to a federal tax on whiskey.

A number of different issues divided the Federalists and the Democratic-Republicans, including the French Revolution, which began in 1789. Initially, many Americans supported the French effort to replace their monarchy and create a republican government. However, the French Revolution quickly became more violent, as thousands of suspected opponents of the revolution were executed during the Reign of Terror. The Federalists, including Washington, were horrified by the violence, while Jefferson and the Democratic-Republicans thought the United States should help its former ally. Washington ensured that the country remain officially neutral.

Washington declined to seek a third term and another Federalist, John Adams, became our second president. Adams signed the Alien and Sedition Acts, which made it a criminal offense to criticize the government, and allowed the president to deport aliens suspected of treason. Adams and the Federalists argued that the laws were necessary in order to improve security as Europe became embroiled in a war against the new French republic. Jefferson and the Democratic-Republicans said the laws restricted free speech. Jefferson made the acts an important topic in 1800 when he successfully ran for president.

Jefferson's victory marked a turning point in the political system because the Democratic-Republicans gained more power while the Federalists went into decline. He repealed the **Alien and Sedition Acts** when he was elected. The Federalists were further weakened when Hamilton was killed in a duel in 1804.

Jefferson accomplished several significant achievements during his presidency, and one of the most important was the **Louisiana Purchase** in 1803. For $15 million, Jefferson bought French territory west of the Mississippi River that doubled the size of the United States. He then appointed Meriwether Lewis and William Clark to lead an expedition to explore the vast new territory and study its geography, vegetation, and plant life. Clark also brought his African-American slave, York, on the journey. York helped hunt and even saved Clark's life during a flood. The expedition was also aided by Sacagawea, a Shoshone woman, who acted as a guide and interpreter. The explorers also established relations with Native American tribes and set the stage for further western expansion in the 1800s.

Several key Supreme Court decisions were also issued during this time. The case of *Marbury vs. Madison* established the policy of judicial review, which declared that the Supreme Court could rule whether or not an act of Congress was constitutional. The case of *McCullough vs. Maryland* affirmed that Congress had the power to pass laws that were "necessary and proper" in order to carry out its other duties. The case also upheld the supremacy of federal laws over state laws when they came into conflict.

War between the United States and Britain broke out in 1812 because the United States was drawn into a conflict between Britain and France. Britain refused to stop interfering with American ships bound for France and had begun forcibly recruiting American citizens into the British navy. Furthermore, the British still occupied several forts near the Great Lakes and continued to encourage Indians to attack American settlements in the Northwest Territories.

This led to war in 1812, and many Native American leaders allied themselves with the British, including the Shawnee warrior Tecumseh. Tecumseh temporarily united several tribes but his confederacy fell apart when he was killed in battle. This further weakened Native American resistance and facilitated American settlement in the Northwest Territory after the war.

The United States also achieved a victory at the **Battle of Lake Erie** where several American ships routed a British squadron. However, an American attempt to invade Canada failed, and the British humiliated the nation by invading Washington D.C. and burning down several public buildings, including the White House. The United States did achieve another victory after hostilities had ceased when future president Andrew Jackson repulsed a British attack at New Orleans. The war did not result in any major territorial gains or losses, but it did reaffirm American independence and gave America its national anthem, the "Star Spangled Banner." It also led to the collapse of the Federalist Party, which had opposed the war. The Democratic-Republicans dominated politics for the next decade, which was known as the "Era of Good Feelings," thus marking the end of the first party system.

Causes and Consequences of Territorial Expansion

Constant immigration meant that land prices in the eastern United States rose, and people sought new economic opportunities on the frontier where land was cheaper. The United States government tried purchasing land from Native Americans, but most refused to relinquish their territories. Native Americans continued to defend their land until Tecumseh was defeated in the **War of 1812.** This defeat helped secure the Northwest Territory, and more settlers began pouring in. After the Louisiana Purchase, Lewis and Clark paved the way for expansion into the Great Plains and further west.

The Cherokee, Chickasaw, Choctaw, Creek, and Seminole tribes of the Southeastern United States were known as the "Five Civilized Tribes" because they had developed a written language and many members had become Christians. Nevertheless, Andrew Jackson signed the **Indian Removal Act** of 1830, which gave him the power to continue buying land from various tribes. The Cherokee filed a lawsuit to protect their territory and won their arguments before the Supreme Court. However, President Andrew Jackson ignored the ruling and eventually used troops to force many tribes off their land and sent them west to Oklahoma during the 1830s. This was known as the "Trail of Tears" because thousands of Native Americans died from starvation, exposure, and disease along the way.

Furthermore, the concept of **Manifest Density** emerged during the 1800s and introduced the idea that God wanted Americans to civilize and control the entire continent. This led to conflict when the province of Texas declared its independence from Mexico and asked to be annexed by the United States. President James K. Polk tried to buy Texas, but when Mexico refused, he sent troops into the disputed territory. Mexican troops responding by attacking an American unit, which led to war in 1846.

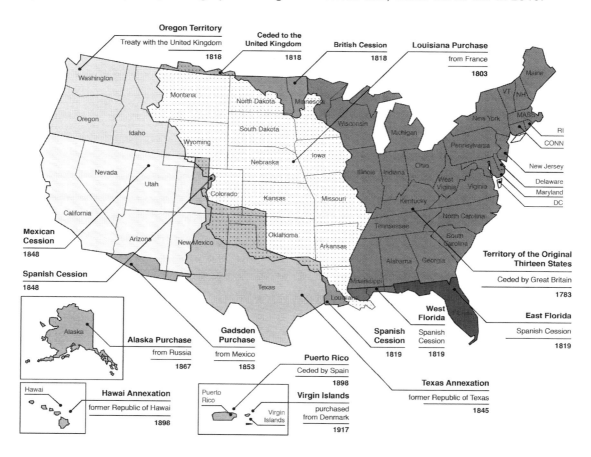

In the **Mexican-American War**, American troops won several battles although the Mexican army usually outnumbered them. The Mexican troops were poorly armed and trained, but, on the other hand, the Americans made use of their highly skilled artillery force. They eventually captured Mexico City and forced the Mexican government to sign the **Treaty of Guadalupe-Hidalgo** in 1848. The treaty recognized American control over Texas and also ceded California, Utah, Colorado, Arizona, New Mexico, and Nevada in exchange for $15 million. Tens of thousands of prospectors flooded into California when gold was discovered there in 1849. The prospectors often encroached on Native American lands, which led to further conflict. In 1854, the United States also acquired additional territories as part of the Gadsden Purchase. The acquisition of so much new territory sparked a debate over whether the land would be open or closed to slavery.

Manifest Destiny also sparked a desire to expand American influence into Central and South America. Adventurers launched several unsuccessful attempts to invade Nicaragua and Cuba.

Several important laws also stimulated western expansion during the second half of the 19th century. Congress passed the Homestead Act in 1862, which allowed citizens to claim 160 acres for only $1.25 per acre. The settler also had to live on the land for five years and make improvements. That same year,

Congress also passed the **Pacific Railroad Act**, which supported the construction of a transcontinental railroad. The United States government provided land and financial support to railroad companies and the first transcontinental link was established in 1869. This facilitated trade and communication between the eastern and western United States.

As Americans poured westward, conflict again broke out between settlers and Native Americans. The discovery of gold in the Black Hills of South Dakota caused prospectors to flood into the area although the U.S. government had recognized the territory belonged to the Sioux. General George Armstrong Custer brought in troops to try and take possession of the Black Hills. This led to disaster when Custer and more than 250 soldiers died at the Battle of Little Big Horn in 1876.

The U.S. government continued its efforts to control Native American tribes. The **Dawes Act** of 1887 encouraged Native Americans to settle on reservations and become farmers in exchange for U.S. citizenship. Chief Joseph was a leader of the Nez Perce tribe who refused to live on a reservation and tried to flee to Canada. However, the U.S. captured Chief Joseph and his tribe and forced them onto a reservation. Reformers also required Native Americans to send their children to boarding schools where they had to speak English and dress like Caucasians instead of maintaining their traditional culture. The schools were often crowded, and students were also subjected to physical and sexual abuse.

In 1890, the Lakota Indians tried to preserve their traditional beliefs by performing a special ceremony called a Ghost Dance. U.S. government officials felt threatened and sent soldiers to try and disarm the Lakota. This led to the **Massacre at Wounded Knee** in 1890 where at least 150 Lakota, including many women and children, were slaughtered. It was the last major conflict between Native Americans and U.S. forces.

The United States purchased Alaska from Russia in 1867 for $7.2 million. At the time, the purchase was unpopular with the public, but seal hunting became very profitable and gold was discovered in 1896. Alaska became a state in 1959.

In 1893, American businessmen launched an armed coup, overthrew the queen of Hawaii, and asked Congress to annex Hawaii. The businessmen owned sugar plantations and feared the queen's attempts to enact reform would threaten their political influence. Hawaii became a U.S. territory in 1898 and a state in 1959.

The last phase of American territorial expansion occurred as a result of the **Spanish-American War** in 1898. New ideas arose in the late 19th century that helped justify further expansion. Some intellectuals applied Charles Darwin's ideas of "survival of the fittest" to the human race and called this new concept Social Darwinism. They used this idea to justify why stronger groups of people colonized and exploited weaker groups. In addition, imperialists also used the idea of the **White Man's Burden** to justify further expansion. They claimed that Caucasians were obligated to civilize and govern groups thought to be less advanced.

These ideas were used to justify America's new status as a colonial power as a result of the Spanish-American War. Although Spain had once been a powerful empire, it had been in decline. The United States went to war against Spain in 1898 when the American battleship USS Maine exploded in Havana Harbor and killed more than 250 sailors. The U.S. Navy defeated the Spanish fleet in several engagements and then the Army followed up with a victory at San Juan Hill, which included the famous charge by Teddy Roosevelt and the Rough Riders.

The war lasted less than four months and made the United States a world power. The U.S. also acquired several Spanish colonies, including Puerto Rico, Guam, and the Philippines. Guam became an important refueling station for American naval forces in the Pacific and remains a U.S. territory today, along with Puerto Rico. When the United States occupied the Philippines, the Filipino people launched a rebellion in order to obtain their independence. The U.S. Army put down the insurrection, but in doing so, they committed many atrocities against the Filipino people. The Philippines would remain an American territory until 1946.

Causes and Consequences of 19th-Century Sectionalism, the Civil War, and Reconstruction

In the early 1800s, political and economic differences between the North and South became more apparent. Politically, a small but vocal group of abolitionists emerged in the North who demanded a complete end to slavery throughout the United States. William Lloyd Garrison edited the abolitionist newspaper *The Liberator* and vehemently denounced the brutality of slavery. His criticism was so vicious that the legislature of Georgia offered a $5,000 bounty to anyone who could capture Garrison and deliver him to state authorities. Other activists participated in the **Underground Railroad**—a network that helped fugitive slaves escape to the Northern United States or Canada.

Economic differences emerged as the North began to industrialize, especially in the textile industry where factories increased productivity. However, the Southern economy remained largely agricultural and focused on labor-intensive crops such as tobacco and cotton. This meant that slavery remained an essential part of the Southern economy. In addition, the North built more roads, railroads, and canals, while the Southern transportation system lagged behind. The Northern economy was also based on cash, while many Southerners still bartered for goods and services. This led to growing sectional tension between the North and South as their economies began to diverge further apart.

These economic differences led to political tension as well, especially over the debate about the expansion of slavery. This debate became more important as the United States expanded westward into the Louisiana Purchase and acquired more land after the Mexican-American War. Most Northerners were not abolitionists. However, many opposed the expansion of slavery into the western territories because it would limit their economic opportunities. If a territory was open to slavery, it would be more attractive to wealthy slave owners who could afford to buy up the best land. In addition, the presence of slave labor would make it hard for independent farmers, artisans, and craftsman to make a living, because they would have to compete against slaves who did not earn any wages. For their part, Southerners felt it was essential to continue expanding in order to strengthen the southern economy and ensure that the Southern way of life survived. As intensive farming depleted the soil of nutrients, Southern slave owners sought more fertile land in the west.

Both the North and South also feared losing political power as more states were admitted to the nation. For example, neither side wanted to lose influence in the United States senate if the careful balance of free and slave state representation was disrupted. Several compromises were negotiated in Congress, but they only temporarily quieted debate. The first such effort, called the Missouri Compromise, was passed in 1820, and it maintained political parity in the U.S. Senate by admitting Missouri as a slave state and Maine as a free state. The **Missouri Compromise** banned slavery in the portion of the Louisiana Purchase that was north of the 36°30' parallel and permitted slavery in the portion south of that line as well as Missouri.

However, the slavery debate erupted again after the acquisition of new territory during the Mexican-American War. The **Compromise of 1850** admitted California as a free state and ended the slave trade, but not slavery itself, in Washington D.C., in order to please Northern politicians. In return, Southern politicians were able to pass a stronger fugitive slave law and demanded that New Mexico and Utah be allowed to vote on whether or not slavery would be permitted in their state constitutions. This introduced the idea of popular sovereignty where the residents of each new territory, and not the federal government, could decide whether or not they would become a slave state or a free state. This essentially negated the **Missouri Compromise** of 1820. The enhanced fugitive slave law also angered many Northerners, because it empowered federal marshals to deputize anyone, even residents of a free state, and force them to help recapture escaped slaves. Anyone who refused would be subject to a $1,000 fine (equivalent to more than $28,000 in 2015).

The debate over slavery erupted again only a few years later when the territories of Kansas and Nebraska were created in 1854. The application of popular sovereignty meant that pro- and anti-slavery settlers flooded into these two territories to ensure that their faction would have a majority when it came time to vote on the state constitution. Tension between pro- and anti-slavery forces in Kansas led to an armed conflict known as "Bleeding Kansas."

John Brown was a militant abolitionist who fought in "Bleeding Kansas" and murdered five pro-slavery settlers there in 1856. He returned to the eastern United States and attacked the federal arsenal at Harper's Ferry, Virginia, in 1859. He hoped to seize the weapons there and launch a slave rebellion, but federal troops killed or captured most of Brown's accomplices and Brown was executed. The attack terrified Southerners and reflected the increasing hostility between North and South.

The sectional differences that emerged in the last several decades culminated in the presidential election of 1860. Abraham Lincoln led the new Republican Party, which opposed slavery on moral and economic grounds. The question of how best to expand slavery into new territories split the Democratic Party into two different factions that each nominated a presidential candidate. A fourth candidate also ran on a platform of preserving the union by trying to ignore the slavery controversy.

Lincoln found little support outside of the North but managed to win the White House since the Democratic Party was divided. Southern states felt threatened by Lincoln's anti-slavery stance and feared he would abolish slavery throughout the country. South Carolina was the first Southern state to secede from the Union and ten more eventually followed. Lincoln declared that the Union could not be dissolved and swore to defend federal installations. The Civil War began when Confederate troops fired on Fort Sumter in Charleston in 1861.

The Civil War

The **First Battle of Bull Run** (also known as the **First Battle of Manassas**) in 1861 was the first major infantry engagement of the Civil War. Both the Northern and Southern troops were inexperienced and although they had equal numbers, the Confederates emerged victorious. Many had thought the war would be short, but it continued for another four years.

The Union navy imposed a blockade on the Confederacy and captured the port of New Orleans in 1862. The Union navy was much stronger than the Confederate fleet and prevented the Southern states from selling cotton to foreign countries or buying weapons.

In 1862, Union forces thwarted a Confederate invasion of Maryland at the Battle of Antietam. This engagement was the single bloodiest day of the war and more than 23,000 men on both sides were

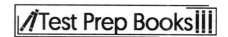

killed or wounded. Union troops forced the Confederates to retreat, and that gave Lincoln the political capital he needed to issue the **Emancipation Proclamation** in 1863. This declaration did not abolish slavery, but it did free slaves in Southern territory. It also allowed African Americans to join the Union navy and about 200,000 did so. The 54th Massachusetts Infantry was a famous unit of African American soldiers who led an assault on Fort Wagner in South Carolina in 1863. Although the attack failed, the 54th Massachusetts witnessed African American troops fighting bravely under fire.

The **Siege of Vicksburg** in 1863 was a major Union victory because they gained control of the Mississippi River and cut the Confederacy in half. This made it difficult the Confederacy to move troops around and communicate with their forces. Grant commanded the Northern forces in the siege and eventually became the Union army's top general.

The **Battle of Gettysburg** in 1863 marked the turning point of the Civil War. Robert E. Lee led Confederate troops into Pennsylvania, but in three days of heavy fighting, the Union army forced them to retreat. The victory bolstered Northern morale and weakened Southern resolve. Never again would Confederate forces threaten Northern territory.

In 1864, Union general William T. Sherman captured Atlanta and then marched more than 200 miles to Savannah. Along the way, he destroyed anything that could support the Southern war effort, such as railroads and cotton mills. At this point, the Southern economy was beginning to collapse. The North had more manpower than the South and could afford to sustain more casualties. The North also had more industrial capacity to produce weapons and supplies and more railroads to transport men and equipment.

Eventually, Robert E. Lee surrendered to Ulysses S. Grant at Appomattox, Virginia, on April 9, 1865. Five days later, John Wilkes Booth assassinated Lincoln in Washington D.C. Vice President Andrew Johnson, a Democrat, succeeded him and soon came into conflict with Republicans in Congress about how to reintegrate Southern states into the nation. This process was known as Reconstruction and lasted from 1865 to 1877.

Reconstruction

Johnson opposed equal rights for African Americans and pardoned many Confederate leaders. However, many Congressional Republicans wanted to harshly punish Southerners for their attempts to secede from the Union. They were known as Radical Republicans because they also wanted to give former slaves equal rights.

Johnson vetoed bills that were designed to protect the rights of freed slaves, but Congress overrode his vetoes. This led to increasing conflict between Johnson and Congress, which eventually caused Radical Republicans to impeach him. Although Johnson was acquitted in 1868, he had very little power, and Radical Republicans took control of the Reconstruction process.

Republicans passed three important constitutional amendments as part of the Reconstruction process. The 13th amendment was ratified in 1865, and it abolished slavery throughout the country. The 14th Amendment was ratified in 1868 and gave equal rights to all citizens. The 15th Amendment was ratified in 1870 and specifically granted all men the right to vote regardless of race.

Southerners resisted these demands and passed laws that prohibited freed slaves from owning weapons or testifying against whites. They also formed militias and vigilante groups, such as the Ku Klux Klan, in order to intimidate African Americans who tried to vote. Congress sent federal troops into Southern states in order to enforce the law and prevent vigilante violence.

After the much-disputed election of 1876, the Democrats offered to let the Republicans have the White House if they agreed to end Reconstruction. After the Republicans agreed, federal troops were withdrawn and African Americans in the South were subjected to discrimination until the Civil Rights movement of the 1960s. Scholars often consider the Reconstruction era the beginning of Jim Crow and a transition into a new form of "institutionalized racism" that still pervades much of modern U.S. society.

Relationships among Industrialization, Urbanization, and Immigration in the Late 19th and Early 20th Centuries

After the end of the Civil War, America experienced a period of intense industrialization, immigration, and urbanization, and all three trends were interrelated. The process of industrialization had begun before the Civil War but expanded into more sectors of the economy in the later part of the century. This era is often called the **Second Industrial Revolution** and included growth in the chemical, petroleum, iron, steel, and telecommunications industries. For example, the Bessemer process made it much easier to produce high quality steel by removing impurities during the smelting process.

The writer Mark Twain called the late 19th century the Gilded Age because the era was also one of extreme social inequality. Some corporations expanded and began to control entire industries. For example, by 1890, the Standard Oil Company produced 88 percent of all the refined oil in the nation. This made a few individuals, such as John D. Rockefeller who owned Standard Oil, extremely wealthy. On the other hand, many workers earned low wages and began to form labor unions, such as the American Federation of Labor in 1886, in order to demand better working conditions and higher pay. Strikes were one of the most common ways workers could express their dissatisfaction, and the Pullman Strike of 1894 was one of the largest such incidents in the 19th century. Workers went on strike after the Pullman Company, which manufactured railroad cars, cut wages by about 25 percent. More than 125,000 workers around the country walked off the job and attacked workers hired to replace them. Federal troops were sent in to end the strike, and more than eighty workers were killed or wounded during confrontations. The strike was unsuccessful, but Congress passed a law making Labor Day a federal holiday in order to placate union members.

Immigration also played an important part in the economic and social changes that occurred during the late 19th century. Immigration patterns changed during this time and immigrants from Southern and Eastern Europe, such as Italy and Poland, began to surpass the number of arrivals from Northern and Western Europe. The immigrants sought economic opportunity in the United States because wages for unskilled workers were higher than in their home countries. Some Americans resented the influx of immigrants because they spoke different languages and practiced Catholicism. In 1924, Congress passed a law that restricted immigration from Southern and Eastern Europe.

Increased **urbanization** was the last factor that contributed to the rapid changes of the Gilded Age. Factories were located near cities in order to draw upon a large pool of potential employees. Immigrants flooded into cities in search of work, and new arrivals often settled in the same neighborhoods where their compatriots lived. Between 1860 and 1890, the urbanization rate increased from about 20 percent to 35 percent. Cities struggled to keep up with growing populations, and services such as sanitation and water often lagged behind demand. Immigrants often lived in crowded living conditions that facilitated the spread of diseases.

Political, Economic, Social, and Cultural Developments from the Progressive Era Through the New Deal

The social inequalities and economic abuses of the Gilded Age did not go unnoticed, and in the 1890s many reformers began to demand change. This period was called the **Progressive Era** and included activists in both the Democratic and Republican parties. The Progressives wanted to use scientific methods and government regulation to improve society. For example, they advocated the use of initiative, referendum, and recall to make government more responsive to its citizens. Progressives also argued that it was necessary to breakup large monopolies (known as trust busting) in order to promote equal economic competition. In 1911, Rockefeller's Standard Oil was split up into thirty-four different companies in order to promote competition, and the Federal Trade Commission was established in 1914 in order to prevent other monopolies from forming. Many Progressives also supported several constitutional amendments that were ratified in early 20th century, including the 17th amendment, which established the direct election of U.S. Senators in 1913 (previously state legislatures had elected senators). They also favored the Prohibition of alcohol that went into effect with the 18th Amendment in 1919. Progressives also advocated for women's rights and backed the 19th Amendment, which gave women the right to vote in 1920.

Many journalists who supported the reform movement of the Progressives were known as **Muckrackers** because they helped expose political corruption and social inequality. Upton Sinclair wrote a novel in 1906 called "The Jungle," which exposed poor working conditions and health violations in the Chicago meatpacking industry. His exposé led to the passages of the **Pure Food and Drug Act** in 1906, which authorized the federal government to inspect the purity of foodstuffs and medicines. Jacob Riis was a photographer who documented the crowded and unhealthy living conditions that many immigrants and poor workers endured.

World War I, from 1914 to 1918, led to a communist revolution in Russia in 1917. Many Americans wanted to prevent political radicals from gaining influence in the United States. A number of strikes and bombings around the country led the federal government to crack down on anarchists, socialists, and communists in an event known as the First Red Scare. In 1919, U.S. Attorney General A. Mitchell Palmer launched a series of raids and arrested resident aliens suspected of belonging to radical groups. About 500 resident aliens were eventually deported.

In 1933, President Franklin D. Roosevelt introduced the **New Deal**, which was a series of executive orders and laws passed by Congress in response to the Great Depression. The programs focused on relief, recovery, and reform, and were enacted until 1938. The second New Deal from 1935-1938 promoted the Social Security Act, labor unions, and aided tenant farmers and migrant workers who were struggling from the economic devastation of the Great Depression.

Social Developments

With the ratification of the 19th Amendment in 1920, women obtained the right to vote. This achievement was partly due to women's contributions on the home front during World War I. Women served as Army nurses and worked in factories to help produce weapons, ammunition, and equipment. As more women entered the workforce, they became more financially independent and began to socialize without being supervised by a chaperone, as was the norm during the 19th century. Overall, women during this period, known as "New Women," took on a more active role in public life, pursued higher education in greater numbers, and sought more sexual freedom. During the 1920s, women, known as "flappers," began to flaunt social conventions by wearing short skirts, bobbing hair, smoking cigarettes, and driving automobiles. Nevertheless, a "glass ceiling" still remains in place decades after women's suffrage in regards to a gender wage gap.

Millions of African Americans also moved north during and after World War I in search of work in a phenomenon known as the Great Migration. This led to increased racial tension as whites and blacks competed for jobs and housing. This culminated in a wave of race riots that swept across the country in the summer of 1919. In Chicago, conflict broke out between whites and blacks at a segregated beach, which led to five days of violence during which thirty-eight people were killed and more than 500 injured. The impact of the Great Migration can still be seen in contemporary, heavily segregated Rust Belt cities such as Gary, IN and Milwaukee, WI.

The invention of the automobile also contributed to social change. Henry Ford applied the method of assembly line construction and scientific management to the automobile manufacturing industry. This made it much cheaper to manufacture cars and allowed more people to purchase them. Automobiles allowed young men and women to socialize and date without adult supervision. Automobiles also improved transportation, increased mobility, and spawned the first suburbs.

Cultural Developments

Motion picture cameras were invented in the late 19th century, and the film industry experienced significant growth in the early 20th century. Because the first movies were silent, dialogue was displayed on intertitles and a live orchestra usually performed during a screening. "Birth of a Nation," by D.W. Griffith, was one of the first major cinema blockbusters, and it portrayed the Ku Klux Klan in a heroic light. Millions of Americans saw the film, which helped the Klan spread throughout the Northern and Western United States.

The **Great Migration** also led to cultural changes during the 1920s known as the **Harlem Renaissance**. The movement was based in the neighborhood of Harlem and led to a rebirth of black literature, art, music, and fashion. Jazz was an important feature of the Harlem Renaissance and challenged musical conventions by emphasizing improvisation and spontaneity. Jazz became very popular with both whites and blacks during the 1920s. Langston Hughes was a poet of the Harlem Renaissance who encouraged his readers to take pride in their black identity.

Economic Developments

The **Panic of 1893** was a worldwide economic depression that devastated the American economy. Businesses went bankrupt, banks collapsed, and unemployment rose to approximately 17 percent. The economy began to recover by 1897, and the beginning of World War I boosted the U.S. economy as European nations bought American goods.

In 1918, the United States emerged from World War I as a major economic power because it had helped finance the Allied war effort and produced large amounts of weapons and equipment. The American agricultural sector also prospered because European farms had been devastated by the war. This sent crop prices up, and farmers used the money to buy more land and equipment. Although the 1920s is usually depicted as an era of economic prosperity, agricultural prices fell after World War I, and farmers were unable to pay back their debts.

Stock market speculation increased during the 1920s, and investors borrowed money in order to purchase shares. This did not cause any concern as long as the stock market went up, but it led to disaster when stock prices fell sharply in October 1929 and investors were unable to repay their loans. The stock market crash may have triggered the **Great Depression**, but it did not cause it. The Great

Depression spread around the globe as nations stopped trading with each other. In the United States, families lost their savings when banks failed because there was no federal insurance. The economy went into a downward spiral because as more people lost their jobs, they had little money to spend, which led to further layoffs and more economic contraction. Unemployment peaked at 25 percent between 1932 and 1933.

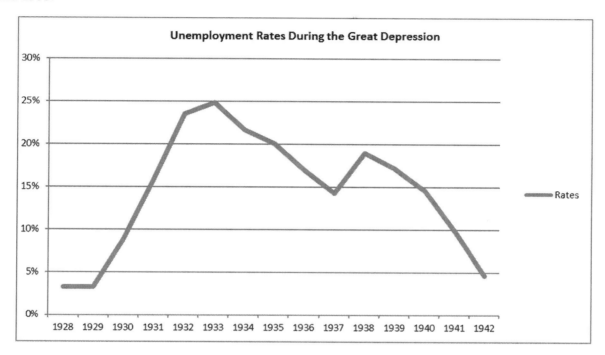

Democratic candidate Franklin D. Roosevelt was elected president in 1932 on his promise to help the economy recover by increasing government spending. After taking office in 1933, Roosevelt introduced a barrage of proposals, called the **New Deal**, that he hoped would boost employment, stimulate demand, and increase government regulation. Some elements of the New Deal were temporary, such as the Civilian Conservation Corps, which put young men to work improving parks between 1933 and 1942. Other New Deal programs endure to this day, such as the Social Security Administration, which has provided pensions to retirees, temporary payments to unemployed workers, and benefits to handicapped individuals since 1935. In addition, the Securities and Exchange Commission was created in 1934 and continues to regulate stock markets and investment companies. The Wagner Act of 1935 was also an important part of the New Deal because it guaranteed the right of workers to unionize and go on strike. The 21st Amendment was ratified in 1933 and repealed Prohibition, which had been hard to enforce and was unpopular. Roosevelt also hoped it would create jobs and stimulate spending. The New Deal helped reduce unemployment, but the economy did not completely recover until America entered World War II and production increased in order to support the war efforts.

Causes and Consequences of American Participation in World War I & II

World War I began in 1914 with the assassination of Franz Ferdinand, the apparent heir of the Austro-Hungarian Empire. A network of secret alliances meant that most European nations were quickly drawn into the conflict, although President Woodrow Wilson initially tried to keep the United States neutral. The war involved two major European alliances: the Triple Entente of Britain, France, and Russia, and the Central Powers, which included Germany and Austria-Hungary. The British implemented a naval blockade that was very successful, and the Germans retaliated by launching submarine attacks. German submarines attacked any ship carrying supplies to the Triple Entente, including the passenger ship RMS Lusitania in 1915. About 1,200 people died, including more than 100 Americans. The Germans temporarily halted their unrestricted submarine campaign, but eventually resumed the attacks in 1917. In addition, in 1917, Germany asked Mexico to attack the United States in a communiqué known as the **Zimmerman telegram**. These events led the United States to join the Triple Entente in 1917, although significant numbers of American troops did not arrive in Europe until 1918. American reinforcement helped the British and French, who had been fighting continuously since 1914, launch a final offensive that defeated Germany in 1918. American forces suffered about 320,000 casualties. As previously noted, World War I also led to significant changes on the home front as women took on new responsibilities, and thousands of African Americans migrated north in search of work. World War I also led to a communist revolution that transformed Russia into the USSR in 1922.

After Germany was defeated in 1918, Wilson made a proposal known as the **Fourteen Points** and argued that the best way to resolve the conflict was by promoting free trade and democracy. For instance, Wilson wanted nations to respect the right to navigate in international waters and create a League of Nations that would resolve future disputes. Some of his suggestions, such as the League of Nations, were adopted, but many were not. In 1919, Germany was forced to sign the Treaty of Versailles, which imposed harsh economic penalties and restricted the German military. Ultimately, the Treaty of Versailles created resentment in Germany that lead to World War II. America emerged as an important player in world affairs after World War I because the American economy had supplied the Triple Entente with arms and equipment and American soldiers helped to achieve victory.

In the period between the world wars, **fascism** became popular in many European countries that were ravaged by the Great Depression. Fascism is a political ideology that advocates for a dictatorship in order to provide stability and unity. Adolf Hitler emerged as a prominent fascist leader in Germany and eventually brought the Nazi party to power in 1933. Germany, Italy, and Japan formed an alliance called the Axis and began to threaten other countries. The League of Nations could not diffuse the conflict. World War II broke out when Germany invaded Poland in 1939. Hitler quickly conquered most of Europe, except for Britain, and attacked the USSR in 1941. The United States sent military equipment and weapons to Britain and the USSR, but did not formally join the war until the Japanese attacked Pearl Harbor on December 7, 1941. Again, women played an important role on the home front by working in factories to build guns, tanks, planes, and ships. African Americans, Native Americans, and Japanese Americans also contributed by fighting on the front lines.

American forces first landed in North Africa where they, along with British and French troops, defeated German and Italian forces in 1942. In 1943, Allied forces invaded Italy, and Soviet troops began to push the German army back out of the USSR. Allied troops landed in France in 1944 and the Soviets began to advance on Germany as well. By May 1945, Hitler had committed suicide and Germany had been defeated.

This also brought about an end to the **Holocaust**. The Holocaust was a genocide committed by Hitler's Nazi Germany and collaborators that resulted in the deaths of more than 6 million Jews and 5 million Romans, gypsies, Afro-Europeans, disabled citizens, and homosexuals. A network of facilities in Germany and its territories were used to house victims for slave labor and mass murder, among other heinous crimes. The **Nuremberg trials** were part of the aftermath of the Holocaust, which served to prosecute important members of Nazi Germany leadership.

In the Pacific theater, American naval forces defeated the Japanese fleet in several key engagements, including the battle of Midway in 1942. American troops began recapturing territory in the Pacific as

well and eventually pushed the Japanese back to their home islands in 1945. The Japanese refused to surrender until American planes dropped atomic bombs on the cities of Nagasaki and Hiroshima in August 1945. Because World War II devastated most of Europe, the United States and the USSR emerged as the only superpowers when it ended. However, the erstwhile allies were suspicious of each other, which led to the Cold War.

Origins, Developments, and Consequences of the Cold War

Although the United States and the USSR worked together to defeat the Axis powers during World War II, the alliance quickly fell apart. As previously discussed, Americans had been afraid of communist influence since the Russian revolution in 1917. The USSR viewed the United States as a capitalist and imperialist power that threatened Soviet security. The USSR and United States divided Europe into spheres of influence, and this mutual hostility led to the Cold War. For example, the Soviets built a wall around the portion of Berlin, Germany that they occupied. The United States formed a military alliance, called the **North Atlantic Treaty Organization**, with its allies in Western Europe. The Soviets responded with their own defensive alliance in Eastern Europe, known as the Warsaw Pact. President Harry Truman announced that the United States would try to contain communist influence and would assist countries threatened by communist aggression. The domino theory predicted that once one country succumbed to communism, neighboring nations would also be at risk (thus following like a stack of dominoes). A resurgence of anti-communist sentiment occurred in the early 1950s when Senator Joseph McCarthy pledged to root out spies within the federal government. Known as **the Second Red Scare** in American history, McCarthy's communist "witch hunts" produced little evidence of his allegations and was censored in 1954 when he refused to reign in his attacks.

The resistance to communism led to a number of indirect conflicts around the globe between the United States and the USSR and China, which had become a communist regime in 1949. For example, Korea was divided into two portions after World War II (at the 38th parallel). When the communist regime in North Korea invaded South Korea in 1950, the United States sent troops to defend South Korea. The USSR and China sent troops, weapons, and equipment to support North Korea. The war ended in a stalemate in 1953, and Korea remains divided to this day. Today, it is divided at a unique demarcation line close to the original 38th parallel separation.

Vietnam was also partitioned into northern and southern regions after World War II. The United States supported South Vietnam and sent troops in ever-increasing numbers. The conflict intensified when North Vietnamese gunboats allegedly attacked a U.S. navy ship in 1964. However, U.S. forces were unable to defeat the Vietnamese communists, who preferred to use guerrilla troops and ambush tactics. The support for the war evaporated in United States as casualties mounted and the draft was unpopular as well. American troops also killed at least 300 civilians in the Vietnamese village of My Lai in 1968. National Guardsmen shot and killed four students and wounded nine others during an anti-war protest at Kent State University in 1970. These scandals made the war extremely unpopular. American troops withdrew in 1973, and South Vietnam was defeated in 1975.

The U.S. and USSR also vied for influence in South America and Africa as well. Some nations refused to pick sides and instead formed the Non-Aligned Movement.

The nuclear arms race and the space race were also important arenas in the **Cold War**. The Soviets took an early lead in the space race and successfully launched the first satellite into orbit in 1957. The Soviets also put the first man, cosmonaut Yuri Gagarin, into space in 1961. However, the United States surpassed the Soviet Union by landing the first man, Neil Armstrong, on the Moon in 1969.

The United States was the only nuclear power for a brief period after World War II, but the Soviets quickly caught up and detonated their own nuclear bomb in 1949. Both sides developed nuclear missiles during the 1950s. This became a very dramatic issue during the **Cuban Missile Crisis**. In 1962, President John F. Kennedy learned that the Soviets were installing nuclear missiles in Cuba, which had become a communist regime. The U.S. navy imposed a blockade, and tensions rose as the Soviets refused to back down. Eventually, negotiations ended the crisis and the Soviets agreed to withdraw the missiles in Cuba if the U.S. agreed to withdraw missiles from Turkey that threatened the USSR. Although tensions remained, the nuclear arms race slowed during the 1970s when the U.S. and USSR entered into negotiations. Both the U.S. and USSR promised to reduce nuclear weapons tests and limit their nuclear arsenals. Throughout the Cold War, a concept known as "Mutually Assured Destruction" (MAD) helped prevent nuclear war. Both the United States and the USSR each had thousands of warheads and MAD predicted that no matter who struck first, the other side would have enough surviving firepower to destroy the aggressor.

During the Cold War, a number of protests and demonstrations occurred in the Soviet satellites in Eastern Europe, including Hungary in 1956 and in Czechoslovakia in 1968. The uprisings were suppressed, but demands for reform continued and, in 1985, Mikhail Gorbachev became the leader of the USSR. He promised to make political and economic reforms, but protestors wanted change to occur more rapidly. Revolutions swept through Eastern Europe in 1989, and the communist regimes in Poland, Romania, and elsewhere crumbled. The **Berlin Wall** fell in 1989, and the Cold War ended when the USSR collapsed in 1991. This left the United States as the sole superpower, although other countries, such as India and China, have become more powerful. Recent scholars, however, are questioning whether the Cold War is once again beginning to heat up with the most recent tensions between Russia and the United States in the early 2000s.

Social, Economic, and Technological Changes Between 1950 and 2000

The post-World War II era led to a number of social, economic, and technological changes in the United States. The counter-culture phenomenon was one of the most powerful social movements in the latter half of the twentieth century in the U.S. The counter-culture movement challenged social norms and rejected traditional authority figures. The movement began in the 1950s with the beatniks, a group of non-conformist writers and artists who were dissatisfied with society. The beatniks sought inspiration in African and Asian cultures and many eschewed materialism.

The **counter-culture movement** became more popular during the 1960s as millions of children from the Baby Boomers generation entered into adulthood. Veterans came home after World War II and started families, and, by the 1960s, many of these young adults also felt disaffected and rebellious. Their parents criticized them because they began to wear colorful clothing and the boys let their hair grow out. Many members of the counter-culture movement, now called hippies, inherited the beatnik's interest in African and Asian cultures. Writer Ken Kesey traveled around the country on a bus encouraging people to experiment with psychedelic drugs, such as LSD. The counter-culture movement influenced musicians and avant-garde artists.

The counter-culture movement was also closely connected to other protest movements during the 1960s, including the Civil Rights movement, which will be discussed later. Many members of the counter-culture movement during the 1960s also opposed the war in Vietnam. The Baby Boomers could be conscripted to fight in Vietnam whether they wanted to or not. In 1965, young men began burning their draft cards, which was a criminal offense, in protest. Massive demonstrations against the war occurred around the country, especially on college campuses, but many other people also refused to

support the war effort, including clergymen and even some veterans who had fought in Vietnam. The counter-culture movement disappeared during the 1970s but had a lasting impact on the social and cultural history of the United States.

Economic Changes

America emerged as one of the most powerful economies in the world after 1945. The US economy, especially manufacturing, was very prosperous during the 1950s and 1960s. The economy successfully switched from wartime production, and consumer demand was very high. During the Great Depression, few families had disposable income. Although most workers earned good wages during World War II, they had little to spend it on because most goods were rationed. Once production of consumer goods resumed, families used their savings to buy cars, household appliances, and televisions. This was good for the economy, and unemployment remained below 5 percent for most of the 1950s and 1960s. However, during the latter part of the 20th century, the manufacturing base in the North and Midwest began to crumble and the area became known as the Rust Belt. Manufacturing jobs began to move from the North and Midwest to states in the South and West, known as the **Sun Belt**, where land was cheap and wages were low.

The world economy also became increasingly interconnected during the post-World War II era. This accelerated the process of globalization, which is the integration of ideas and products from different cultures. This benefitted the United States economically because businesses, such as McDonald's and Coca-Cola, found many consumers around the world who were eager to consume American goods. However, the process works both ways, and many aspects of foreign culture, such as Japanese cartoons and animation, have become very popular in the United States. Many critics also point out that globalization has hurt the American economy in recent decades because manufacturing jobs have gone overseas to countries in South America and Asia where wages are low.

Technological Changes

Several technological changes have had a significant impact on the U.S. economy as well. The Cold War led to advances in nuclear power and aerospace engineering. The development of computers, in particular, has helped accelerate the transition to a post-industrial economy where information technology and other services have replaced traditional manufacturing jobs. The first computers were used to break coded messages during World War II and had very limited computing power. The invention of transistor technology in 1947 made computers cheaper, smaller, and more reliable. The invention of integrated circuits in the 1960s and 1970s increased computing power and gave birth to the first personal computers. The Internet was created in 1969, but widespread use in business and academia did not begin until the 1980s. These developments have made it much easier to share information and have increased economic opportunities. But, the increasing use of robots, especially in the manufacturing industry, have also made the economy more efficient while also causing layoffs.

Political Realignments from the New Deal and Great Society Through the Rise of Conservatism

President Franklin D. Roosevelt created the New Deal in order to stimulate the economy and improve government regulation. The New Deal also marked an important shift in American politics because the Democratic Party began to favor government intervention while Republicans opposed it. This was a reversal of the parties' previous platforms. The Democratic Party relied on a coalition of labor unions, Catholics, African Americans, and other minorities. The Republican Party included conservatives, evangelicals, and business leaders.

The Great Society was another major government program that the Democratic Party supported. President Lyndon B. Johnson sought to end poverty and improve education. For example, he raised the minimum wage and created programs to provide poor Americans with job training. The Great Society also implemented a number of Civil Rights laws that will be discussed in greater detail later.

The presidential election of 1980 was another watershed moment. Republican nominee Ronald Reagan carried forty-three states, and the Republicans won a majority in the U.S. Senate after twenty-eight years of Democratic control. Reagan presented an optimistic message and broadcasted a television advertisement that proclaimed "It's morning again in America." He promised to restore America's military power, cut government regulations, and reduce taxes. Reagan enjoyed the support of resurgent conservative Christian evangelicals, who wanted to restore morality to American society. They were particularly concerned about issues such as abortion. The **Moral Majority**, founded by Baptist minister Jerry Falwell in 1979, was one key group that helped Reagan win the election. This coalition helped realign party loyalties, as more liberal Republicans and conservative Democrats shifted their allegiance.

Impact of Race, Gender, and Ethnicity Throughout American History

Race, gender, and ethnicity have been important themes in American history from the colonial era to the present. Individuals from different races, ethnicities, and genders have had very different experiences throughout the same historical events or eras. It is important to distinguish between race and ethnicity: race refers to a group of people with common ancestry, while ethnicity refers to cultural background, such as language and tradition.

Race played an important part in colonial America because both Caucasians and Africans occupied positions of servitude. White immigrants who could not purchase passage to the New World sometimes agreed to become indentured servants. Their employers paid for their passage across the Atlantic, and in exchange, the indentured servant agreed to work without wages for at least five years. However, African slaves were rarely able to free themselves. The strong connection between slavery and race meant that all blacks, whether free or enslaved, were viewed as inferior. After the American Revolution, most free blacks, even those living in northern states, were denied the right to vote. Although the Civil Rights Movement in the 1960s made great gains, many activists claim more must be done in order to overcome the legacy of racial inequality in America. Japanese Americans experienced discrimination during and after World War II, specifically with the implementation of Japanese internment camps in America. This forced 110,000 to 120,000 Japanese Americans into camps, 62 percent of which were United States citizens.

Discrimination against certain ethnicities is also prominent throughout American history as well. For example, many Americans resented the arrival of German and Irish immigrants during the 1800s because they spoke a different language or practiced different religions. Hispanics were also subject to discrimination, and in 1943, a number of Hispanic youths were attacked during the **Zoot Suit Riots**.

Gender differences in the United States have also been impossible to ignore. For example, until the 1840s, most married women in the United States were unable to enter into contracts, own property, or retain their own wages. As previously discussed, women were unable to vote until the 19th Amendment was ratified in 1920. The **Women's Rights Movement** in the U.S. ranged from 1848 to 1920. This movement called for a woman's right to vote, the right to bodily autonomy, freedom from sexual violence, the right to hold public office, the right to work, the right to fair wages and equal pay, and the right to own property and obtain an education. Women continue to demand change during the 21st century for reasons such as the gender wage gap, better resources for women's health, female

reproductive rights, and for protection of basic human rights, such as bringing greater awareness to rape culture, violence against women, and protection against female sex trafficking.

How Participants in the Political Process Engage in Politics and Shape Policy

The Constitution provides for a series of checks and balances between the legislative, executive, and judicial branches of the federal government. Members of Congress debate and vote on legislation, although the president may request that legislators consider a certain proposal. The president may veto legislation that he or she disagrees with, but Congress can override the veto with a two-thirds majority in both chambers. The Supreme Court may review legislation and declare it unconstitutional. The president selects nominees to the Supreme Court but the Senate must confirm them.

Branch	Role	Checks & Balances on Other Branches	
Executive	Carries out the laws	Legislative Branch • Proposes laws • Vetoes laws • Calls special sessions of Congress • Makes appointments • Negotiates foreign treaties	Judicial Branch • Appoints federal judges • Grants pardons to federal offenders
Legislative	Makes the laws	Executive Branch • Has the ability to override a President's veto • Confirms executive appointments • Ratifies treaties • Has the ability to declare war • Appropriates money • Has the ability to impeach and remove President	Judicial Branch • Creates lower federal courts • Has the ability to impeach and remove judges • Has the ability to propose amendments to overrule judicial decisions • Approves appointments of federal judges
Judicial	Interprets the laws	Executive Branch • Has the ability to declare executive actions unconstitutional	Legislative Branch • Has the ability to declare acts of Congress unconstitutional

Constituents, businesses, professional associations, civic organizations, and other interest groups may lobby members of Congress and ask them to propose legislation or support a certain proposal. Interest groups may also lobby the president and other executive branch officials in order to try and influence policies. The president appoints fifteen men and women to head a variety of executive departments, including defense, transportation, education, and many others. These officials are known as the president's cabinet, and they advise the president on various matters. These discussions are used to help formulate policies, and the cabinet members are responsible for putting the decisions into action. The president may also issue executive orders that instruct federal employees how to enforce certain policies. For example, in 1948, President Harry S. Truman issued an executive order that established racial integration within the armed forces.

Elections are an essential part of our democracy. Although members of Congress are prohibited from accepting gifts, individuals and interest groups can make campaign contributions during elections, which occur every two years in the House, four years for the White House, and six years in the Senate. Individuals can give a maximum of $2,700 per year to individual candidates. Since 2010, individuals, corporations, labor unions, and other contributors can give an unlimited amount to organizations known as **super PACs** (Political Action Committees). Super PACs may not give money to candidates or parties but can air commercials that support a specific issue.

Although President George Washington warned against creating political parties in his farewell address in 1796, they have been a part of American politics since the founding of our country. The Republicans and Democrats hold conventions to nominate candidates for state and national elections. The parties also create a platform, which is a set of goals and principles, at each convention in order to set priorities and inform members about issues. During elections, parties also organize rallies and urge voters to support their candidates and issues. Some Americans belong to either the Republican or Democratic Party, but voters who belong to neither party are known as independents.

Emergence of the United States as a World Power and the Evolving Role of United States in the World

The United States gained prestige and international status after the Spanish-American War of 1898, because the United States defeated Spain and acquired several colonies. American participation in World War I made the United States an economic and financial leader as well. The United States loaned money to Britain and France and supplied weapons and equipment that helped the Triple Entente achieve victory. The United States and USSR emerged from World War II as the only surviving superpowers because so much of the rest of the world had been devastated. This system was described as bipolar because there were two centers of power.

The United States was the leader of the free world during the Cold War and formed military and economic alliances with other nations. With the collapse of the USSR in 1991, the United States was the only surviving world power. This era was a unipolar system because there were no other major powers that could rival the United States.

Although the U.S. was still considered a world power, September 11, 2001 demonstrated that America was nevertheless vulnerable to attacks. The attacks on 9/11 killed more than 3,000 civilians and first-responders. Although American troops have been stationed in Iraq and Afghanistan for more than a decade, both nations are still unstable, and the lack of progress has damaged American prestige. Furthermore, new economic and military powers have risen to challenge American dominance in world affairs. The Chinese economy has grown significantly, and the Chinese government has expanded and improved its armed forces. India has also boosted its manufacturing industry and has purchased new fighter jets and an aircraft carrier. The European Union, Russia, and Brazil have also emerged as

potential rivals that might create a multipolar environment. However, the United States is still the world's largest economy and remains a cultural leader.

The Influence of Religion Throughout American History

Religion has also been an important, albeit divisive, theme in American history since the colonial period. The British colonies in North America attracted settlers from many different religions, including Catholics in Maryland, Quakers in Pennsylvania, Puritans in New England, and Anglicans in Virginia. This led to conflict and tension. For example, Puritans in New England expelled dissenters and even executed four Quakers between 1659 and 1661.

In the 1730s and 1740s, a religious revival known as the First Great Awakening swept through the British colonies in North America. This movement emphasized a more personal connection to Christ, and some Protestant preachers, such as Jonathan Edwards, began to present their sermons in a more passionate and emotional style. This "fire and brimstone" form of religious dissemination became the cornerstone of the **First Great Awakening**. These passionate sermons—and the emotions that they stirred—caused divisions within Protestant congregations. Those who supported the Great Awakening were known as **New Lights** while those who opposed it were called **Old Lights**. The Baptists and Methodists became more popular during the revival because they embraced this new style of preaching.

The **Second Great Awakening** occurred in the early 1800s and urged Protestants to work not only for their own salvation but for the salvation of others as well. This helped fuel a social reform movement that promoted the abolition of slavery, temperance, and prison reform. The question of slavery caused schisms in the Baptist and Methodist churches during the 1840s. The Second Great Awakening, much like the First Great Awakening, inaugurated the creation of several **New Religious Movements** (NRMs) in the United States, especially in the southern states.

A third revival occurred in the late 1800s that emphasized temperance. As previously discussed, the religious right emerged after World War II and began to play an important part in American politics, especially during the election of President Ronald Reagan in 1980.

Although Catholics were a minority during the colonial period of American history, Catholicism has become the largest religious denomination in the United States. Many colonial governments actually banned Catholicism, but the American Revolution brought more toleration. However, anti-Catholic sentiment renewed in the 1800s as immigrants from Ireland and Germany, many of whom were Catholic, arrived in ever-increasing numbers. The arrival of Italian immigrants in the late 1800s and early 1900s also increased Protestant-Catholic tension in America. Many Americans feared that Catholic immigrants would be more loyal to Pope than they would be to the Constitution. This led to the creation of the **Know-Nothings,** who sought to limit immigration and physically attacked Catholics. Anti-Catholic sentiment remained an issue even until the presidential election of 1960 when John F. Kennedy, a Catholic, won the Democratic nomination. Kennedy helped allay fears by promising to respect the separation of church and state. Since then, anti-Catholicism has largely disappeared.

Small numbers of Jews immigrated to the U.S. during the colonial period, but large numbers of Jews from Eastern Europe began to arrive in the late 19th and early 20th centuries. Jews contributed to the American economy in many different ways but drew criticism from anti-Semites because of their prominence in the financial industry. The **Anti-Defamation League** was founded in 1913 to combat anti-Semitic sentiments. In the 1920s, the resurgent Ku Klux Klan revived anti-Semitism. The Anti-Defamation League sponsored events after World War II to commemorate the Holocaust and repudiate Holocaust deniers. Anti-Semitism has declined, but the Anti-Defamation League reported that more than 900 anti-Semitic incidents occurred across the country in 2014.

Muslim immigration in the 1800s remained modest. The first mosque was not constructed in the United States until the 20th century. In the latter part of the 20th century, more Muslims, especially from Pakistan, began arriving in the United States. In the wake of the 9/11 attacks, Islamophobic incidents increased, and Muslims were victims of harassment, intimidation, and assaults. The United States' current battle with ISIS in the Middle East, North Africa, and Europe has also increased Islamophobia.

Major Economic Transformations in the United States

The American economy has changed dramatically since the 1700s, a century where agriculture was the main economic activity. The First Industrial Revolution began in the early 1800s when steam-powered machines were used to increase productivity, especially in the textile industry. The invention of steamboats and railroads made it much cheaper and faster to ship goods across the country in the mid-1800s as well. After the Civil War, the Second Industrial Revolution led to increased productivity and efficiency in the many industries, including metallurgy, chemicals, telecommunications, and energy. This led to significant social changes as immigration and urbanization increased. Workers began to form labor unions in order to demand better wages and working conditions, which led to strikes and conflict with law enforcement officials.

In the early 20th century, Henry Ford introduced the moving assembly line to the automobile manufacturing industry, which made it easier for middle- and working-class families to buy cars. Other industries adopted Ford's methods, which led to lower prices for many consumer goods. The stock market crash in 1929 helped trigger the Great Depression, which resulted in a vicious downward economic spiral. Franklin D. Roosevelt introduced the New Deal to try and boost the economy, but only the outbreak of World War II led to full employment. The United States emerged from World War II as the world's largest economy and pent-up consumer demand fueled prosperity during the post-war era.

The development of computers in the latter part of the 20th century improved communications and led to greater economic efficiency. However, it also marked the beginning of the post-industrial economy in the United States. Traditional manufacturing jobs began to disappear as robots replaced unskilled workers. On the other hand, careers in the information technology sector grew and became a key component of the new high-tech economy. Many unskilled manufacturing jobs also went overseas to countries in Asia and South America as the global economy became more interconnected.

The Causes and Consequences of Changing Immigration Patterns to the United States and Internal Immigration within the United States

Immigration has always played an important role in American history, although patterns have changed over time. Most immigrants came to the United States in search of better economic opportunities, although some have sought religious and political freedom as well. During the colonial period, most immigrants came from Britain, but during the mid-1800s that began to change. In the 1840s, a fungus destroyed the potato crop in Ireland, which led to widespread famine. Many Irish starved, but millions also emigrated, with many heading to the United States. A wave of revolutions also swept across Europe in 1848 and many participants, especially Germans, fled to the United States when the movements failed. This shift led to tension and conflict between immigrants and those born in America. Native-born Americans resented the immigrants' cultural differences and the increased competition for jobs. In the

late 1800s, immigration patterns changed again and arrivals from Southern and Eastern Europe began to increase. In 1924, Congress passed an act that limited immigration from these areas.

In 1965, immigration patterns changed again when Congress passed the Immigration and Nationality Act of 1965 act that changed the way immigration quotas were calculated. Immigrants from Central and South America, as well as Asia, became very numerous. Like their predecessors, most came in search of jobs. Recently, many refugees are fleeing violence that erupted from the revolutions that swept through the Middle East between 2010 and 2012. Cultural differences and economic competition between immigrants and native-born American citizens continue.

Internal Migration

Migration within the United States has also been an important theme since the colonial period. During the 18th and 19th centuries, the focus was on westward expansion because settlers sought cheap land. This often led to violence between settlers and Native Americans who refused to relinquish their territory.

The 19th century also saw an increase in migration from rural areas to cities as individuals sought employment in factories. This trend continued until the late 20th century and more than 80 percent of Americans now live in cities. World War I led to the Great Migration of African Americans from the South to the North. This influx also created more competition for jobs and housing and resulted in a wave of race riots in 1919.

The Struggles and Achievements of Individuals and Groups for Greater Political and Civil Rights Throughout American History

Although the Declaration of Independence declared "all men are created equal," blacks, women, and other minorities struggled for more than a century to make this dream a reality. The U.S. Constitution legalized slavery, and it was not abolished until the 13th Amendment was ratified in 1865. The 14th Amendment, ratified in 1868, granted African Americans citizenship, and the 15th Amendment, ratified in 1870, explicitly granted them the right to vote. However, white Southerners passed laws, known as the **Jim Crow system**, that prevented blacks from exercising their rights and, when that failed, they relied on violence and intimidation to oppress African Americans. For example, many Southern states required voters to pass literacy tests and used them to prevent blacks from casting a ballot. Whites were either exempt from the test or were held to much lower standards. Blacks who protested their oppression could be assaulted and even killed with impunity. In the 1896 decision *Plessy vs. Ferguson*, the U.S. Supreme Court ruled that "separate but equal" schools for white and black students were permissible. In reality, black schools were almost always inferior to white schools.

The emergence of the Civil Rights Movement after World War II finally destroyed the Jim Crow system. In the 1954 decision *Brown vs. Board of Education*, the Supreme Court reversed the "separate but equal" doctrine and declared that separate schools were inherently unequal because they stigmatized African American students. In 1957, President Dwight D. Eisenhower used federal troops to force the high school in Little Rock, Arkansas, to integrate and accept nine black students. This encouraged civil rights activists to demand additional reforms. In 1955, Rosa Parks refused to give up her seat on a bus in Montgomery, Alabama, which led to a boycott. Martin Luther King Jr. led the bus boycott and became a national leader in the Civil Rights Movement. In 1960, four students in Greensboro, North Carolina, launched a peaceful sit-in at a segregated lunch counter, which sparked similar protests around the country. White activists from the North went south to help blacks register to vote, and in 1964 three activists were murdered in Mississippi. That same year, King led 250,000 protesters in a march on Washington D.C. where he delivered his famous "I Have a Dream Speech."

Although King advocated for peaceful protests, many other civil rights activists disagreed with him. For example, Malcolm X believed that blacks should use violence to defend themselves. Furthermore, King worked with white activists while Malcolm X rejected any cooperation. Malcolm X was assassinated in 1965, and, despite his reputation as a non-violent leader, King was also gunned down in 1968.

Under mounting pressure, Congress passed several important pieces of legislation. The 1964 Civil Rights Act banned discrimination based on race, color, religion, sex, or national origin. The Voting Rights Act of 1965 prohibited the use of poll taxes or literacy tests to prohibit voting. The Civil Rights Act of 1968 banned housing discrimination. In 1967, Carl Stokes became the first black mayor of a major American city, Cleveland. That same year, Thurgood Marshall became the first African American to serve on the Supreme Court. President Gerald Ford declared February to be black history month. In 1989, Colin Powell became the first black chairman of the Joint Chiefs of Staff. Despite these reforms, activists claim institutional racism is still a problem in the 21st century. The Civil Rights movement inspired women, Latinos, and other groups to make similar demands for equal rights.

Women

In 1776, Abigail Adams urged her husband, founding father John Adams, to advocate for women's rights, but it would take more than a century before women could vote. In 1848, activists organized a convention in Seneca Falls, New York, to organize the women's suffrage movement, and their efforts slowly gained momentum. The ratification of the 19th Amendment in 1920 finally gave women the right to vote.

Although women had achieved political equality, they continued to demand reform throughout the 20th century. In the early 1900s, Margaret Sanger provided women with information about birth control, which was illegal at the time. Women entered the industrial workforce in large numbers during World War II, but when the war ended, they were fired so that veterans would have jobs when they came home. Many women were frustrated when told they had to return to their domestic lives. Simone de Beauvoir, a French writer, published her book "The Second Sex" after World War II, and an English translation was published in 1953. It highlighted the unequal treatment of women throughout history and sparked a feminist movement in the United States. In 1963, Betty Friedan published a book, called "The Feminine Mystique," that revealed how frustrated many suburban wives were with the social norms that kept them at home. During the 1960s, women participated in the sexual revolution and exerted more control over their own sexuality. In 1972, Congress passed Title IX, which prohibited sexual discrimination in education and expanded women's sports programs. In the 1970s, women's rights activists also pushed for greater access to birth control, and in 1973 the Supreme Court issued the controversial decision *Roe vs. Wade* which removed many barriers to abortion services. Women also demanded greater protection from domestic abuse and greater access to divorce.

During the 20th century, many American women made notable achievements, including Amelia Earhart, who was the first woman to cross the Atlantic in an airplane in 1928. In 1981, Sandra Day O'Connor became the first woman to serve on the Supreme Court. In 1983, Sally Ride became the first female astronaut. In 1984, Geraldine Ferraro became the first woman to run for vice-president, although she was unsuccessful. However, many activists continue to demand reform in the 21st century. For example, women only account for 20 percent of the U.S. Senate and House of Representatives. Furthermore, women only earn 79 percent of what men in similar jobs are paid. In 1980, President Jimmy Carter declared March to be women's history month.

Hispanics

After World War II, many Hispanics also began to demand greater equality. In 1949, veterans protested a refusal by a Texas town to bury a Mexican American soldier, who died during World War II, in the local cemetery, because only whites could be buried there. Activists called themselves Chicanos, a term that previously was used as a pejorative to describe Mexican Americans. Cesar Chavez was a labor union activist who organized transient Hispanic agricultural workers in an effort to obtain better working conditions in the 1960s and 1970s. Activists encouraged a sense of pride in Chicano identity, especially in arts and literature. In 1968, President Lyndon B. Johnson declared National Hispanic Heritage Month would run from mid-September to mid-October.

In 1959, biochemist Severo Ochoa became the first Hispanic to win a Nobel Prize. Franklin Chang-Diaz became the first Hispanic astronaut in 1986, and he flew a total of seven space shuttle missions. In 1990, Oscar Hijuelos became the first Hispanic American to win the Pulitzer Prize. Sonja Sotomayor became the first Hispanic to serve on the Supreme Court in 2009.

Native Americans

Native Americans suffered centuries of oppression at the hands of European colonists, and later American settlers as they pushed further west. Native Americans resisted attempts to encroach on their lands but were pushed onto smaller and smaller reservations. The **Massacre at Wounded Knee** in 1890 was the last major conflict between Native Americans and U.S. forces. However, American officials continued to try and force Native Americans to assimilate into white culture.

In 1968, a group of Native Americans formed the **American Indian Movement** in order to combat racism and demand greater independence. Between 1969 and 1971, a group of Native American activists occupied the federal prison on Alcatraz Island near San Francisco, although it had been closed since 1963. The activists offered to buy back the island for $9.40 in order to draw attention to how the federal government had forced tribes to sell their lands at low prices. Other activists disrupted Thanksgiving Day ceremonies aboard a replica of the Mayflower in Boston in 1970. In 1971, Native American activists also occupied Mount Rushmore, which is located on ground the Native Americans consider sacred. Violence broke out between activists and law enforcement officials in 1973 when Native Americans occupied the town of Wounded Knee, sight of the famous massacre.

In 1970, President Richard Nixon granted Native American tribes more autonomy. In 1978, Congress passed the American Indian Religious Freedom Act, which guaranteed Native Americans' rights to practice their religious ceremonies and visit sacred sites. In 1990, President George H.W. Bush declared November Native American History Month. In 1969, Navarre Scott Momaday became the first Native American to win a Pulitzer Prize for his book "House Made of Dawn." In 2014, Diane Humetewa, a member of the Hopi tribe, became the first Native American woman to serve as a federal judge. However, many Native American communities still suffer from high rates of unemployment, alcoholism, and domestic abuse.

Asian Americans

Asian Americans also faced discrimination throughout American history and in 1882, Congress passed a law banning all Chinese immigrants. During World War II, more than 100,000 Japanese Americans were interned in concentration camps. In 1982, two American autoworkers beat Vincent Chin to death with a baseball bat because his assailants blamed him for the loss of jobs in the automotive manufacturing industry.

In the 1960s, activists demanded that the term "Asian American" replace the word "oriental," because it carried a stigma. Asian Americans also promoted a sense of pride in their cultural identity and successfully pushed for the creation of ethnic studies programs. Ellison Onizuka became the first Asian American astronaut in 1985, although he perished in the space shuttle Challenger disaster. In 1990, President George H.W. Bush declared May Asian Pacific American Heritage Month and Sheryl WuDunn became the first Asian American to win a Pulitzer Prize that same year.

World History

Ancient Civilizations

Ancient Greece

Ancient Greece formed from scattered farming communities between 800 BCE and 500 BCE. In this early era of Greece, the polis, or city-state, held all of the political power locally. City-states were self-ruling and self-sufficient. The idea of a self-governing state had an enduring effect on the government of Greece and would result in the demokratia (rule by the people), which would spread and influence the world. As farming villages grew and marketplaces were built, a government with laws, an organized army, and tax collection took shape.

Each city-state was different from one another, but some unifying traits included a common language, a shared belief system, an agriculturally based economy, and rule by several wealthy citizens instead of rule by a king or queen. However, these few aristocratic rulers, known as *oligarchs,* often owned the best and most land, which created tension as the population grew. As a result, many citizens moved to less populated or newly conquered areas. By 800 BCE, there were over 1500 city-states, each with its own rulers and rules. Greek city-states were concentrated on the coast, resulting in greater contact with other civilizations through trade. City-states' governments and culture continued to diverge as time progressed. For example, in the fifth century BCE, Athens became the first direct democracy in the world, and Athenian citizens would vote directly on legislation. Only adult, male, landowning citizens could vote, but it was a remarkable departure from all contemporary forms of government to provide for direct democracy, especially relative to other city-states' oligarchies. Another world-renowned example is Sparta, which based its entire social system and constitution on military training and ability.

The Greek religion was polytheistic. Every city-state had a temple dedicated to a particular god or goddess; however, the whole of ancient Greece believed that Zeus, residing in Mount Olympus, was the most powerful of the gods. The physical presence of the temple, the rituals and festivals that dotted the Greek year, and the widespread belief in the gods controlling every aspect of human life heavily influenced their agricultural economy, government, and interactions with other ancient civilizations.

The ancient Greeks were known for their citizen-soldiers, known as *hoplites.* No ancient civilization could field a professional military due to economic restraints, such as a lack of a banking system and the need for agricultural laborers, but the hoplites were famous in ancient times for their tactics and skill. Hoplites were armed with spears and large shields, and they would fight in a phalanx formation. The Romans would later adopt many of the Greek military principles. Greek city-states fought numerous wars among each other, the largest being the Peloponnesian War, as well as wars against Persia. Fought between 499 BCE and 449 BCE, the Greco-Persian Wars pitted the Greek city-states against the mighty Persian Empire after the latter invaded. Although ancient sources are difficult to authenticate, it is certain that the Persian forces vastly outnumbered the Greeks who historically struggled to unite, even against a common enemy. This conflict included the legendary Battle of Thermopylae where three hundred

Spartans, led by the Spartan king Leonidas, held off the elite contingent of the Persian army, the Immortals, for two days. After several setbacks and disastrous turns, the Greeks eventually defeated the Persian fleet at the naval Battle of Mycale and forced the Persians out of Europe. Greek unification did not last beyond this victory, and by 404 BCE, Sparta crushed Athens in the Peloponnesian War. Athens would never again attain its status as the leading city-state.

Roman Empire

Although already one of the world's most powerful civilizations, Rome began to strain under political pressure and domestic unrest in the mid-first century BCE. In 48 BCE, Gaius Julius Caesar seized power over the republic, but his assassination in 44 BCE on the Ides of March threw the republic back into turmoil. Caesar's great-nephew turned son adopted by will, Octavian, eventually emerged as the sole leader of Rome, and historians define this point as the beginning of the Roman Empire. Octavian would serve as the first emperor under the name Augustus. His rule would be one of the most peaceful and prosperous in Roman history, often referred to as the *Pax Romana* or *Pax Augusta*. Although the Roman Empire did not adhere to the republic's democratic principles and separations of powers, the Roman Empire would be the vehicle that enabled Rome to conquer and administer enormous territory.

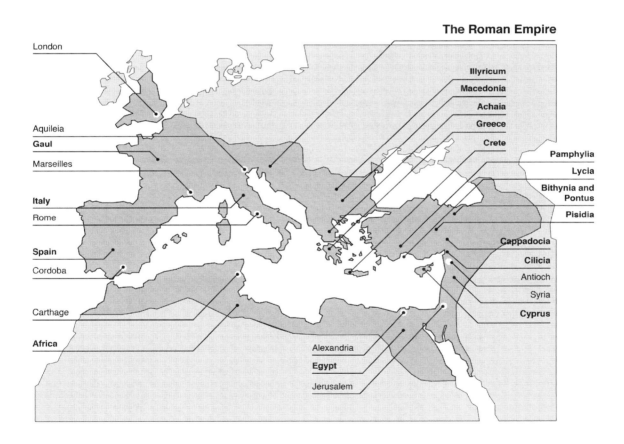

As Rome became an empire, its influence both in the ancient world and in the modern world began to take shape. Rome's ability to absorb and adapt the cultural achievements of Greece and push them on conquered cultures was a key to their success. Rome was highly influenced by Greek culture, religions, ideas, literature, and politics but kept at its roots the Roman ideals of simplicity, honesty, and loyalty. Rome was able to hold together a government that included multiple races, languages, and cultures in peace through the successful use of these ideas. In addition, Rome applied concepts developed by the Persians in the administration and political organization of its territories. By the time Rome became an empire, the government was highly structured with a complex civil service that addressed and administrated localized affairs.

Rome's decline began well before its eventual fall. There are many aspects to Rome's demise, including social, political, moral, religious, and economic. Each took their toll on the strength of the empire, and by 400 A.D., Rome collapsed under public unrest and religious discord, along with the invasion of the Huns of Mongolia and Germanic tribes. Although ultimately defeated, Rome's legacy extends all the way through to the present day. The Roman Republic's democratic elements and robust civil service would be the model for much of the West, especially the United States. That is to say nothing of the advancements in literature, technology, architecture, urban planning, and hygiene across the empire that influenced every future Western civilization.

Persian Empire

The Persian Empire consisted of multiple countries, religions, languages, and races governed by a central government. Cyrus the Great was known for his social and political acumen. He was able to navigate the empire's diversity with his "carrot or stick" approach. Cyrus the Great would offer foreign civilizations some degree of home rule, as long as they paid tribute to Persia and adopted some of its norms, or else the might of the legendary Persian military would crush them. As long as the citizens of Persia paid taxes and were peaceful, they were allowed to keep their own religious customs, local culture, and local economies. It was not until his successors that this political policy began to wan with the onset of multiple rebellions that weakened the centralized government.

The government of Persia delegated power among four governing capitals. Each state had a satrap, or governor. The satrapy government allowed for regional self-governance with a military leader and an official record-keeper that reported to the central government. The empire was also innovative in its road construction and postal systems. By allowing some degree of regional autonomy, Persia was able to rule over an unprecedented territory in ancient history. For example, Babylon even requested to be part of Persia because of its unique policies. The empire's enormity and vast scope influenced world history for centuries. Persian scholars and political philosophers would later influence rulers in the Renaissance and Enlightenment eras.

Maurya Empire

The Maurya Empire established a centralized government to govern its vast territories, and it specialized in tax collection, administration, and the military. It was modeled after the Greek and Persian governments, who, through trade and invasion, had influenced Chandragupta's government layout. Previously, regional chieftains and small armies governed India, which led to continuous skirmishes and wars. Chandragupta cleared out the chieftains and imposed regulated laws and tax reforms. The centralized form of government allowed for a period of peace, scientific advancement, and religious growth.

The centralized government was made up of four provinces organized under one capital. Each emperor had a cabinet of ministers known as a *Mantriparishad,* or Council of Ministers, to help guide him—an idea that is still used in governments across the world. Princes, or Kumaras, likewise oversaw each province, with a council of ministers called the *Mahamatyas*. A civil service was developed to govern many aspects of life and infrastructure, including waterworks, roads, and international trade. The army was expanded into one of the largest in the world at the time. Trade became a major source of revenue as other empires sought spices, food, and cloth from India.

India's three main religions flourished in this period. Hinduism, a blend of multiple beliefs, appeared in the Epic Age and became a central religion. Buddhism appeared as a consequence of the harsh social structure that had left a wide gap in the social and economic freedoms of the people. Chandragupta later accepted Jainism, a religion of total peace and unity with the world. Overall, the Maurya Empire featured a balance of religions that promoted peace as foundational and sought social harmony. The centralized government discouraged the infamous Indian caste system, which organized society by social status and led to discrimination against the lower castes.

Chinese Empire

Between 1000 BCE and 500 A.D., ancient China was unified under three successive dynasties: the Zhou Dynasty, the Qin Dynasty, and the Han Dynasty, in respective chronological order. The Zhou Dynasty was the longest dynasty in Chinese history and began after the fall of the Shang Dynasty. Originally, the Zhou Dynasty had moved away from the Shang Empire, created their own government, and formed alliances with enemies of the Shang. When war eventually broke, the people of Shang, so angered by their own government's foolishness, put up little resistance against the rebellion.

Under the Zhou Dynasty, the kingdom's ruler legitimized their power through the Mandate of Heaven, meaning they believed the rulers of the land were put in place by a higher being that could not be disposed. The Zhou claimed that the Shang Dynasty had forfeited their claim due to their mismanagement of the kingdom. This would be a common theme for dynasty takeovers. A centralized government was established, but the Zhou Dynasty never achieved complete centralized control across the kingdom. The economy was heavily agricultural and organized based on feudalism, an economical system in which a wealthy, landowning class rules the peasant class. These aristocratic rulers retained considerable power and regularly rebelled against the central government.

The Qin Dynasty was the first imperial dynasty, originally organized under Emperor Qin Shi Huangdi. The imperial state had a more centralized government, which limited the aristocratic landowners' power, stabilized the economy, and boosted the army. The Qin Dynasty formed a political structure that allowed China to start major building projects like the Great Wall of China. Its form of government would be adopted by many dynasties in China's history. The Qin Dynasty was short-lived and ended when Emperor Qin Shi Huangdi died prematurely, and rebel leaders fought for control of the kingdom. Liu Bang of Han defeated Xiang Yu of Chu at the Battle of Gaixia in 202 BCE, establishing the Han Dynasty.

Like the previous imperial dynasty, power was consolidated under a single emperor who dominated the Han Dynasty's centralized government. Under the emperor, a cabinet of ministers and chosen nobility acted as advisors who retained limited power. The Han dynasty was a golden era of Chinese innovation and technology, all driven by the tremendous growth in commerce and trade. To facilitate commerce, the Han Dynasty issued one of the world's earliest currencies under a money economy. Han coinage would remain the dominant currency from its introduction in 119 BCE until the Tang Dynasty in 618 A.D. A uniform currency was an essential part of the legendary Silk Road, which began under the Han Dynasty.

World History in Western Civilization

Development of Early Western Civilization

There were a number of powerful civilizations during the classical period. Mesopotamia was home to one of the earliest civilizations between the Euphrates and the Tigris rivers in the Near East. The rivers provided water and vegetation for early humans, but they were surrounded by desert. This led to the beginning of irrigation efforts to expand water and agriculture across the region, which resulted in the area being known as the Fertile Crescent.

The organization necessary to initiate canals and other projects led to the formation of cities and hierarchies, which would have considerable influence on the structure of later civilizations. For instance, the new hierarchies established different classes within the societies, such as kings, priests, artisans, and workers. Over time, these city-states expanded to encompass outside territories, and the city of Akkad became the world's first empire in 2350 BCE. In addition, Mesopotamian scribes developed systemized drawings called pictograms, which were the first system of writing in the world; furthermore, the creation of wedge-shaped cuneiform tablets preserved written records for multiple generations.

Later, Mesopotamian kingdoms made further advancements. For instance, Babylon established a sophisticated mathematical system based on numbers from one to sixty; this not only influenced modern concepts, such as the number of minutes in each hour, but also created the framework for math equations and theories. In addition, the Babylonian king Hammurabi established a complex set of laws, known as the Code of Hammurabi, which would set a precedent for future legal systems.

Meanwhile, another major civilization began to form around the Nile River in Africa. The Nile's relatively predictable nature allowed farmers to use the river's water and the silt from floods to grow many crops along its banks, which led to further advancements in irrigation. Egyptian rulers mobilized the kingdom's population for incredible construction projects, including the famous pyramids. Egyptians also improved pictographic writing with their more complex system of hieroglyphs, which allowed for more diverse styles of writing. The advancements in writing can be seen through the Egyptians' complex system of religion, with documents such as the *Book of the Dead* outlining not only systems of worship and pantheons of deities but also a deeper, more philosophical concept of the afterlife.

While civilizations in Egypt and Mesopotamia helped to establish class systems and empires, other forms of government emerged in Greece. Despite common ties between different cities, such as the Olympic Games, each settlement, known as a polis, had its own unique culture. Many of the cities were oligarchies, in which a council of distinguished leaders monopolized the government; others were dictatorships ruled by tyrants. Athens was a notable exception by practicing an early form of democracy in which free, landholding men could participate, but it offered more freedom of thought than other systems.

Taking advantage of their proximity to the Mediterranean Sea, Greek cities sent expeditions to establish colonies abroad that developed their own local traditions. In the process, Greek merchants interacted with Phoenician traders, who had developed an alphabetic writing system built on sounds instead of pictures. This diverse network of exchanges made Greece a vibrant center of art, science, and philosophy. For example, the Greek doctor Hippocrates established a system of ethics for doctors called the Hippocratic Oath, which continues to guide the modern medical profession. Complex forms of literature were created, including the epic poem "The Iliad," and theatrical productions were also developed. Athens in particular sought to spread its vision of democratic freedom throughout the world, which led to the devastating Peloponnesian War between allies of Athens and those of oligarchic Sparta from 431 to 404 BCE.

Alexander the Great helped disseminate Greek culture to new regions, also known as *diffusion*. Alexander was in fact an heir to the throne of Macedon, which was a warrior kingdom to the north of Greece. After finishing his father's work of unifying Greece under Macedonian control, Alexander successfully conquered Mesopotamia, which had been part of the Persian Empire. The spread of Greek

institutions throughout the Mediterranean and Near East led to a period of Hellenization, during which various civilizations assimilated Greek culture; this allowed Greek traditions, such as architecture and philosophy, to endure into the present day.

Greek ideas were later assimilated, along with many other concepts, into the Roman Empire. Located west of Greece on the Italian peninsula, the city of Rome gradually conquered its neighbors and expanded its territories abroad; by 44 BCE, Rome had conquered much of Western Europe, northern Africa, and the Near East. Romans were very creative, and they adapted new ideas and innovated new technologies to strengthen their power. For instance, Romans built on the engineering knowledge of Greeks to create arched pathways, known as aqueducts, to transport water for long distances and devise advanced plumbing systems.

One of Rome's greatest legacies was its system of government. Early Rome was a republic, a democratic system in which leaders are elected by the people. Although the process still heavily favored wealthy elites, the republican system was a key inspiration for later institutions such as the United States. Octavian "Augustus" Caesar later made Rome into an empire, and the senate had only a symbolic role in the government. The new imperial system built on the examples of earlier empires to establish a vibrant dynasty that used a sophisticated legal code and a well-trained military to enforce order across vast regions. Even after Rome itself fell to barbarian invaders in fifth century A.D., the eastern half of the empire survived as the Byzantine Empire until 1453 A.D. Furthermore, the Roman Empire's institutions continued to influence and inspire later medieval kingdoms, including the Holy Roman Empire; even rulers in the twentieth century called themselves Kaiser and Tsar, titles which stem from the word "Caesar."

In addition, the Roman Empire was host to the spread of new religious ideas. In the region of Israel, the religion of Judaism presented a new approach to worship via monotheism, which is the belief in the existence of a single deity. An offshoot of Judaism called Christianity spread across the Roman Empire and gained popularity. While Rome initially suppressed the religion, it later backed Christianity and allowed the religious system to endure as a powerful force in medieval times.

Middle Ages

Early Middle Ages

The Middle Ages refers to the period from the fifth century to the fifteenth century, beginning with the fall of the Roman Empire and ending with the Renaissance and Age of Exploration. Sharp population decline, intensely localized governance, frequent invasions, famine, and disease defined the early Middle Ages and explain why it is sometimes referred to as the *Dark Ages*. Manorialism and feudalism were the dominant economic systems of the period. Peasants would rent patches of land to farm on enormous manors of aristocrats, while knights and lower nobles would exchange military service with the aristocracy in exchange for control over a manor. In addition, much of the knowledge gained during the Age of Antiquity was lost during this period.

High Middle Ages

During the High Middle Ages, signs of revival began to emerge. Christians began to see the need to live out the fundamental convictions of Christianity and also saw the need for the clergy to exemplify Christ. After several reforms, religious orders developed, such as the Franciscans and Dominicans. The orders protected the knowledge and texts of the church, becoming a strong intellectual body. As a consequence, there was a revival in learning in the monasteries that trickled out to the cathedrals and then to schools.

Around 570 A.D., the Islamic prophet Muhammad was born in Mecca. Muhammad was a trade merchant who, coming into contact with Christian and Jewish traders, blended their religions with his own religious experience in which he believed that Allah was the one true god. He believed that Allah had called him to preach the Islamic religion. At first, he met with little success, as most Arabs believed in many and differing gods. However, in a few years, he was able to unite the nomadic tribes under Islam.

After Muhammad's death, his successors, known as *caliphs,* developed the religion of Islam into a system of government and spread the faith and government control into the Middle East, North Africa, Spain, and southern France. At one point, the Islamic Empire was larger than the Roman Empire. With invasion, Islam spread the Arabic language and embraced Greek science and Indian mathematics. From 900 A.D. to 1100 A.D., Islam experienced a golden age.

In 1095, European Christians launched military strikes against Muslims in the Holy Land, and the entire series of armed religious conflicts is known as the *Crusades*. During the Crusades, Italy's trade flourished because the movement of people facilitated commerce and communication with the Middle East and Africa. In the High Middle Ages, Italy expanded trade into Europe, and merchants across Europe began to settle in areas with good trade routes. Others who had a trade to sell settled in these areas, forming towns and local governments. The development of commerce would be the impetus for the Renaissance.

Age of Exploration
The traveling merchants, the Crusades, the conquests of foreign lands, and the writings of ancient Greece expanded the known world of Europeans to include Europe, northern Africa, the Middle East, and Asia. Early explorers such as Marco Polo brought back amazing stories and exotic goods from Asia, while ports in the Middle East and around the Mediterranean spread cultures through trade. However, the very existence of America and Australia was unknown to the ancient and medieval world. Likewise, there was very little knowledge of sub-Saharan Africa until the late Renaissance era.

In an effort to find better trade routes to China, explorers discovered unknown lands that would change the world in dramatic fashion. Over a two hundred year period from 1450 to 1650, the great explorers of the age would discover new lands, unknown people, and better trade routes to the silks, spices, precious metals, and other sought-after goods Europe was eager to own.

Portugal and Spain funded the first explorations and, along with Italy, dominated the discovery of new lands and trade routes for the first one hundred years of exploration. In 1488, Portuguese explorer Bartolomeu Dias became the first European to sail around the Cape of Good Hope in South Africa and the first European to sail from the Atlantic Ocean to the Indian Ocean. On a voyage lasting from 1497 to 1499, Vasco da Gama, another Portuguese explorer, followed the route of Dias and became the first European to reach India by sea.

Portuguese explorers' success led to Portugal's dominance over trade with Africa and Asia. In West Africa, the Portuguese traded for slaves, and in east Africa, they captured city-states and opened trading posts. The coastal trading posts were utilized to launch further exploration and trade farther east with China and Japan. During a voyage launched in 1500, Dias went on to reach Brazil after his ship was blown off course to Africa. Brazil would later become Portugal's most lucrative colony due to the sugar plantations farmed by African slaves and indigenous people.

By the 1530s, France, England, the Netherlands, and Scotland were beginning to send explorers on their own expeditions. In 1534, the king of France sent Jacques Cartier to discover a western passage to the Asian markets, and during his voyage in 1534, Cartier became the first European to travel down the Saint Lawrence River and explore Canada, which Cartier named after Iroquois settlements that he encountered. Englishman Francis Drake was the first European to successfully circumnavigate the world, completing the three-year voyage in 1580. Another Englishman, Henry Hudson, was hired by the Dutch East India Company to find a northwest passage to India, and he explored the modern New York metropolitan area in the early seventeenth century. The Dutch would use this knowledge to colonize the area around the Hudson River, including New Amsterdam.

Even more devastating than the loss of their land, contact by Europeans exposed the indigenous people of America to devastating new diseases. Without any type of immunization, mild European diseases decimated the populations of the natives. Often the illness and death of natives made conquering the areas swift, and with it the loss of the culture, traditions, and languages of the native people. However, diseases such as syphilis and cholera were brought back from expeditions, ravaging European countries. The high death toll from disease, coupled with the deaths from native-born slave labor, caused a labor shortage that the Spanish replenished with slaves from their trade deals in West Africa. These slaves were mainly brought to the Caribbean Islands, though they were shipped to other Spanish colonies. The British colonies would later import millions of slaves to the modern-day American South to harvest cotton.

The Italian Renaissance

The Renaissance, meaning *rebirth*, began in the fourteenth century in Italy and spread throughout Europe during the fifteenth century. Its philosophy was humanism, or the study of man and his relationship with the world. It was a time when reason and knowledge were highly valued. The Roman Catholic Church kept pace with Europe's focus on mankind and nature, instead of heaven and heavenly beings. Popes sponsored educational enhancements and were, in some instances, as is the case with Pope Pius II, trained as classical scholars. In the early 1500s, Julius II had masters such as Michelangelo create artistic masterpieces that celebrated humans. Indeed, the arts moved toward a more realistic and proportional style with Italian painters such as Leonardo da Vinci, Raphael, and Titian leading the way.

The literary greats of the age were writing in their own vernacular, or language, instead of Latin. This was one of the greatest leaps forward; it not only built their native language, but it also allowed the Italian people to learn and grow in literacy. This and the advances in printing made the written word more accessible and widely dispersed than ever before. The dissemination of knowledge to larger groups of people would change the world, especially as the Renaissance spread to other European nations.

Renaissance in Northern Europe

The ideas in Italy began to spread northward, allowing the arts to flourish in Germany, such as Albrecht Dürer, and in the Netherlands, like Johannes Vermeer and Rembrandt. England began a long history of great literature with Geoffrey Chaucer's *The Canterbury Tales* and Edmund Spenser's *The Faerie Queene*. The highest literary achievements of the Renaissance came from two English playwrights, Christopher Marlowe and his better-known contemporary, William Shakespeare. But the Renaissance did not stray far from religious themes; instead, they humanized them, as is the case of Italian works of art. It was also an early changing point in Christianity, as theologians like Meister Eckhart, Thomas à Kempis, and Sir Thomas More began to use humanism to question the need for priestly intervention, favoring instead direct worship of God.

The invention of movable type by Johannes Gutenberg in 1439 started a revolution in printing that saw the expansion of books go from approximately 100,000 laboriously hand-copied books in Europe to over 9 million by 1500. The literacy rate in Europe improved vastly, as did the printing of religious writings. News could now travel to distant places, allowing for unprecedented communication both locally and globally. Movable type would be one of the major inventions of the Renaissance, heavily influencing the Reformation and Enlightenment.

The Reformation

In 1517, Martin Luther, a German monk and professor of theology, nailed his famous *Ninety-Five Theses,* or *Disputation on the Power of Indulgences,* to the door of the cathedral in Wittenberg, Germany. Pope Leo X demanded Luther to rescind, but Luther stood his ground, which launched the Protestant Reformation. There were serious problems in the Catholic Church, including clergy accepting simony, or the sale of church offices; pluralism, or having multiple offices; and the violation of vows. In addition, the worldly behavior of the church leaders and the biblical ignorance of the lower clergy prompted Luther to ignite a fire that could not be swept away or cleared up. The Roman Catholic Church could not weather this call for reform like it had done before. This was instead a call to cast off the Catholic faith for Protestantism. Many church denominations were formed under Protestantism, the first being Lutheranism, which gained strength in Germany and Switzerland.

Shortly after, the Roman Catholic Church issued a Counter Reformation in an attempt to quell the spread of Protestantism by addressing some of the complaints. In its initial stages, the Counter Reformation had little effect, and many Germans adopted Lutheranism as its officially recognized religion. By 1555, the Catholic Church recognized Lutheranism under the Peace of Augsburg, which allowed rulers to decide on which religion their kingdom would follow. In Germany there was peace, but civil wars broke out in France and the Netherlands. The Spanish-ruled Netherlands' struggle was as political as it was social, with other countries joining the fight against Catholic Spain.

In the 1600s, the peace in Germany faded as the country allied itself with either the Protestant Union or the Catholic League. The Thirty Years' War broke out in 1618 and became one of the most destructive wars in European history. It was a war of political and religious hostility that would involve Germany, Denmark, France, Austria, Spain, and Sweden, to some degree. Though it ended in 1648 with the Peace of Westphalia, France and Spain would wage war until 1659. The Treaty of Westphalia emphasized national self-determination, which directly led to the development of the nation-state. For the first time in human history, local people controlled the right to build a nation-state with the accompanying legitimacy to control their region. The new states, most of which were carved out of the Holy Roman Empire, were allowed to determine their religion, including Catholicism, Lutheranism, and Calvinism.

The Enlightenment

In the Enlightenment, also known as the *Age of Reason,* that followed the Renaissance, Europe began to move toward a view that men were capable of improving the world, including themselves, through rational thinking. The Enlightenment placed a heavy emphasis on individualism and rationalism. During the Renaissance, scholars looked at the Middle Ages as a lost period and considered their own time as modern and new. The Enlightenment, building on the foundations of humanism, began a prolific era of literature, philosophy, and invention.

By the 1700s, Europe had entered the High Enlightenment Age, where events started to take place as a result of the rational thought promoted by the first half of the age. The idea that everything in the universe could be reasoned and cataloged became a theme that set Diderot to work at the first

encyclopedia and inspired Thomas Paine and Thomas Jefferson during the initial political unrest in the American colonies.

In the later years of the Enlightenment, the ideal vision that society could be reborn through reason was tested in the French Revolution of 1789. Instead of becoming a leader in rational thinking and orderly government, the revolution turned into the Reign of Terror that saw the mass execution of French citizens and opened the way for the rise of Napoleon.

American Revolution

The American Revolution occurred as a result of changing values in the Thirteen Colonies that broke from their traditional relationship with England. Early on in the colonization of North America, the colonial social structure tried to mirror the stratified order of Great Britain. In England, the landed elites were seen as intellectually and morally superior to the common man, which led to a paternalistic relationship. This style of governance was similarly applied to the colonial system; government was left to the property-owning upper class, and the colonies as a whole could be seen as a child dutifully serving "Mother England."

However, the colonies' distance from England meant that actual, hereditary aristocrats from Britain only formed a small percentage of the overall population and did not even fill all the positions of power. By the mid-eighteenth century, much of the American upper class consisted of local families who acquired status through business rather than lineage. Despite this, representatives from Britain were appointed to govern the colonies. As a result, a rift began to form between the colonists and British officials.

Uncertain about whether they should remain loyal to Britain, representatives from twelve colonies formed the First Continental Congress in 1774 to discuss what they should do next. When Patriot militiamen at Lexington and Concord fought British soldiers in April 1775, the Revolutionary War began. While the rebel forces worked to present the struggle as a united, patriotic effort, the colonies remained divided throughout the war. Thousands of colonists, known as Loyalists or Tories, supported Britain. Even the revolutionaries proved to be significantly fragmented, and many militias only served in their home states. The Continental Congress was also divided over whether to reconcile with Britain or push for full separation. These issues hindered the ability of the revolutionary armies to resist the British, who had superior training and resources at their disposal.

Even so, the Continental Army, under General George Washington, gradually built up a force that utilized Prussian military training and backwoods guerrilla tactics to make up for their limited resources. Although the British forces continued to win significant battles, the Continental Army gradually reduced Britain's will to fight as the years passed. Furthermore, Americans appealed to the rivalry that other European nations had with the British Empire. The support was initially limited to indirect assistance, but aid gradually increased. After the American victory at the Battle of Saratoga in 1777, France and other nations began to actively support the American cause by providing much-needed troops and equipment.

In 1781, the primary British army under General Cornwallis was defeated by an American and French coalition at Virginia, which paved the way for negotiations. The Treaty of Paris in 1783 ended the war, recognized the former colonies' independence from Great Britain, and gave America control over territory between the Appalachian Mountains and Mississippi River. However, the state of the new nation was still uncertain. The new nation's government initially stemmed from the state-based structure of the Continental Congress and was incorporated into the Articles of Confederation in 1777.

French Revolution

Unlike the United States' revolution against a ruler across the ocean, the French Revolution was an internal fight. In 1789, tension between the lower class (peasants) and middle class(bourgeois) and the extravagant wealthy upper class of France came to a head. The Old Regime, headed by the monarchy, was overthrown, and the Third Estate, made up of the bourgeois class, seized power. The American Revolution, overtaxation, years of bad harvests, drastic income inequality, and the Enlightenment influenced the French Revolution. In August 1789, the National Constituent Assembly, a democratic assembly formed during the French Revolution, passed the Declaration of the Rights of Man and of the Citizen, which defined the natural right of men to be free and equal under the law.

Napoleon

France radically changed the government from a monarchy to a democracy with provisions for civil rights, religious freedoms, and decriminalization of various morality crimes, like same-sex relationships. Two political powers emerged: liberal republicans called *Girondist*s and radical revolutionaries, known as *Jacobins*. Conflict between the parties resulted in the Reign of Terror—a period of mass executions— and eventually the rise of Napoleon who set up a nationalist military dictatorship. During the revolution, Napoleon Bonaparte consolidated power after becoming famous for his heroism during the revolutionary wars against Britain, Austria, and other monarchies that sought to retain their right of royal rule. However, by 1804, Napoleon declared himself emperor and remilitarized France, and he conquered most of Europe in a series of global conflicts collectively known as the *Napoleonic Wars,* starting in 1803 and continuing until Napoleon's defeat at the Battle of Waterloo in 1815.

After the chaos sparked by the French Revolution that fanned across Europe during the revolutionary wars, European powers met at the Congress of Vienna in November 1814 to June 1815 to rebalance power and restore old boundaries. The Congress of Vienna carved out new territories, changing the map of Europe. France lost all of its conquered territories, while Prussia, Austria, and Russia expanded their own. With the restoration of a balance of power, Europe enjoyed nearly fifty years of peace.

Latin American Wars of Independence

Fueled by the successful American Revolution, Napoleon's rise to power, and the writings of the Enlightenment, a spirit of revolution swept across the Americas. The French colony in Haiti was the first major revolution occurring in 1791. The Haitian Revolution was the largest slave uprising since the Roman Empire, and it holds a unique place in history because it is the only slave uprising to establish a slave-free nation ruled by nonwhites and former slaves. In 1804, the Haitians achieved independence and became the first independent nation in Latin America. When Napoleon conquered Spain in 1808, Latin American colonies refused to recognize his elder brother, Joseph Bonaparte, as the new Spanish monarch and advocated for their own independence. Known as the *Latin American Wars of Independence,* Venezuela, Colombia, Ecuador, Argentina, Uruguay, Paraguay, Chile, Peru, and Bolivia all achieved independence between 1810 and 1830. In 1824, Mexico declared itself a republic when, after several attempts by the lower classes of Mexico to revolt, the wealthier classes joined and launched a final and successful revolt. When Napoleon overtook Portugal, King John VI fled to Brazil and set up court. Later he left his son Pedro behind to rule. Pedro launched a revolution that saw him crowned emperor.

By the mid-1800s, the revolutions of Latin America ceased, and only a few areas remained under European rule. The U.S. President James Monroe issued the Monroe Doctrine, which stated that the Americas could no longer be colonized. It was an attempt to stop European nations, especially Spain, from colonizing areas or attempting to recapture areas. England's navy contributed to the success of the

doctrine, as they were eager to increase trade with the Americas and establish an alliance with the United States.

First World War

The onset of World War I began with the precarious balance of power and the geographic divisions written by the Napoleonic Wars' Vienna Congress.

Austria-Hungary's large empire was diverse in culture and included various peoples of several nationalities, languages, and beliefs. However, minorities in their lands in the Balkans grew tired of foreign control. This was especially true in Bosnia, which was all but under control by the nationalistic secret military society, the Black Hand. This nationalistic sentiment grew until, in 1914, Gavrilo Princip, a Serb patriot and member of the Black Hand, assassinated Archduke Franz Ferdinand, heir presumptive to the throne of Austria-Hungary. In response, Emperor Franz Joseph I of Austria-Hungary declared war on the kingdom of Serbia, officially launching the First World War.

Europe had tied itself into a tangled web of alliances and mutual protection pacts. Germany and Austria-Hungary were allies. Russia promised protection to France and Serbia, and England maintained a tacit support to its past allies throughout the mainland. Each of the Allies soon mobilized to support each other. Germany had already planned for declarations of war, however, and was nervous about fighting a two-border war against both France and Russia, so it developed the Schlieffen Plan—a strategy to quickly demolish French resistance before turning around to fight Russia on the Eastern Front. However, this plan relied on the neutrality of England; after Germany invaded Belgium to attack France, England's declaration of war ensured that a long war would be inevitable.

The Great War lasted from 1914 to 1918 and was the deadliest war in European history until World War II, with approximately 16 million combatants and civilians dying in the conflict. The carnage was largely a result of technological innovation outpacing military tactics. World War I was the first military conflict to deploy millions of soldiers and the first war to involve telephones, tanks, aircrafts, chemical weapons, and heavy artillery. These twentieth-century technological innovations were deployed alongside outdated military tactics, particularly trench warfare. As a result, hundreds of thousands of troops would die during battles without achieving any meaningful strategic gains. Countries were devastated by the loss of the male population and struggled to cope with a depleted workforce, and widows and orphans struggled to regain any degree of normalcy.

Due to the high death tolls, the Allies' need of the financial support, and the anger associated with the war, the Treaty of Versailles harshly punished Germany, who the Allies blamed for the war. The Allies coerced Germany into signing the treaty that was a death sentence to their country's economy. It contained a "guilt clause," which, unlike the Congress of Vienna's terms for the similarly belligerent France, made oppressive demands on Germany. The treaty took German lands, enforced a heavy reparations debt that was impossible to pay, and stripped Germany of its colonies. After suffering enormous losses during the war itself, the Treaty of Versailles ensured that no national recovery would be possible.

In the aftermath, Russia, Italy, and Germany turned to totalitarian governments, and colonies of Europe started to have nationalistic, anticolonial movements. The Russian Revolution of 1917 led to a civil war in which the Bolsheviks, or Communists, took control under the guidance of Communist revolutionary Vladimir Lenin and established the Soviet Union. The Communist government turned into a dictatorship when Stalin emerged as leader in 1924. Stalin ruled with an iron fist and executed all of his political opponents, including the Bolsheviks. Dissatisfaction with the treaty in Italy led to the rise of fascist leader Benito Mussolini. Germany suffered through several small revolutions, splintering political parties, and class division; this, combined with wartime debt and hyperinflation—a result of the Treaty of Versailles—caused many to become desperate, especially during the throes of the Great Depression. Adolf Hitler, a popular leader in the National Socialist German Workers' Party (Nazi Party), organized street violence against Communists. In the 1932 parliamentary elections, the Nazis emerged as the largest party in the *Reichstag* (German Parliament), but the Nazis did not have enough votes to name Hitler as chancellor. The street violence against Communists and Jews continued unabated, and on January 30, 1933, political pressure led to President von Hindenburg naming Adolf Hitler the chancellor of Germany. Hitler immediately expelled Communists, the second most popular political party, from the *Reichstag*, and coerced the *Reichstag* to pass the Enabling Act of 1933, effectively creating a dictatorship.

Second World War

Nazi Germany had risen to power through the 1920s and 1930s, with Hitler's belief that Germany would only recover its honor if it had a resounding military victory over Europe. Nazi ideology adhered to an extreme nationalism advocating for the superiority of the German people and the necessity of expanding their lands into an empire. Jews, Communists, and other nonconformists were banned from political and social participation.

In 1936, German troops violated the Treaty of Versailles by moving outside Germany's borders, with a remilitarization of the Rhineland. The Rome-Berlin Axis, an alliance between Germany and Italy, was forged in the same year. Germany was the only European power to support Italy's invasion and annexation of Ethiopia, and in exchange, Italy supported Germany's annexation of Austria. In 1936, a civil war broke out in Spain between Spanish nationalist rebels and the government of the Spanish Republic. Mussolini and Hitler supported the Spanish nationalist general Francisco Franco and used the Spanish Civil War as a testing ground for their new alliance. The Allies did not respond to these actions, and when Germany demanded the return of the Sudetenland, a territory in Czechoslovakia, France and Great Britain agreed in hopes of an appeasement despite the protests of the Czech government. Hitler then moved into more areas farther afield, which prompted the Soviet Union to sign a nonaggression pact with Germany. On September 1, 1939, Germany invaded Poland, and on September 3, 1939, France and Great Britain declared war on Germany, jumpstarting the deadliest conflict in world history.

Although less discussed than the Holocaust, the Japanese military committed similar war crimes across Asia, executing between three and ten million Chinese and Koreans, among others, between 1937 and 1945. In one event, the Rape of Nanking, Japanese soldiers captured Nanking and brutally murdered 300,000 civilians. An additional twenty thousand women, children, and elderly were raped during the massacre. Japanese newspapers closely covered a contest between two Japanese officers to see who could kill more people with a sword during the Rape of Nanking. Stalin also committed heinous war crimes during World War II, with estimates ranging from four to ten million deaths as a result of executions and sentences to the Gulag. The United States has also faced criticism for its decision to drop two nuclear bombs on the Japanese cities of Hiroshima and Nagasaki, killing more than 129,000 civilians, leveling both cities, and ending the war. The American government justified the use of nuclear weapons as the only way to avoid a ground invasion of Japan that would have cost more Japanese and American lives than the bombs.

Towns and cities had been leveled, civilian and soldier death tolls were crippling to economies, and countries struggled well into the 1950s to recover economically. It became a breeding ground for Communism, and in China, the end of the war meant a reprisal of the civil war between Mao Zedong's Communists and nationalists that had been interrupted by world war. Another result of the war was a

changed map of the world, as countries were divided or newly formed, and the end of most of Britain's colonialism occurred as a result of the empire's economic and military losses. Following the war, Great Britain, France, Portugal, Belgium, Italy, the Netherlands, and Japan had either granted freedom to colonies or lost areas during the war. Many African and Middle Eastern countries would be granted their independence; however, the newly formed countries' borders were drawn according to those of the former colonies, creating ethnic and religious tensions that still exist today.

In an effort to stop a world war from occurring again, the Allies created the United Nations to be a safeguard and upholder of peace. This proved especially important, yet difficult, as the world was divided between a capitalist Western bloc and a Communist Eastern bloc. Germany was divided between the United States and Soviet Union to maintain peace and to better control the reconstruction of Germany; occupation zones were established, with East Germany occupied by the Soviet Union and West Germany occupied by Great Britain, France, and the United States.

Industrialization, Nationalism, Immigration, and Globalization in Modern World History

Industrialization

In the modern world, industrialization is the initial key to modernization and development. For developed nations, the process of industrialization took place centuries ago. England, where the *Industrial Revolution* began, actually began to produce products in factories in the early 1700s. Later, the United States and some nations of Western Europe followed suit, using raw materials brought in from their colonies abroad to make finished products. For example, cotton was spun into fabric on elaborate weaving machines that mass-produced textiles. As a result, nations that perfected the textile process were able to sell their products around the world, which produced enormous profits. Over time, those nations were able to accumulate wealth, improve their nation's infrastructure, and provide more services for their citizens. Similar to the events of the eighteenth and nineteenth centuries, nations throughout the world are undergoing the same process in today's world. China exemplifies this concept. In China, agriculture was once the predominant occupation, and although it is true that agriculture is still a dominant sector of the Chinese economy, millions of Chinese citizens are flocking to major cities like Beijing, Shanghai, and Hangzhou, due to the availability of factory jobs that allow its workers a certain element of *social mobility*, or the ability to rise up out of one's socioeconomic situation.

Nationalism and Imperialism

With most revolutions, nationalism, or the devotion to one's country, plays a central role. The American and French revolutions, along with the revolutions of Latin America, were fought with the desire to improve the prosperity and position of its civilians. After the Napoleonic Wars and the Congress of Vienna, the years of undisturbed peace resulted in a buildup of nationalism. Countries like Italy and Poland resented Austrian and Russian rule as much as they had disliked French occupation. A rise in nationalism in Germany was a constant threat to Austria, as it tried to govern multiple cultures and languages across a wide geographic area. The precarious situation would remain hostile to some degree until the outbreak of World War I. The Industrial Revolution had made the lower and middle classes restless for change and improvements. By 1848, uprisings began to spring up all over Europe, beginning with France. Many who had nationalistic leanings toward a country that was either no longer in existence or had been forced into another country were able to separate from other nations. The Hungarians broke with Vienna, though they were forced back soon after, the Romans split from papal power, and the Italians threatened rebellion.

The development of imperialism began in the mid-nineteenth century and lasted until the twentieth century, with much of the imperialized world gaining freedom after World Wars I and II. The spread of imperialism that was to follow the revolutions of the eighteenth and nineteenth centuries can be traced, in part, to the idea of nationalism. Some countries believed they were doing a good, and even a moral, thing by conquering and colonizing new territory to spread their culture, traditions, religion, and government. However, a darker side of nationalism—the feeling of superiority and right—caused the takeover of areas and the enforcement of foreign rules and laws.

Globalization

The world economy also became increasingly interconnected during the post-World War II era. This accelerated the process of globalization, which is the integration of ideas and products from different cultures. This benefitted the United States economically because businesses, such as McDonald's and Coca-Cola, found many consumers around the world who were eager to consume American goods. However, the process works both ways, and many aspects of foreign culture, such as Japanese cartoons and animation, have become very popular in the United States. Many critics also point out that globalization has hurt the American economy in recent decades because manufacturing jobs have gone overseas to countries in South America and Asia where wages are low.

Immigration

Immigration has changed the demographics of countries and can have positive and negative effects. Migration and immigration have occurred due to famine, warfare, and lack of economic prospects. Immigration can aid countries struggling to maintain a workforce, and it can also bring in needed medical professionals, scientists, and others with special training. However, immigration also puts strain on developed economies to support migrants who arrive without the necessary education and training to thrive in the advanced economies. Until recently, immigrants were encouraged, or in some cases, forced to assimilate and take on the customs and culture of their new country. For example, in the United States, legislation was passed to force German immigrants to learn English. More recently, developed countries have struggled to assimilate new arrivals to their countries, such as the recent surge of refugees into Europe. Unfortunately, the failure to adequately assimilate immigrants has created greater inequality and prevalence of radical behavior.

Civics and Government

Political Theory

Major Political Concepts

Politics is the process of governance, typically exercised through the enactment and enforcement of laws, over a community, most commonly a state. Political theory involves the study of politics, especially concerning the efficacy and legitimacy of those responsible for governance. The major concepts in political theory include **power** and **authority**. The concepts of power and authority are closely related but possess certain significant distinctions.

Power is the ability of a ruling body or political entity to influence the actions, behavior, and attitude of a person or group of people; in short, power implies a degree of control over a human community. In order to possess **authority**, the ruling body or political entity must be recognized as having the right and justification to exercise power. This is commonly referred to as **legitimacy**. In representative governments, authority is garnered from the citizens through democratic processes, but in more autocratic regimes, influential elites grant that authority. In some cases, a ruling body or political entity

may possess authority recognized by its citizens or influential elites but lack the power to influence those citizens and political entities or effect change within the system of governance. When power and authority are not properly aligned, governments are extremely weak, often deadlocked, and at risk of collapse or revolution.

Government is the result of the decisions made by a society during the political process and is a physical manifestation of the political entity or ruling body. The government determines and enforces the power of the state. A government includes the formal institutions of a society with the power and authority to enact, enforce, and interpret laws. The many different forms of government are determined based on this delegation of power between those institutions. Government encompasses the functions of **law**, **order**, and **justice** and is responsible for maintaining the society.

Sovereignty refers to a political entity's right and power to self-govern, including enacting and enforcing its own taxes and laws without interference from external forces. A political entity may possess varying degrees of sovereignty, as some sovereign states may still be subject to influence by outside political entities. For example, the members of the European Union cede some sovereignty in order to enjoy membership. A state's sovereignty is legitimate when outside political entities recognize the right of the state to self-govern. Both sovereignty and legitimacy are requirements to form a state.

The terms *nation* and *state* are often used interchangeably, but in political theory, they are two very distinct concepts. **Nation** refers to a people's cultural identity, while **state** refers to a territory's political organization and government.

Unlike states, there are no definitive requirements to be a nation; the nation just needs to include a group that is bound together by some shared defining characteristics such as the following:

- Language
- Culture and traditions
- Beliefs and religion
- Homeland
- Ethnicity
- History
- Mythology

The term **state** is commonly used to reference a nation-state, especially in regard to their government. There are four requirements for a political entity to be recognized as a state:

- **Territory:** a clearly defined geographic area with distinct borders

- **Population:** citizens and noncitizens living within the borders of the territory with some degree of permanence

- **Legitimacy:** legal authority to rule that is recognized by the citizens of the state and by other states

- **Sovereignty:** a political entity's right and power to self-govern without interference from external forces

Nation-state is the term used to describe a political entity with both a clearly defined nation and state. In a nation-state, the majority population of the state is a nation that identifies the territory as their

homeland and shares a common history and culture. It is also possible to have several nations in the same nation-state. For example, there are Canadians in Canada and nations of Aboriginal peoples. The presence of multiple nations raises issues related to sovereignty.

Example of a nation: Sikhs in India

Example of a state: Vatican City

Example of a nation-state: Germany

Liberty and Freedom

Governments provide different **liberties** and **freedoms**, depending on the form of government that, in turn, is influenced by the states' shared history and culture. Philosophers started articulating defined rights and liberties and arguing for their adoption and protection by the government during the Enlightenment. Liberalism advocates for increasing liberty and freedom in society and economics, while authoritarian regimes offer considerably less protection. The terms liberty and freedom both reference the same thing in society. Examples of liberty and freedom relate to speech, religion, press, economic pursuits, etc. The degrees and breadth of protections are context dependent, and they might be fluid and influenced by the political process. Many states delineate liberty and freedoms in their constitution. Rights and privileges are different because they involve something that is legal to do or possible according to the government sanctioning the action; thus, rights and privileges are typically more easily revoked from both specific individuals and larger communities.

Rights and Privileges

A **right** is a freedom, protection, or entitlement afforded to a person. In political theory, a right or **privilege** specifically refers to that which is granted to the citizens of a political territory according to the legal system or the social conventions of the society. Rights and privileges typically cannot be removed without due process under the law, although this depends on the form of government. Like the distinction between liberty and freedom, the difference between rights and privileges is indistinguishable. They are often the concrete representations of liberty and freedom in a society; as rights and privileges grow in number and strength, the society enjoys more liberty and freedom. Rights are the privileges provided by the government; however, the term right is commonly used to refer to a privilege of higher status that receives more scrutiny and protection from the state, usually requiring more due process to revoke or alter.

Example of a right: the right to vote

Example of a privilege: the privilege to receive a driver's license

Law, Order, and Justice

Governments must establish law, order, and justice through the government's political institutions. Protecting citizens' rights under the systems of laws, as well as effectively punishing the violators, is one of government's foundational responsibilities. Establishing law, order, and justice is an exercise in sovereignty—states that do not impose law and order under a reasonable standard of justice risk public unrest and challenges to the regime's legitimacy.

Law is the system of rules, policies, and regulations that govern behavior that is enacted by a political entity and enforced by the formal institutions created by the political entity. **Order** refers to each citizen,

institution, and political entity acting as the law. Political entities secure order through the execution of laws through its enforcing institutions, like the police and justice system.

The term **justice** often refers to the fairness in the political entity's treatment of its citizens. When discussing its relation to law and order, justice refers to the necessary, agreed-upon consequences for failure to act according to the law. In political theory, justice is the social contract between the political entity and its citizens in which all parties involved have agreed upon what is considered "fair" in a society and what the consequences of acting contrary to the law should be.

Citizenship and Nationality

Citizenship is the political term referring to the legal status of an individual recognized by the state or government as a protected member. **Citizens** are subject to the laws of their state and granted the rights and liberties afforded by that state, including the right to vote and be elected to office.

Citizenship does not necessarily denote **nationality**, which refers to the relationship between an individual and a nation, rather than a legal status. Nationality is a component of citizenship, but nationality does not afford an individual any of the rights associated with citizenship of a state without official sanction from the state.

Major Political Theorists

Aristotle, 384 BC to 322 BC

A student of Plato, Aristotle was a Greek philosopher best known for his theories of rationality and ethics. In *Politics*, Aristotle asserted that "man is a political animal" and that man must actively participate in politics to further the political well-being of the city-state. He believed that man could only attain the happy life that nature intended through noble acts that bettered the community in which they lived. To this end, he envisioned an ideal society in which the most virtuous and knowledgeable men ruled and the citizens lived virtuous lives in the service of their city-state. Aristotle is one of the leading contributors to Western philosophy and, later, Liberalism. The Roman Republic applied much of his theories on ethics and civics to their government.

Niccolo Machiavelli, 1469-1527

Machiavelli was an Italian diplomat, politician, and historian during the Italian Renaissance. He is most known for his infamous political treatise, *The Prince,* which inspired the political philosophy known as Machiavellianism. In *The Prince*, Machiavelli instructed rulers that it is "better to be feared than to be loved" and asserted that a ruler must be willing to commit any acts necessary to maintain power and establish the stability of their society. He advocated violence, deceit, and immoral acts and advised rulers to seek out and eliminate any potential political rivals within the community. Machiavellianism prioritizes maintaining and consolidating power over all else.

Thomas Hobbes, 1588-1679

Thomas Hobbes was an English philosopher most notable as being the founder of social contract theory. In his 1651 book, *Leviathan,* Hobbes stated that without a strong central authority, the people would live in constant fear; they would go without industry, knowledge, or commodities if there was no system for enforcing good behavior As such, he asserted that the people enacted social contracts with ruling bodies, in which the people agreed to submit to the laws and limitations imposed by the ruling body in exchange for the protection and quality of life afforded to them. Hobbes was responsible for founding many of the fundamentals of liberal thought, stating that all men are naturally equal and that a ruling body may only gain legitimacy through the consent of the people.

John Locke, 1632-1704

John Locke was an English philosopher influenced by Thomas Hobbes' social contract theory. Though he agreed with the theory on some points, he argued against Hobbes' assertion that man requires protection from a ruling body in order to maintain a civilized society, Instead, Locke believed that men are rational and tolerant beings by nature. Locke is considered the "Father of Liberalism." His works on political theory include the famous *Two Treatises on Government*, which contributed significantly to modern ideals of liberal theory, including espousing the need to protect the right to life, liberty, health, and possessions. Locke's philosophy on rights and self-government influenced the writers of the Declaration of Independence.

Jean-Jacques Rousseau, 1712-1778

Jean-Jacques Rousseau was a philosopher, writer, and composer during the Enlightenment in France whose political writings, *Discourse on Equality* and *The Social Contract,* influenced aspects of the French Revolutionary War and many modern political theories. Rousseau thought man was free in the more primitive stage, and this freedom could only be maintained in a state if the people remained sovereign through representative government. He advocated for religious equality and argued that the people are sovereign, rather than a divinely empowered monarch, and that the people should have the right to rule themselves. His work was banned in France, but his ideals influenced the people and inspired many of the political reforms that led to the Revolutionary War.

Immanuel Kant, 1724-1804

A German philosopher, Kant is best known for the set of ideas known as Kantianism, which states that individuals possess dignity and deserve respect. Although more famous for his philosophy, Kant discussed political theory in his essay, *Perpetual Peace: A Philosophical Sketch,* where he argued that world peace may be achieved through universal democracy and international cooperation. In this work, he explained that, in order to end all wars and create lasting peace, all states must form constitutional republics in which elected officials adhere to the rule of the constitutional law to govern the state.

John Stuart Mills, 1806-1873

John Stuart Mills, an English philosopher and political economist, was considered the "most influential English-speaking philosopher of the nineteenth century" and was best known for being the first member of Parliament to advocate women's suffrage. His book *On Liberty* promoted utilitarianism, which advocates that people should always make decisions based on what would be the most net positive. In his work, Mills sought to limit the power exercised upon the individual by any ruling body and stated that moral actions are those that promote utility and increase individuals' and society's well-being. He called for limited constraints upon individual behavior that only restrict those actions that cause harm to others.

Karl Marx, 1818-1883

Karl Marx, a philosopher, social scientist, historian, and revolutionary, is considered one of the most influential Socialist thinkers of the nineteenth century. His ideas became known as Marxism, and Marx heavily influenced powerful Socialist and Communist leaders, such as Vladimir Lenin, with his 1848 pamphlet *The Communist Manifesto.* In this pamphlet, he explained that, in a capitalist society, perpetual "class struggle" exists in which a ruling class (**bourgeois**) controls the means of production and exploits the working class (proletariat), who are forced to sell labor for wages. He advocated for the working class to rebel against the ruling class and establish a classless society with collective ownership of the means of production. He envisioned world history as a series of stages in which capitalism eventually collapses into Communism.

Vladimir Lenin, 1870-1924

As a leading figure in the Russian Revolution and eventual founding of the Soviet Union, Lenin was one of the most controversial and influential political figures in the international Communist movement. Strongly influenced by Karl Marx, Lenin established the one-party system of government, which advocates pure Communism: a classless, egalitarian society in which the people abide by one rule—from "each according to his ability, to each according to his needs," a slogan coined by Marx. Lenin led the Bolsheviks into power after the Russian Revolution, and his Communist Party dominated the centralized one-party system. Lenin redistributed land among the peasants and nationalized most private industries, and he suppressed all political opposition with aggression, most notably during the Red Terror. Lenin advocated for Communist revolution across the world. Joseph Stalin, who further consolidated power of the increasingly authoritarian centralized government, succeeded Lenin as head of the Soviet Union.

Political Orientations

A **political orientation** is a set of ideals, principles, and doctrines that inform how government should be organized, specifically the balance between individual rights, government power, and economic policy. Thus, the country's dominant orientation will largely shape the state form of government. Orientations are also known as political ideologies and are commonly plotted on a political system to provide a visual representation for comparative purposes. The simplest political spectrum orders political orientations on a left-to-right axis. The far left and far right represent the most extreme political orientations, while the center is more moderate.

Left Axis

In general, the orientations on the left emphasize social and economic equality and advocate for government intervention to achieve it. Examples of "leftist" ideologies include Communism, Socialism, Progressivism, and Liberalism.

Communism

Communism is a radical political ideology that seeks to establish common ownership over production and abolish social status and money. Communists believe that the world is split between two social classes—capitalists and the working class (often referred to as the proletariat). Communist politics assert that conflict arises from the inequality between the ruling class and the working class; thus, Communism favors a classless society. Political philosophers Karl Marx and Friedrich Engels argued that Communism is society's destiny since capitalism would ultimately collapse. The collapse of the Soviet Union weakened Communism across the world.

Socialism

Socialists prioritize the health of the community over the rights of individuals, seeking collective and equitable ownership over the means of production. Consequently, **Socialism** is closely tied to an economic system. Under the political philosopher Karl Marx, Socialism is the transitional stage between capitalism and Communism. In general, Socialism is less extreme than Communism. Socialists are more willing to work within the framework of democracy to elect Socialist policies, like social security, universal health care, unemployment benefits, and other programs related to building a societal safety net. Most modern countries include some degree of Socialism, such as progressive taxation and welfare systems. The Scandinavian countries of Denmark, Finland, Sweden, and Norway are examples of countries with a socialized welfare system.

Progressivism

Progressive ideals maintain that progress in the form of scientific and technological advancement, social change, and economic development improve the quality of human life. Progressive ideals include the view that the political and economic interests of the ruling class suppress progress, which results in perpetual social and economic inequality.

Liberalism

Liberalism developed during the Age of Enlightenment in opposition to absolute monarchy, royal privilege, and state religion. In general, Liberalism emphasizes liberty and equality, and liberals support freedom of speech, freedom of religion, free markets, civil rights, gender equality, and secular governance. Liberals support government intervention into private matters to further social justice and fight inequality; thus, liberals often favor social welfare organizations and economic safety nets to combat income inequality. A popular form of Liberalism is **Populism**, which advocates for incorporating advances in technology, social organization, science, and economics into the government to benefit society.

Right Axis

Orientations on the right of the spectrum generally value the existing and historical political institutions and oppose government intervention, especially in regard to the economy. Examples include Conservatism, Libertarianism, and Fascism.

Conservatism

Conservatism is a political ideology that prioritizes traditional institutions within a culture and civilization. In general, Conservatives oppose modern developments and value stability. Since Conservatism depends on the traditional institution, this ideology differs greatly from country to country. Conservatives often emphasize the traditional family structure and the importance of individual self-reliance. Fiscal Conservatism is one of the most common variants, and in general, the proponents of fiscal Conservatism oppose government spending and public debt.

Libertarianism

Libertarianism opposes state intervention on society and the economy. Libertarians advocate for a weak central government, favoring more local rule, and seek to maximize personal autonomy and protect personal freedom. Libertarians often follow a conservative approach to government, especially in the context of power and intervention, but favor a progressive approach to rights and freedom, especially those tied to personal liberty, like freedom of speech.

Fascism

Fascism is a form of totalitarianism that became popular in Europe after World War I. Fascists advocate for a centralized government led by an all-powerful dictator, tasked with preparing for total war and mobilizing all resources to benefit the state. This orientation's distinguishing features include a consolidated and centralized government. In addition, Fascists show more willingness to use violence as a means to accomplish state goals. Popular support for Fascism fell drastically after World War II. There is widespread disagreement among historians, political theorists, and other commentators as to whether governments or political movements qualify as Fascist regimes. Kim Jong-un's status as the supreme leader of North Korea is often cited as a Fascist regime due to the extreme consolidation of totalitarian power. Similarly, Vladimir Putin's Russia is often characterized as Fascist. There are also many European political parties—typically, nationalist parties that oppose immigration and international governance—that advocate for policies widely considered to be Fascist, like the Golden Dawn in Greece.

However, few modern movements and regimes would describe themselves as Fascist. Fascism implies some comfort with totalitarian government, and it is most closely associated with the Fascist regimes during World War II, especially Adolf Hitler's Nazi Party.

Off the Spectrum

Some political orientations do not fit within the political spectrum because they can fit alongside other orientations and ideologies. For example, Nazism is both a Fascist and nationalist ideology. Examples include nationalism, internationalism, and anarchism.

Nationalism

Nationalism is an ideology that prioritizes a loyalty and devotion to the home country. Typically, nationalists exalt their country over all others; thus, nationalism prioritizes promoting its own culture over all other cultures and foreign interests. Nationalism developed alongside the growth of the European nation-state during the nineteenth century. Nationalists can be on the far right or far left. Nationalism is a powerful unifying force within a country or region. In modern times, nationalist groups oppose globalism, such as free trade agreements and supranational political entities, viewing global cooperation as a threat to their national culture.

Internationalism

Internationalist ideals support strong social, political, and economic unity among the states, nations, and people of the world. Internationalists value political institutions on an international level, commonly referred to as supranational political entities. Member states cede some degree of sovereignty to the international entity and receive the associated benefits. Internationalism advocates for free trade and globalism. Nongovernmental organizations play an important role in internationalism's approach to global problems. Examples of supranational organizations include the United Nations, European Union, North Atlantic Treaty Organization (NATO), and World Trade Organization (WTO).

Anarchism

Anarchism is a radical political orientation that advocates for self-government and abolishment of existing political institutions; instead, society would be based on voluntary institutions. Anarchists oppose the state, arguing that political institutions organize society hierarchically, which creates the conflict. Thus, law and order is then accomplished by freeing the people from the state's corruption.

United States Government and Politics

Constitutional Underpinnings

The role of government is to maintain a society and provide public services through its formal institutions, protect the citizens of the state, and regulate the economic system. To determine how a government should perform these functions and to protect the rights and liberties of the citizens, states enact a **constitution**, a written document that typically establishes the form of government and delegation of powers within the government, delineates limits on government, and defines protected rights, liberties, and privileges.

The many underpinnings, or foundations, upon which the *Constitution* of the United States was founded include:

Articles of Confederation (1781-1789)
The Articles of Confederation established a formal agreement or confederation between the original thirteen states. The Articles of Confederation established a central government composed of a unicameral legislative assembly in which each state was granted a single representative. Passing a bill required votes from nine of the thirteen representatives. Under the Articles of Confederation, the centralized government, the Continental Congress, was granted very limited powers, rendering it largely ineffective. Those powers included:

- Borrowing money from states or foreign government
- Creating post offices
- Appointing military offices
- Declaring war
- Signing treaties with foreign states

The weak central government established under the Articles of Confederation lacked the power to impose taxes, enforce a draft to staff the new army and navy, regulate trade, or enforce the laws enacted in Congress. As such, the sovereignty remained primarily with the states. Under the Articles, the states reserved the powers to impose taxes upon each other and the citizens of their states, regulate trade within their states, coin and print money, and sign treaties with foreign states. The states also often ignored the laws enacted by the Congress because there was no executive branch to enforce the law.

This imbalance of power between the central government and the states led to crisis within the states, resulting in economic difficulties and violence. The lack of common currency and investment in interstate infrastructure greatly hindered economic growth. In the years 1786 and 1787, farmers in several states staged a series of protests over local tax and debt collection imposed on struggling farms, commonly known as **Shay's Rebellion**.

Constitutional Convention, 1787
The failures of the Articles of Confederation to effectively govern on a national level directly led to the Constitutional Convention, and those experiences influenced the founders' decision to include a more robust federal government in the United States Constitution. The Constitutional Convention faced several challenges, including disputes over representation between large and small states, tension between the southern and northern states over slavery, differing visions of how power would be delegated within the government, and opposition to ceding states' sovereignty to a national federal government.

New Jersey Plan
Led by William Patterson, the New Jersey Plan called for a unicameral legislature that would grant each state a single vote. It proposed a plural executive power selected by the legislature, which would possess no veto power over the legislature, as well as judges appointed by the executive power for the duration of their lives.

The Virginia Plan
Drafted by James Madison, the Virginia Plan featured a bicameral legislature with two houses. The representatives of the lower house were to be selected by the people, and then the lower house would

elect the upper house. The number of representatives of each house would be based upon population or the amount of money contributed to the federal government by each state; thus, large states supported the Virginia Plan. In this plan, the legislature could appoint judges and select a single executive with veto power.

Compromises
Connecticut Compromise
This compromise included aspects of both the New Jersey Plan and Virginia Plan in a bicameral legislature. Representation in the House of Representatives was proportional to a state's population, and in the Senate, states enjoyed equal representation with two senators per state.

Slavery Compromises
Several other compromises were made during the Convention, including the *Three-Fifths Compromise*, which, in an effort to appease both the South states who wanted slaves to be counted as part of the population for the purpose of representation but not counted for the purpose of taxes, and the North, who demanded slaves be counted for taxes but not representation, the framers of the Constitution determined that three-fifths of the slave population of each state would be counted for the purpose of both taxes and representation. In an additional compromise called the *Commerce and Slave Trade Compromise*, Congress agreed not to tax exports from states or ban the trading of slaves for twenty years. This eased Southerners' fears that if the Northern states controlled the federal government, then they could enforce antislavery policies.

Constitution vs. Articles of Confederation
The Constitution addressed the weaknesses of the Articles of Confederation in the following ways:

- Unlike the sovereign states under the Articles of Confederation, the people are now the sovereign, and they bestow sovereignty to both the states and federal government, according to principles of Federalism.

- The Constitution creates a robust central government with many specific and implied powers.

- The Constitution empowers the federal government to levy taxes against the states.

- The Constitution establishes an executive branch of the federal government to enforce the laws; it is led by a president who serves as the commander-in-chief.

- The Constitution establishes a federal judiciary branch with a Supreme Court and lower courts to interpret the laws enacted by the legislative branch.

- The Constitution removes the states' power to coin and print money and establishes a national currency; Congress may regulate interstate and international commerce.

- The Constitution specifies representation in Congress based on population and equal representation for each state in the Senate.

- The Constitution requires a simple majority in both houses to enact laws rather than a vote of at least nine out of thirteen, as specified in the Articles of Confederation. In addition, senators vote separately under the Constitution, while states vote as a single bloc in the Constitution.

- The Constitution requires a two-thirds majority vote in the House of Representatives and a two-thirds majority in the Senate to amend the Constitution, while the Articles of Confederation required a unanimous vote.

Federalism

To strengthen the central government, while still appeasing the individual states who preferred to remain sovereign over their territories, the framers of the Constitution based the new government upon the principles of **Federalism**—a compound government system that divides powers between a central government and various regional governments. The Constitution clearly defined the roles of both the state governments and the new federal government, specifying the limited power of the federal government and reserving all other powers not restricted by the Constitution to the states in the Tenth Amendment to the Constitution, commonly referred to as the Reservation Clause.

The Constitution establishes the specific powers granted to the federal and state governments.

- Delegated powers: the specific powers granted to the federal government by the Constitution
- Implied powers: the unstated powers that can be reasonably inferred from the Constitution
- Inherent powers: the reasonable powers required by the government to manage the nation's affairs and maintain sovereignty
- Reserved powers: the unspecified powers belonging to the state that are not expressly granted to the federal government or denied to the state by the Constitution
- Concurrent powers: the powers shared between the federal and state governments

The Constitution would delegate the following expanded powers to the federal government:

- Coin money
- Declare war
- Establish lower federal courts
- Sign foreign treaties
- Expand the territories of the United States, and admit new states into the union
- Regulate immigration
- Regulate interstate commerce

The following powers were reserved for the states:

- Establish local governments
- Hold elections
- Implement welfare and benefit programs
- Create public school systems
- Establish licensing standards and requirements
- Regulate state corporations
- Regulate commerce within the state

The **concurrent** powers granted to both the federal and state governments in the Constitution include:

- The power to levy taxes
- The power to borrow money
- The power to charter incorporations

Ratifying the Constitution

The framers of the Constitution signed the Constitution on September 17, 1787, but the Articles of Confederation required nine of the thirteen states to ratify the document. Conventions were held in all thirteen states and sparked heated debates between those who supported and those who opposed the new system of government. The Federalists supported the expansion of the federal government, and the anti-Federalists feared that a stronger central government would weaken the states. The anti-Federalists also sought additional protection for civil liberties. The debates between these two parties continued for two years and inspired a series of essays known as the *Federalist Papers* and *Anti-Federalist Papers* authored anonymously by leaders of their respective party.

Notable Federalists and authors of the *Federalist Papers* include:

- Alexander Hamilton: founder of the Federalist Party and advocate for a centralized financial system

- George Washington: commander-in-chief of the Continental Army and future first president of the United States

- James Madison: one of the primary drafters of the Constitution and the future fourth president of the United States

- John Jay: president of the Continental Congress and future first chief justice of the United States

- John Adams: future second president of the United States

Notable anti-Federalists and authors of the *Anti-Federalist Papers* include:

- Thomas Jefferson: primary author of the Declaration of Independence and future third president of the United States

- Patrick Henry: governor of Virginia (1776-1779, 1784-1786)

- Samuel Adams: governor of Massachusetts (1794-1797), lieutenant governor of Massachusetts (1789-1794), and president of the Massachusetts Senate (1782-1785, 1787-1788)

- George Mason: one of only three delegates who did not sign the Constitution at the Constitutional Convention and author of Objections to This Constitution of Government (1787) and the Virginia Declaration of Rights of 1776, which served as the basis for the Bill of Rights

The first state to ratify the Constitution was Delaware in a unanimous vote on December 7, 1787. Several other states followed, and eventually, after ten months, New Hampshire became the ninth state to ratify the Constitution in June 1788. However, some states still remained divided between Federalist and anti-Federalist sentiments and had yet to approve the document, including the two most populous states, Virginia and New York. To reconcile their differing views, the Federalists agreed to include a bill of rights if anti-Federalists supported the new Constitution. Federalist sentiment prevailed, and the

remaining states approved the document. On May 29, 1790, the last holdout, Rhode Island, ratified the Constitution by two votes. As promised, the Bill of Rights—the first 10 amendments to the Constitution—was added in 1791, providing expanded civil liberty protection and due process of law.

Powers, Structure, and Processes of National Political Institutions

A **political institution** is an organization created by the government to enact and enforce laws, act as a mediator during conflict, create economic policy, establish social systems, and carry out some power. These institutions maintain a rigid structure of internal rules and oversight, especially if the power is delegated, like agencies under the executive branch.

The Constitution established a federal government divided into three branches: legislative, executive, and judicial.

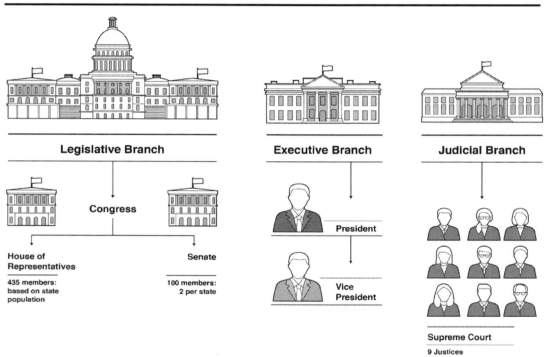

Executive Branch

The executive branch is responsible for enforcing the laws. The executive branch consists of the president, the vice president, the president's cabinet, and federal agencies created by Congress to execute some delegated.

The president of the United States:

- Serves a four-year term and is limited to two terms in office
- Is the chief executive officer of the United States and commander-in-chief of the armed forces
- Is elected by the Electoral College

- Appoints cabinet members, federal judges, and the heads of federal agencies
- Vetoes or signs bills into law
- Handles foreign affairs, including appointing diplomats and negotiating treaties
- Must be thirty-five years old, a natural-born U.S. citizen, and have lived in the United States for at least fourteen years

The vice president:

- Serves four-year terms alongside and at the will of the president
- Acts as president of the Senate
- Assumes the presidency if the president is incapacitated
- Assumes any additional duties assigned by the president

The cabinet members:

- Are appointed by the president
- Act as heads for the fifteen executive departments
- Advise the president in matters relating to their departments and carry out delegated power

Note that the president can only sign and veto laws and cannot initiate them himself. As head of the executive branch, it is the responsibility of the president to execute and enforce the laws passed by the legislative branch.

Although Congress delegates their legislative authority to agencies in an enabling statute, they are located in the executive branch because they are tasked with executing their delegating authority. The president enjoys the power of appointment and removal over all federal agency workers, except those tasked with quasi-legislative or quasi-judicial powers.

Legislative Branch

The legislative branch is responsible for enacting federal laws. This branch possesses the power to declare war, regulate interstate commerce, approve or reject presidential appointments, and investigate the other branches. The legislative branch is **bicameral**, meaning it consists of two houses: the lower house, called the **House of Representatives**, and the upper house, known as the **Senate**. Both houses are elected by popular vote.

Members of both houses are intended to represent the interests of the constituents in their home states and to bring their concerns to a national level. Ideas for laws, called **bills**, are proposed in one chamber and then are voted upon according to the body's rules; should the bill pass the first round of voting, the other legislative chamber must approve it before it can be sent to the president.

The two houses (or **chambers**) are similar though they differ on some procedures such as how debates on bills take place.

House of Representatives

The **House of Representatives** is responsible for enacting bills relating to revenue, impeaching federal officers including the president and Supreme Court justices, and electing the president in the case of no candidate reaching a majority in the Electoral College.

In the House of Representatives:

- Each state's representation in the House of Representatives is determined proportionally by population, with the total number of voting seats limited to 435.

- There are six nonvoting members from Washington, D.C., Puerto Rico, American Samoa, Guam, Northern Mariana Islands, and the U.S. Virgin Islands.

- The **Speaker of the House** is elected by the other representatives and is responsible for presiding over the House. In the event that the president and vice president are unable to fulfill their duties, the Speaker of the House will succeed to the presidency.

- The representatives of the House serve two-year terms.

- The requirements for eligibility in the House include:

  o Must be twenty-five years of age
  o Must have been a U.S. citizen for at least seven years
  o Must be a resident of the state they are representing by the time of the election

Senate
The **Senate** has the exclusive powers to confirm or reject all presidential appointments, ratify treaties, and try impeachment cases initiated by the House of Representatives.

In the Senate:

- The number of representatives is one hundred, with two representatives from each state.
- The vice president presides over the Senate and breaks the tie, if necessary.
- The representatives serve six-year terms.
- The requirements for eligibility in the Senate include:
  o Must be thirty years of age
  o Must have been a U.S. citizen for the past nine years
  o Must be a resident of the state they are representing at the time of their election

Legislative Process
Although all members of the houses make the final voting, the senators and representatives serve on committees and subcommittees dedicated to specific areas of policy. These committees are responsible for debating the merit of bills, revising bills, and passing or killing bills that are assigned to their committee. If it passes, they then present the bill to the entire Senate or House of Representatives (depending on which they are a part of). In most cases, a bill can be introduced in either the Senate or the House, but a majority vote of both houses is required to approve a new bill before the president may sign the bill into law.

Judicial Branch
The **judicial branch**, though it cannot pass laws itself, is tasked with interpreting the law and ensuring citizens receive due process under the law. The judicial branch consists of the **Supreme Court**, the highest court in the country, overseeing all federal and state courts. Lower federal courts are the district courts and court of appeals.

The Supreme Court:

- Judges are appointed by the president and confirmed by the Senate.
- Judges serve until retirement, death, or impeachment.
- Judges possess sole power to judge the constitutionality of a law.
- Judges set precedents for lower courts based on their decisions.
- Judges try appeals that have proceeded from the lower district courts.

Checks and Balances

Notice that a system of **checks and balances** between the branches exists. This is to ensure that no branch oversteps its authority. They include:

- Checks on the Legislative Branch:
    - The president can veto bills passed by Congress.
    - The president can call special sessions of Congress.
    - The judicial branch can rule legislation unconstitutional.
- Checks on the Executive Branch:
    - Congress has the power to override presidential vetoes by a two-thirds majority vote.
    - Congress can impeach or remove a president, and the chief justice of the Supreme Court presides over impeachment proceedings.
    - Congress can refuse to approve presidential appointments or ratify treaties.
- Checks on the Judicial Branch:
    - The president appoints justices to the Supreme Court, as well as district court and court of appeals judges.
    - The president can pardon federal prisoners.
    - The executive branch can refuse to enforce court decisions.
    - Congress can create federal courts below the Supreme Court.
    - Congress can determine the number of Supreme Court justices.
    - Congress can set the salaries of federal judges.
    - Congress can refuse to approve presidential appointments of judges.
    - Congress can impeach and convict federal judges.

The three branches of government operate separately, but they must rely on each other to create, enforce, and interpret the laws of the United States.

Checks and Balances

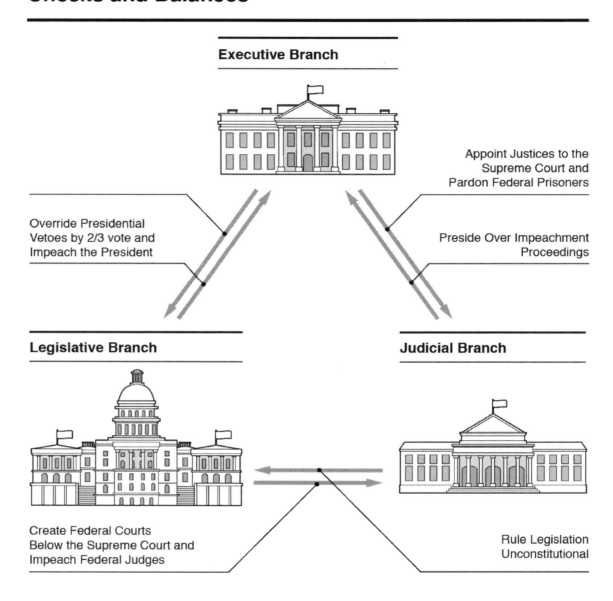

Executive Branch

Appoint Justices to the Supreme Court and Pardon Federal Prisoners

Preside Over Impeachment Proceedings

Override Presidential Vetoes by 2/3 vote and Impeach the President

Legislative Branch

Judicial Branch

Create Federal Courts Below the Supreme Court and Impeach Federal Judges

Rule Legislation Unconstitutional

How Laws are Enacted and Enforced

To enact a new law:

- The bill is introduced to Congress.
- The bill is sent to the appropriate committee for review and revision.
- The approved bill is sent to the Speaker of the House and the majority party leader of the Senate, who places the bill on the calendar for review.

- The houses debate the merits of the bill and recommend amendments.

  o In the House of Representatives, those who wish to debate about a bill are allowed only a few minutes to speak, and amendments to the bill are limited.

  o In the Senate, debates and amendments are unlimited, and those who wish to postpone a vote may do so by filibuster, refusing to stop speaking.

- The approved bill is revised in both houses to ensure identical wording in both bills.
- The revised bill is returned to both houses for final approval.
- The bill is sent to the president, who may

  o Sign the bill into law

  o Veto the bill

  o Take no action, resulting in the bill becoming law if Congress remains in session for ten days or dying if Congress adjourns before ten days have passed

The Role of State Government

While the federal government manages the nation as a whole, state governments address issues pertaining to their specific territory. In the past, states claimed the right, known as **nullification**, to refuse to enforce federal laws that they considered unconstitutional. However, conflicts between state and federal authority, particularly in the South in regard to first, slavery, and later, discrimination, have led to increased federal power, and states cannot defy federal laws. Even so, the Tenth Amendment limits federal power to those powers specifically granted in the Constitution, and the rest of the powers are retained by the states and citizens. Therefore, individual state governments are left in charge of decisions with immediate effects on their citizens, such as state laws and taxes.

In this way, the powers of government are separated both horizontally between the three branches of government (executive, legislative, and judicial) and vertically between the levels of government (federal, state, and local).

Like the federal government, state governments consist of executive, judicial, and legislative branches, but the exact configuration of those branches varies between states. For example, while most states follow the bicameral structure of Congress, Nebraska has only a single legislative chamber. Additionally, requirements to run for office, length of terms, and other details vary from state to state. State governments have considerable authority within their states, but they cannot impose their power on other states.

Separation of Powers

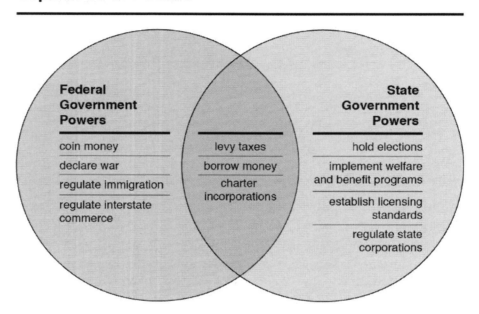

Federal Government Powers

- coin money
- declare war
- regulate immigration
- regulate interstate commerce

(overlap)
- levy taxes
- borrow money
- charter incorporations

State Government Powers

- hold elections
- implement welfare and benefit programs
- establish licensing standards
- regulate state corporations

Civil Liberties and Civil Rights

The protection of **civil liberties** is one of the most important political values upon which American society is based. Though the terms **civil liberties** and **civil rights** are commonly used interchangeably, they describe two very distinct types of protections. Civil liberties refer to the legal protections afforded to U.S. citizens against government action, while civil rights refer to equal treatment under the law, especially in relation to minority groups, like women, African Americans, and Hispanics.

Civil Liberties

A **civil liberty** is a protection from legal action by the government. Civil liberties are granted by the Constitution in the first ten amendments, collectively known as the Bill of Rights, which were added to the Constitution in 1791. Civil liberties are conditional and do not afford protection from government action in every scenario. They can be restricted when they infringe on the rights of others; for example, with defamation, child pornography, or "fighting words." They also may be suspended with just cause, such as in the case of limiting the freedom of press to protect national security.

The Bill of Rights

The first ten amendments of the Constitution are called the **Bill of Rights**. They were passed to win over anti-Federalists during the ratification of the Constitution. Anti-Federalists wanted assurances that the federal government would protect certain fundamental civil liberties. The Bill of Rights includes:

- Amendment I: Establishes freedom of religion, speech, and press; the right to assemble in peaceful protest; and the right to petition the government without fear of reprisal

- Amendment II: Establishes the right to bear arms

- Amendment III: Establishes the right to refuse to quarter, or house, soldiers in time of war

- Amendment IV: Establishes protection against unreasonable search and seizure and requires a warrant based on probable cause supported by specific information

- Amendment V: Protects against self-incrimination in criminal trials, except in cases of military court martial; protects against being tried more than once for the same crime, known as double jeopardy; and protects against seizure of private property for public use without compensation

- Amendment VI: Establishes extensive set of rights to protect defendants in a criminal trial—the right to a speedy and timely trial before a judge and impartial jury of peers, the right to be informed of criminal accusations, the right to present and compel witnesses in defense of the accused, the right to confront witnesses against the accused, and the right to assistance of counsel

- Amendment VII: Protects the right to a trial by jury in civil cases exceeding a dollar amount of $20

- Amendment VIII: Protects against cruel and unusual punishment and excessive fines

- Amendment IX: Establishes the existence of additional fundamental rights unnamed in the Constitution; protects those rights that are not enumerated

- Amendment X: Reserves all powers that are not specified to the federal government or prohibited to the states or the people, establishing the principles of separation of powers and Federalism

Civil Rights

Civil rights concern who is protected, while *civil liberties* concern what is protected. Civil rights refer to protection against unfair treatment based on characteristics such as gender, race, ethnicity, religion, sexual orientation, and disability. The struggle for civil rights has a long history in the United States. Following the Civil War, the ratification of three amendments—Thirteenth, Fourteenth, and Fifteenth, collectively known as the Reconstruction Amendments—expanded the constitutional protection of equal civil rights.

The Thirteenth Amendment abolished slavery and involuntary servitude, except as punishment for a crime. The issue of slavery was no longer in the states' hands. Although the Emancipation Proclamation freed slaves in the Confederacy, the status of former slaves remained uncertain as the war neared its conclusion. Many Northerners did not hold strong views on slavery, but most wanted to punish the South and resolve the primary cause of the bloody Civil War. The Northern states all immediately ratified the amendment, and in December 1865, enough reconstructed Southern states ratified the amendment for it to be adopted into law.

The Fourteenth Amendment prohibited states from depriving life, liberty, or property without due process and from violating equal protection based on race, color, or previous condition of servitude. Now, all persons born or naturalized in the United States were considered legal citizens. Although revolutionary for the theoretical rights of all American citizens, newly freed or otherwise, the Fourteenth Amendment did not provide actual federally enforced equal protection until the Civil Rights Act of 1964.

The Fifteenth Amendment prohibits the government from denying a citizen the right to vote for reasons of race, color, or previous condition of servitude. Adopted in 1870, the last of the Reconstruction Amendments, the Fifteenth Amendment sought to protect newly freed slaves' right to vote. As discussed below, most states interpreted the amendment to only apply to male suffrage. In addition, Southern states passed a series of laws to systematically disenfranchise African Americans, like poll taxes, literacy tests, and residency rules. The use of violence and intimidation for political purpose was also common. Meaningful change did not occur until the Civil Rights Movement, nearly one hundred years later. In 1964, the Twenty-Fourth Amendment prohibited the states and federal government from charging a poll tax or fee to vote. Later, the Voting Rights Act of 1965 empowered the federal government to enforce the Fifteenth Amendment on the states for the first time.

Women's Suffrage

The Fourteenth Amendment specified equal treatment for all citizens; however, it did not establish women's right to vote in elections, known as women's suffrage. Although landowning women were allowed to vote in New Jersey in the late eighteenth century, the right was removed in 1807. The fight for women's suffrage continued in the middle of the nineteenth century. Famous women's rights activists include Susan B. Anthony, Lucy Stone, and Elizabeth Cady Stanton, who authored the *Declaration of Rights and Sentiments*, which demanded access to the civil liberties granted to all men. Women gained the right to vote in 1869 in Wyoming and 1870 in Utah.

The women's suffrage movement gained momentum in the early twentieth century after their increased participation in the economy during World War I when much of the workforce went overseas to fight. The National Women's Party picketed outside the White House and led a series of protests in Washington, resulting in the imprisonment of the party's leader, Alice Paul. In 1918, Woodrow Wilson declared his support for women's suffrage despite earlier opposition, and in 1920, Congress passed the Nineteenth Amendment, which made it illegal for states to withhold voting rights based on gender.

Jim Crow Laws

Southern states circumvented the Fourteenth Amendment and imposed what were referred to as Jim Crow laws, which established racial segregation of public facilities. These "separate but equal" facilities included the military, workplaces, public schools, restaurants, restrooms, transportation, and recreational facilities. Despite the label of "separate but equal," most facilities reserved for African Americans were considerably inferior.

In 1896, the Supreme Court handed down a decision in the case of *Plessy vs. Ferguson*, in which Homer Plessy, a Louisiana man of mixed race, attempted to board a railway car reserved for "whites only" and was charged for violating the separate car law. Plessy subsequently filed suit against the state, claiming they violated his Fourteenth Amendment rights. The Supreme Court decided in favor of the state, ruling that the law was not unconstitutional. The Supreme Court upheld separate but equal laws until the 1954

case of *Brown vs. the Board of Education of Topeka* where the Supreme Court ruled that racial segregation of public schools violated the Fourteenth Amendment.

Civil Rights Movement

Brown vs. the Board of Education prohibited segregation in 1954, but the Civil Rights Movement, led by the National Association for the Advancement of Colored People (NAACP) and such famous activists as Martin Luther King Jr. and Malcolm X, did not secure the enforcement of the Fourteenth Amendment until the passage of the Civil Rights Act of 1964, which outlawed discrimination based on gender, race, ethnicity, and religion. African American and Native American women, however, did not gain the right to vote until the Voting Rights Act of 1965, which enforced the voting rights articulated in the Fourteenth Amendment and Fifteenth Amendment. Section 5 of the Voting Rights Act prevented states with a history of discrimination from altering their voting laws without getting approval from the attorney general or a federal district court.

Political Beliefs and Behaviors

Political beliefs are the beliefs held by the citizens of a nation about the government, leaders, policies, and the related political issues of their state. Political beliefs differ among individual citizens, but in America, a strong basis of democracy shapes the political beliefs, behaviors, and attitudes.

Democratic Values

The foundation of democratic values upon which the United States is based include:

- The people are sovereign, and they elect a representative government to exercise that sovereignty.
- The citizens of the nation are equal under the law.
- The peaceful transition of power is valued regardless of election results.
- The private property of individuals cannot be taken by force by the government without due process or fair compensation.
- The civil liberties of the citizens of the state cannot be abridged or violated by the government without due process.
- The government should be accountable to the citizenry.

Political Socialization

American citizens undergo a process of **political socialization** from early childhood to adulthood during which they develop their individual sense of political identity and civic pride. Children learn about politics in the home from an early age, whether from the views, opinions, and facts of family and friends, or through the media to which they are exposed.

In school, they learn about the nation's political history, basic politics, and democratic values, as well as the ideals of patriotism and the processes of government. As they grow older, they join interest groups, labor unions, religious groups, and political organizations that further influence their political beliefs. This socialization shapes not only the political beliefs and values of individual citizens and groups but the political ideals of the nation and public opinion.

Public Opinion

Public opinion is the shared political ideals, opinions, and attitudes of the people of a state regarding the politics, current events, and social issues that influence policy and shape the political atmosphere of a state. Public opinion is the result of political beliefs, socialization, and current events. Political scientists measure public opinion through:

- Distribution of opinion across demographics such as age, race, gender, and religion
- Strength of the opinion
- Stability of the opinion over time

Public opinion refers to the majority opinion in a democratic state. Citizens express public opinion through the interest groups they join, the media they consume and interact with, and the leaders they elect. To measure public opinion, scientists use polls to gather data. Accurate polling requires:

- Random sampling of representative populations
- Unbiased questions
- Clear instructions for how to answer questions
- Controlled procedures such as the use of telephone, mail, Internet, or in-person interviews with an unbiased pollster
- Accurate reporting of the results, including information about methods, inconsistencies, respondents, and possible sources and degree of error

Political Participation

Citizens express their political beliefs and public opinion through participation in politics. The conventional ways citizens can participate in politics in a democratic state include:

- Obeying laws
- Voting in elections
- Running for public office
- Staying interested and informed of current events
- Learning U.S. history
- Attending public hearings for information and to express their opinions on issues, especially on the local level
- Forming interest groups to promote their common goals
- Forming political action committees (PACs) that raise money to influence policy decisions
- Petitioning government to create awareness of issues
- Campaigning for a candidate
- Contributing to campaigns
- Using mass media to express political ideas, opinions, and grievances

Voting

In a democratic state, the most common way to participate in politics is by voting for candidates in an election. Voting allows the citizens of a state to influence policy by selecting the candidates who share their views and make policy decisions that best suit their interests, or candidates who they believe are most capable of leading the country. In the United States, all citizens—regardless of gender, race, or religion—are allowed to vote unless they have lost their right to vote through due process, such as felons.

Since the Progressive movement and the increased social activism of the 1890s to the 1920s that sought to eliminate corruption in government, direct participation in politics through voting has increased. Citizens can participate by voting in the following types of elections:

- **Direct primaries:** Citizens can nominate candidates for public office.

- **National, state, and municipal elections:** Citizens elect their representatives in government.

- **Recall elections:** Citizens can petition the government to vote an official out of office before their term ends.

- **Referendums:** Citizens can vote directly on proposed laws or amendments to the state constitution.

- **Voter initiatives:** Citizens can petition their local or state government to propose laws that will be approved or rejected by voters.

Electoral Process, Political Parties, Interest Groups, and Mass Media
Electoral Process
During the **electoral process**, the citizens of a state decide who will represent them at the local, state, and federal level. Different political officials that citizens elect through popular vote include but are not limited to:

- City mayor
- City council members
- State representative
- State governor
- State senator
- House member
- U.S. Senator
- President

The Constitution grants the states the power to hold their own elections, and the voting process often varies from city to city and state to state.

While a popular vote decides nearly all local and state elections, the president of the United States is elected by the **Electoral College,** rather than by **popular vote**. Presidential elections occur every four years on the first Tuesday after the first Monday in November.

The electoral process for the president of the United States includes:

Primary Elections and Caucuses
In a presidential election, *nominees* from the two major parties, as well as some third parties, run against each other. To determine who will win the nomination from each party, the states hold **primary elections** or **caucuses**.

During the primary elections, the states vote for who they want to win their party's nomination. In some states, primary elections are closed, meaning voters may only vote for candidates from their registered party, but other states hold **open primaries** in which voters may vote in either party's primary.

Some states hold **caucuses** in which the members of a political party meet in small groups, and the decisions of those groups determine the party's candidate.

Each state holds a number of delegates proportional to its population, and the candidate with the most delegate votes receives the domination. Some states give all of their delegates (**winner-take-all**) to the primary or caucus winner, while some others split the votes more proportionally.

Conventions

The two major parties hold national conventions to determine who will be the nominee to run for president from each party. The **delegates** each candidate won in the primary elections or caucuses are the voters who represent their states at the national conventions. The candidate who wins the most delegate votes is given the nomination. Political parties establish their own internal requirements and procedures for how a nominee is nominated.

Conventions are typically spread across several days, and leaders of the party give speeches, culminating with the candidate accepting the nomination at the end.

Campaigning

Once the nominees are selected from each party, they continue campaigning into the national election. Prior to the mid-1800s, candidates did not actively campaign for themselves, considering it dishonorable to the office, but campaigning is now rampant. Modern campaigning includes, but is not limited to:

- Raising money
- Meeting with citizens and public officials around the country
- Giving speeches
- Issuing policy proposals
- Running internal polls to determine strategy
- Organizing strategic voter outreach in important districts
- Participating in debates organized by a third-party private debate commission
- Advertising on television, through mail, or on the Internet

General Election

On the first Tuesday after the first Monday in November of an election year, every four years, the people cast their votes by secret ballot for president in a **general election**. Voters may vote for any candidate, regardless of their party affiliation. The outcome of the popular vote does not decide the election; instead, the winner is determined by the Electoral College.

Electoral College

When the people cast their votes for president in the general election, they are casting their votes for the **electors** from the Electoral College who will elect the president. In order to win the presidential election, a nominee must win 270 of the 538 electoral votes. The number of electors is equal to the total number of senators and representatives from each state plus three electoral votes for Washington D.C. which does not have any voting members in the legislative branch.

The electors typically vote based on the popular vote from their states. Although the Constitution does not require electors to vote for the popular vote winner of their state, no elector voting against the popular vote of their state has ever changed the outcome of an election. Due to the Electoral College, a nominee may win the popular vote and still lose the election.

For example, let's imagine that there only two states, Wyoming and Nebraska, in a presidential election. Wyoming has three electoral votes and awards them all to the winner of the election by majority vote. Nebraska has five electoral votes and also awards them all to the winner of the election by majority vote. If 500,000 people in Wyoming vote and the Republican candidate wins by a vote of 300,000 to 200,000, the Republican candidate will win the three electoral votes for the state. If the same number of people vote in Nebraska, but the Republican candidate loses the state by a vote of 249,000 to 251,000, the Democratic candidate wins the five electoral votes from that state. This means the Republican candidate will have received 549,000 popular votes but only three electoral votes, while the Democratic candidate will have received 451,000 popular votes but will have won five electoral votes. Thus, the Republican won the popular vote by a considerable margin, but the Democratic candidate will have been awarded more electoral votes, which are the only ones that matter.

	Wyoming	Nebraska	Total # of Votes
Republican Votes	300,000	249,000	549,000
Democratic Votes	200,000	251,000	451,000
Republican Electoral Votes	3	0	3
Democratic Electoral Votes	0	5	5

If no one wins the majority of electoral votes in the presidential election, the House of Representatives decides the presidency, as required by the Twelfth Amendment. They may only vote for the top three candidates, and each state delegation votes as a single bloc. Twenty-six votes, a simple majority, are required to elect the president. The House has only elected the president twice, in 1801 and 1825.

Here how many electoral votes each state and the District of Columbia have:

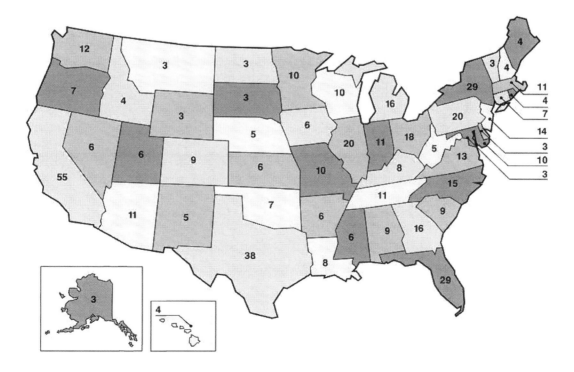

Political Parties

A **political party** is an organized group of voters who share the same political values and support or oppose the same policies. Members of a political party vote for the candidates from their party who they believe share their values and will approve or reject the policies they support or oppose. Political parties often determine the positions party members take on issues of policy, such as the economy, taxation, and social services.

The **Founding Fathers** of the United States opposed the divisiveness they associated with political parties, and President George Washington railed against the evil of political parties in his Farewell Address. However, the ratification of the Constitution led to the creation of the first two American political parties, the Federalists and the anti-Federalist Democratic-Republican Party. Andrew Jackson, the seventh president of the United States, is considered the country's first official Democratic president. During his candidacy, his opposition organized under the Whig Party. The Whigs asserted Congress' supremacy over the president and primarily focused on economic concerns, like banking and violations.

Slavery divided the nation and created unrest among the political parties, as members took opposing views and splintered into separate sects of the party or started new parties with members who shared their views. The Whig Party, so divided by the differing views of the members, collapsed. Former Whigs joined or formed the following parties:

- **Constitutional Union Party:** Devoted itself to a single-issue platform of preserving the Union and recognizing the Constitution as the supreme rule of law. The party did not take a firm issue on slavery, but vigorously opposed secession.

- **Democratic Party:** Divided into northern and southern factions over slavery, but the Democrats sought to compromise and remain unified.

- **Know-Nothing Party:** Advocated for an anti-immigration single-party platform, especially immigrants from Catholic countries.

- **Republican Party:** Formed in response to the Kansas-Nebraska Act, which threatened to extend slavery into new territories, called for the abolition of slavery and argued for a more modernized economy.

Modern Political Parties

The defeat of the South in the Civil War resulted in the Republicans holding power until the 1930s, when Franklin D. Roosevelt, a Democrat, was elected president. Roosevelt instituted the New Deal, which included many social policies that built an expansive social welfare program to provide financial support to citizens during the Great Depression. The Republican Party opposed this interference by the government, and the two parties became more strongly divided. The political landscape again shifted during the Civil Rights Movement, as Southern Democrats fled to the Republican Party over their opposition to enforcing federal civil rights onto states. This strengthened the modern coalition between economic conservatives and social conservatives.

Today, the Democrats and Republicans are still the two major parties, though many third parties have emerged. The Republicans and Democrats hold opposing views on the degree of state intervention into private business, taxation, states' rights, and government assistance.

The ideals of these parties include:

Republican (or the Grand Old Party [GOP])

- Founded by abolitionists
- Support capitalism, free enterprise, and a policy of noninterference by the government
- Support strong national defense
- Support deregulation and restrictions of labor unions
- Advocate for states' rights
- Oppose abortion
- Support traditional values, often based on Judeo-Christian foundations, including considerable opposition to same-sex marriage

Democrat

- Founded by anti-Federalists and rooted in classical Liberalism
- Promote civil rights, equal opportunity and protection under the law, and social justice
- Support government-instituted social programs and safety nets
- Support environmental issues
- Support government intervention and regulation, and advocate for labor unions
- Support universal health care

Some prominent third parties include:

- **Reform Party:** support political reform of the two-party system
- **Green Party:** support environmental causes
- **Libertarian Party:** support a radical policy of nonintervention and small, localized government

Interest Groups

An **interest group** is an organization with members who share similar social concerns or political interests. Members of political interest groups work together to influence policy decisions that benefit a particular segment of society or cause. Interest groups might include:

- Activist groups, like the NAACP, American Civil Liberties Union (ACLU), or People for the Ethical Treatment of Animals (PETA)
- Corporations, like pharmaceutical companies or banks
- Small-business advocates
- Religious groups, like the Concerned Women PAC and the Muslim Public Affairs Council
- Unions, such as the Association of Teacher Educators and International Brotherhood of Electrical Workers

Lobbyists

To promote their causes and influence policy in their favor, many interest groups employ **lobbyists**, paid advocates who work to influence lawmakers. Lobbying is a controversial practice, but it is sanctioned and protected as free speech. Lobbying from interest groups has a powerful impact on many policy decisions made in the United States. Examples of lobbyist groups include American Israel Public Affairs Committee (AIPAC) and Pharmaceutical Research and Manufacturers of America.

Mass Media

Mass media refers to the various methods by which the majority of the general public receives news and information. Mass media includes television, newspapers, radio, magazines, online news outlets, and social media networks. The general public relies on mass media for political knowledge and cultural socialization, as well as the majority of their knowledge of current events, social issues, and political news.

Evolution of Mass Media

- Until the end of the nineteenth century, print media such as newspapers and magazines was the only form of mass communication.

- In the 1890s, after the invention of the radio, broadcast media become a popular form of communication, particularly among illiterate people.

- In the 1940s, television superseded both print and broadcast media as the most popular form of mass media.

- In 1947, President Harry Truman gave the first political speech on television.

- In 1952, Dwight Eisenhower was the first political candidate to air campaign ads on television.

- Today, the Internet is the most widespread mass media technology, and citizens have instant access to news and information, as well as interactive platforms on which they can communicate directly with political leaders or share their views through social media, blogs, and independent news sites.

Influence of Mass Media on Politics

Mass media has a powerful effect on public opinion and politics. Mass media:

- Shapes public interests
- Enables candidates to reach voters wherever they are
- Determines what is and is not considered important in society based on how it prioritizes events and issues
- Provides the context in which to report events
- Is paid for by advertisers who may pressure news outlets to suppress or report information in their own interests

Comparative Politics and International Relations

Forms of Government and Types of Regimes

Government is the physical manifestation of the political entity or ruling body of a state. It includes the formal institutions that operate to manage and maintain a society. The form of government does not determine the state's **economic system**, though these concepts are often closely tied. Many forms of government are based on a society's economic system. However, while the form of government refers to the methods by which a society is managed, the term **economy** refers to the management of resources in a society. Many forms of government exist, often as hybrids of two or more forms of government or economic systems. Forms of government can be distinguished based on protection of civil liberties, protection of rights, distribution of power, power of government, and principles of Federalism.

Regime is the term used to describe the political conditions under which the citizens live under the ruling body. A regime is defined by the amount of power the government exerts over the people and the number of people who comprise the ruling body. It is closely related to the form of government because the form of government largely creates the political conditions. Regimes are governmental bodies that control both the form and the limit of term of their office. For example, authoritarianism is an example of a form of government and type of regime. A regime is considered to be ongoing until the culture, priorities, and values of the government are altered, ranging from the peaceful transitions of power between democratic political parties to the violent overthrow of the current regime.

The forms of government operated by regimes of government include:

Aristocracy

An **aristocracy** is a form of government composed of a small group of wealthy rulers, either holding hereditary titles of nobility or membership in a higher class. Variations of aristocratic governments include:

- **Oligarchy:** form of government where political power is consolidated in the hands of a small group of people

- **Plutocracy:** type of oligarchy where a wealthy elite class dominates the state and society

Though no aristocratic governments exist today, it was the dominant form of government during ancient times, including the:

- Vassals and lords during the Middle Ages, especially in relation to feudalism
- City-state of Sparta in ancient Greece

Authoritarian

An **authoritarian** state is one in which a single party rules indefinitely. The ruling body operates with unrivaled control and complete power to make policy decisions, including the restriction of denying civil liberties such as freedom of speech, press, religion, and protest. Forms of authoritarian governments include **autocracy**, **dictatorship**, and **totalitarian**—states or societies ruled by a single person with complete power over society.

Examples of states with authoritarian governments:

- Soviet Union
- Nazi Germany
- Modern-day North Korea

Democracy

Democracy is a form of government in which the people act as the ruling body by electing representatives to voice their views. Forms of democratic governments include:

- **Direct democracy:** democratic government in which the people make direct decisions on specific policies by majority vote of all eligible voters, like in ancient Athens

- **Representative democracy:** democratic government in which the people elect representatives to vote in a legislative body. This form of soft government providing for the election of representatives is also known as representative republic or indirect democracy. Representative democracy is currently the most popular form of government in the world.

The presidential and parliamentary systems are the most common forms of representative democracy. In the **presidential** system, the executive operates in its own distinct branch. Although the executive and legislative branches might enjoy powers checking each other, as in the American presidential system, the two functions are clearly separated. In addition, the president is typically both the head of state and head of government. Examples of presidential systems include Brazil, Nigeria, and the United States.

In the **parliamentary** system, the prime minister serves as the head of the government. The legislative branch, typically a parliament, elects the prime minister; thus, unlike in the presidential system, the parliament can replace the prime minister with a vote of no confidence. This practically means that the parliament has considerable influence over the office of prime minister. Parliamentary systems often include a president as the head of state, but the office is mostly ceremonial, functioning like a figurehead. Examples of parliamentary systems include Germany, Australia, and Pakistan.

The presidential system is a form of government better designed to distribute power between separate branches of government. This theoretically provides more stability. The president serves for a limited term of years, while prime ministers serve until replaced after receiving a vote of no confidence.

In the parliamentary system, the interdependence and interconnectedness between the parliament and prime minister facilitates efficient and timely governance, capable of adjusting to developing and fluid situations. In contrast, the presidential system is more prone to political gridlock because there is no direct connection between the legislative and executive branches. The legislature in a presidential system cannot replace the executive, like in the parliamentary system. The separation of powers in a presidential system can lead to disagreement between the executive and legislature, causing gridlock and other delays in governance.

Federalism is a set of principles that divides power between a central government and regional governments. Sovereign states often combine into a federation, and to do so, they cede some degree of sovereignty to establish a functional central government to handle broad national policies. The United States Constitution structures the central government according to principles of Federalism. Canada is another example of a form of government with a Federalist structure.

Monarchy

Monarchy is a form of government in which the state is ruled by a *monarch*, typically a hereditary ruler. Monarchs have often justified their power due to some divine right to rule. Types of monarchies include:

- **Absolute monarchy:** a monarchy in which the monarch has complete power over the people and the state

- **Constitutional monarchy:** a type of monarchy in which the citizens of the state are protected by a constitution, and a separate branch, typically a parliament, makes legislative decisions. The monarch and legislature share power.

- **Crowned republic:** a type of monarchy in which the monarch holds only a ceremonial position and the people hold sovereignty over the state. It is defined by the monarch's lack of executive power.

Examples of monarchies:

- Kingdom of Saudi Arabia is an absolute monarchy.
- Australia is a crowned republic.

Major Types of Electoral Systems

An **electoral system** defines the procedures by which voters make decisions regarding who will rule a state and how those elected officials will make policy decisions for the state. Electoral systems dictate how the members of the ruling body are selected, how votes translate into positions, and how seats are filled in the political offices at each level of government. States use three types of electoral systems to determine the outcomes of elections: plurality, majority, and proportional representation.

Plurality

In a **plurality** electoral system, a candidate must receive a plurality, or highest percent, of votes in order to win a political seat. For example, if three candidates are running for election to a single-seat political office and Candidate A receives 37 percent, Candidate B receives 39 percent, and Candidate C receives 24 percent, Candidate B will win the seat, even though the majority of the electorate actually voted *against* them.

One benefit of a plurality electoral system is that additional rounds of voting are not required if none of the candidates receive a majority vote. However, a plurality does not reflect the majority opinion, as in the example above, where most voters voted against the winner. Plurality electoral systems often lead to a wide variety of political parties because it is possible to win without a majority.

Majority

In a **majority** electoral system, a candidate must receive a majority of votes in order to be awarded a seat. If none of the candidates hit the mark, another round of voting is called, also referred to as a **runoff**. Some states only include the two candidates with the most votes to participate in the second round, while other states only require a candidate to reach a certain threshold of votes (for example, 15 percent) in order to be included as a candidate in the second round of voting.

A benefit of majority electoral systems is that they are more representative of public opinion than a plurality, as no candidate may win if the majority has voted against them. However, until a candidate receives a vote of 50 percent plus 1, additional rounds of voting must occur, which can lead to confusion. In addition, the second round of voting can lead to the less popular candidate winning. The Twelfth Amendment of the United States mandates that if no presidential candidate wins a majority of electoral delegates, then the House determines the winner by majority vote. The House could then elect a candidate that received considerably less support in the first round, such as a third-party candidate.

Proportional Representation

A **proportional representation** electoral system is often used for elections in which more than one seat in a legislative body is open. Proportional representation awards seats in proportion to the percent of votes each party receives. It is most common in parliamentary systems. For example, if there are ten seats and Party A receives 40 percent, Party B receives 20 percent, Party C receives 30 percent, and Party D receives 10 percent, Party A is awarded four seats, Party B is awarded two, Party C is awarded three, and Party D is awarded one.

A benefit of proportional representation is that it is the most representative of the people. However, it requires complex procedures and formulas to allocate votes and award seats, causing some confusion in the process. There are two types of proportional representation electoral systems—party list systems and single transferable systems.

Party List Systems: This method employs the **highest average method**, in which the votes received by each party are divided based on a formula (for example, the number of seats), and the highest quotient values are used to determine which parties receive a seat. In some party list systems, the voters directly select a political party, while others pool the total vote of the candidates from the party. The political parties typically issue lists of candidates that would serve in the government if their political party wins the seats.

Single Transferable Systems: In this method, voters rank the candidates in order of preference. To determine the number of votes to qualify, a candidate must receive a number of votes equal to $[v/(s + 1)] + 1$ in which v represents the number of votes and s represents the number of seats.

The number 1 rankings are counted first, and the first candidate to receive the required votes is awarded a seat. Any further votes that ranked that candidate number 1 will instead be awarded to the candidate ranked number 2. When a second candidate receives the required votes, any votes for them will be transferred to the next candidate on the list, and so on, until all the seats are filled. In this system, if no one receives the required number of votes, the candidate with the lowest number of #1 ranking votes is dropped, and the process begins again.

Countries that use proportional representation electoral systems:

- Ireland
- Belgium

Foreign Policy
Foreign policy refers to a state's international policy governing and informing their interactions with other states. A state's foreign policy typically defines the methods they employ to safeguard the state against foreign states, the social and economic goals of the state, and how the state will achieve these goals in the global arena through their relations with foreign states. Foreign policy is typically an executive function, either through the head of state or delegated to the foreign minister.

A state's foreign policy is influenced by several factors, including:

- Public opinion
- Economic and domestic stability
- Current events
- Social and humanitarian interests

The foreign policy of the United States has changed dramatically since the Founding Fathers established a policy of isolationism, which persisted well into the nineteenth century. After World War II, the United States emerged alongside the Soviet Union as the lone remaining global support powers. The United States led Europe and her allies during the Cold War's fight against the spread of Communism. When the Cold War ended after the collapse of the Soviet Union, the United States was left as the only true superpower, enjoying an unrivaled military and one of the world's most productive economies. American foreign policy shifted toward funding sustainable development economic projects in struggling countries and supporting democracy across the globe. The United States remains the leader of the NATO and has entered into several free trade agreements, most notably the North American Free Trade Association between Mexico, America, and Canada.

The president and secretary of state of the United States determine and enforce the U.S. foreign policy. The goals of U.S. foreign policy include:

- Maintain national security
- Promote world peace
- Promote civil rights and democracy
- Ally with other states to solve international problems
- Promote global cooperation and trade

To accomplish these goals, the State Department:

- Employs foreign diplomats to meet and talk with officials from foreign countries
- Maintains U.S. embassies in foreign nations from which to practice diplomacy
- Joins and supports international organizations such as:
    - NATO
    - WTO
    - United Nations

Theories of International Relations

The study of international relations involves analyzing the methods and effectiveness of different countries' approaches to international relations. Those approaches are informed by theories of international relations. The foundational theories of international relations are Realism and Liberalism.

Realism

Realism believes that states' foreign policy is guided by principles of "realpolitik"—the idea that states' politics and foreign relations are based primarily on the specific circumstances and factors. Realism advocates for studying international relations through the inevitable conflict between states pursuing power.

There are four basic tenets common to all schools of Realism.

1. States are the central actors, not individuals or international institutions.
2. There is no supreme authority guiding international relations, so the system is anarchic.
3. States act rationally to advance their self-interest.
4. All states are interested in maintaining or expanding their power as a means to self-preservation.

Liberalism

Liberalism has roots dating back to the Enlightenment, advocating for equality between states of the world. Liberals assert that global cooperation and interdependence is a force of world peace. Liberals seek to form international organizations to respond to international issues. There are three foundational principles common to all schools of Liberalism.

- Rejects "realpolitik" and the assumption that conflict is the inevitable outcome for international relations

- Supports international cooperation for the states' mutual benefit

- Seeks to establish international organizations and nongovernmental organizations to influence states' policy decisions and respond to international issues

International Relations in Practice

Examples of practices related to the execution of states' foreign policy, and the subject of the study of international relations, include diplomacy, conflict, treaties, and cooperation.

Diplomacy is the act of meeting with officials from foreign states to:

- Discuss matters of international interest
- Foster harmonious relationships
- Organize international partnerships
- Broker agreements
- Resolve disputes

States conduct diplomacy to negotiate compromises between states; accomplish militaristic, economic, environmental, and humanitarian goals; and garner allies in case of future crisis or conflict.

Conflict

International **conflict** refers to disputes between states. If not defused with international intervention and diplomacy, conflicts often lead to violence and armed conflict. The causes of international conflict might include:

- Territorial disputes
- Economic interests
- Religious or cultural tensions
- Social or humanitarian interests
- Political differences
- Civil rights and liberties
- Struggle for power

Treaties

To promote harmonious international relations, solve disputes, or protect against foreign invasions, states enter into **treaties** with other states. Treaties resolve armed conflict, establish trade agreements, or form defense alliances by:

- Creating compromises between the states for mutual benefit because treaties are invalid when signed under coercion
- Outlining the responsibilities and benefits of membership
- Limiting the actions of the states
- Describing the consequences of breaking the treaty agreement

Cooperation

To achieve common goals, states *cooperate*, or work together. When states cooperate, they form military alliances to protect each other against other states and make trade agreements that outline how trade is conducted fairly between the states to benefit both states' economies. States form

international organizations that regulate military action and trade between states. Some of the most powerful of these include:

- United Nations: a global organization created after World War II in an effort to prevent another world war, foster cooperation among the nations of the world, and protect human rights across the world

- WTO: a global organization created to encourage, regulate, and promote fair trade among the nations of the world

- NATO: an organization of states from North America and Europe, led by the United States, formed after the end of World War II to counter the Soviet Union and Warsaw Pact's efforts to spread Communism across the world

Power and Problems of International Organizations and International Law

States form **international organizations** to foster cooperation among the nations of the world. These include organizations that are dedicated to regulating trade, outlining rules for dealing with conflict among states, settling territory disputes, solving global problems, and promoting social or humanitarian efforts around the world. International organizations also develop the body of **international laws**—the rules and regulations for interactions between the states. International organizations are important for the enforcement of international agreements and treaties. Without these organizations, there would be no way to enforce obligations and rights, effectively rendering those agreements useless.

Benefits of international organizations:

- States gain a forum to discuss matters of international politics and peacefully settle disputes.

- States increase their political power and wealth by combining their power with other powerful nations.

- States collaborate to solve global issues.

- States protect each other against more powerful foreign enemies; conflict is less likely due to the threat of stronger collective action.

Problems of International Law Enforcement

International laws enacted by international organizations are difficult to enforce, as no international enforcement agency exists to ensure states abide by the rules and regulations agreed upon by the organization. As such, states are bound only by the understanding that it is in their best interest to comply with the rules of the organization, rather than out of fear of retaliation from other states, except in the most extreme cases. There are also issues related to the balance of sovereignty between international organizations and states.

The methods by which organizations can punish state members who fail to follow the rules of the organization include:

- **Reciprocal action:** If a state acts against another state, the offended state may reciprocate the act.

- **Economic boycott:** If a state fails to comply with the organization's rules, the other states may refuse to trade with the offending state.

- **Damaged reputation:** If a state fails to follow the rules, the organization speaks out publicly to shame the offending state; thus, other states are less likely to do future business with the violating state.

- **Collective military action:** If a state violates the rules of the organization to an extreme extent, such as by committing government-sanctioned genocide or interfering with the territorial integrity of another state, states and international organizations may use military force to punish the offending state.

Geography

Concepts and Terminology of Physical and Human Geography

Geographers utilize a variety of different maps in their study of the spatial world. Projections are maps that represent the entire world (which is spherical) on a flat surface. *Conformal projections* preserve angles locally, maintaining the shape of a small area in infinitesimal circles of varying sizes on a two-dimensional map. Conformal projections tend to possess inherent flaws due to their two-dimensional nature. For example, the most well-known projection, the *Mercator projection*, drastically distorts the size of land areas at the poles. In this particular map, Antarctica, one of the smallest continents, appears massive, almost rivaling the size of North America. In contrast to the poles, the areas closer to the central portion of the globe are more accurate. Other projections attempt to lessen the amount of distortion; the *Equal-area projection*, for example, attempts to equally represent the size of landforms on the globe.

Nevertheless, equal-area projections like the *Lambert projection* also inherently alter the size of continents, islands, and other landforms, both close to Earth's center and near the poles. Other projections are a hybrid of the two primary models. For example, the *Robinson projection*, also referred to as the *Goode's homolosine projection*, tries to balance form and area in order to create a more visually accurate representation of the spatial world. Despite the efforts to maintain consistency with shapes, projections cannot provide accurate representations of the Earth's surface, due to their flat, two-dimensional nature. In this sense, projections are useful symbols of space, but they do not always provide the most accurate portrayal of spatial reality.

Unlike projections, *topographic maps* display contour lines, which represent the relative elevation of a particular place and are very useful for surveyors, engineers, and/or travelers. Hikers of the Appalachian Trail or Pacific Crest Trail, for example, may call upon topographic maps to calculate their daily climbs.

Thematic maps are also quite useful to geographers because they use two-dimensional surfaces to convey complex political, physical, social, cultural, economic, or historical themes.

Thematic maps can be broken down into different subgroups: *dot-density maps* and *flow-line maps*. A *dot-density map* is a type of thematic map that illustrates the volume and density in a particular area. Although most dots on these maps represent the number of people in an area, they don't always have to do that. Instead, these maps may represent the number of events, such as lightning strikes, that have taken place in an area. *Flow-line maps* are another type of thematic map, which utilize both thin and thick lines to illustrate the movement of goods, people, or even animals between two places. The thicker the line, the greater the number of moving elements; a thinner line would, of course, represent a smaller number.

Similar to topographic maps, an *isoline map* is also useful for calculating data and differentiating between the characteristics of two places. In an *isoline map*, symbols represent values, and lines can be drawn between two points in order to determine differences. For example, average temperature is commonly measured on isoline maps. Point A, which is high in the mountains, may have a value of 33 degrees, while point B, which is in the middle of the Mojave Desert, may have a value of 105 degrees. Using the different values, it is easy to determine that temperatures in the mountains are 72 degrees cooler than in the desert. Additionally, isoline maps help geographers study the world by creating questions. For example, is it only elevation that is responsible for the differences in temperature? If not, what other factors could cause such a wide disparity in the two values? Utilizing these, and other sorts of maps, is essential in the study of geography.

Using Mental Maps to Organize Spatial Information

Mental maps are exactly what they sound like—maps that exist within someone's mind. The cognitive image of a particular place may differ from person to person, but the concept of remembering important places does not. For example, the commonalities usually emerge relative to the knowledge of one's workplace, school, home, or favorite restaurants. Furthermore, mental maps also embody the means of travelling from point A to point B. One may know the best route on public transit, the least hilly bike path, or the roadways that have the least amount of traffic. In places where someone has very little interaction, mental maps usually tend to be minimally informative, due to the absence of any personal experience in a particular place.

Maps are also organized through scale. Scale is simply the ratio of a distance on the ground to the corresponding distance on paper. Geographers and cartographers attempt to make the image on paper representative of the actual place. For example, the United States Geological Survey (USGS) utilizes the mathematical ratio of 1/24,000 in all of its topographical maps. This scale means that one inch on the map is equivalent to 24,000 inches—or nearly two-thirds of a mile—on the ground. The two primary types of maps, *large scale* and *small scale*, essentially serve the same purpose, but for two different types of places. Large-scale maps represent a much smaller area with greater detail, while small-scale maps are representative of much larger areas with less detail.

Recognizing and Interpreting Spatial Patterns Presented at Different Scales from Local to Global

Two primary realms exist within the study of geography. The first realm, *physical geography*, essentially correlates with the land, water, and foliage of the Earth. The second realm, *human geography*, is the study of the Earth's people and how they interact with their environment. Like land and water on Earth, humans are also impacted by different forces such as culture, history, sociology, technological advancement and changes, and access to natural resources. For example, human populations tend to be higher around more reliable sources of fresh water. The metropolitan area of New York City, which has abundant freshwater resources, is home to nearly 20 million people, whereas Australia, both a continent and a country, has almost the same population. Although water isn't the only factor in this disparity, it certainly plays a role in a place's *population density*—the total number of people in a particular place divided by the total land area, usually square miles or square kilometers. Australia's population density stands at 8.13 people per square mile, while the most densely populated nation on Earth, Bangladesh, is home to 2,894 people per square mile.

Population density can have a devastating impact on both the physical environment/ecosystem and the humans who live within the environment/ecosystem of a particular place. For example, Delhi, one of India's most populated cities, is home to nearly five million gasoline-powered vehicles. Each day, those vehicles emit an enormous amount of carbon monoxide into the atmosphere, which directly affects the quality of life of Delhi's citizens. In fact, the problem of the smog and pollution has gotten so severe that many drivers are unable to see fifty feet in front of them. Additionally, densely populated areas within third-world nations, or developing nations, struggle significantly in their quest to balance the demands of the modern economy with their nation's lack of infrastructure. For example, nearly as many automobiles operate every day in major American cities like New York and Los Angeles as they do in Delhi, but they create significantly less pollution due to cleaner burning engines, better fuels, and governmental emission regulations.

Although it's a significant factor, population density is not the only source of strain on the resources of a particular place. Historical forces such as civil war, religious conflict, genocide, and government corruption can also alter the lives of a nation's citizens in a profound manner. For example, the war-torn nation on the Horn of Africa, Somalia, has not had a functioning government for nearly three decades. As a result, the nation's citizens have virtually no access to hospital care, vaccinations, or proper facilities for childbirth. Due to these and other factors, the nation's *infant mortality rate*, or the total number of child deaths per 1,000 live births, stands at a whopping 98.39/1000. When compared to Iceland's 1.82/1000, it's quite evident that Somalia struggles to provide basic services in the realm of childbirth and there is a dire need for humanitarian assistance.

Literacy rates, like the infant mortality rate, are also an excellent indicator of the relative level of development in a particular place. Like Somalia, other developing nations have both economic and social factors that hinder their ability to educate their own citizens. Due to radical religious factions within some nations like Afghanistan and Pakistan, girls are often denied the ability to attend school, which further reduces the nation's overall literacy rate. For example, girls in Afghanistan, which spent decades under Taliban control, have a 24.2 percent literacy rate, one of the lowest rates of any nation on Earth that keeps records (Somalia's government is so dysfunctional records don't exist).

Although literacy rates are useful in determining a nation's development level, high literacy rates do exist within developing nations. For example, Suriname, which has a significantly lower GDP than Afghanistan, enjoys a nearly 96 percent literacy rate among both sexes. Utilizing this and other data, geographers can create questions regarding how such phenomena occur. How is Suriname able to educate its population more effectively with fewer financial resources? Is it something inherent within their culture? Demographic data, such as population density, the infant mortality rate, and the literacy rate all provide insight into the characteristics of a particular place and help geographers better understand the spatial world.

Locating and Using Sources of Geographic Data

Geographic data is essential to fully understanding both the spatial and human realms of geography. In reference to the human population, different factors affect the quality of life one experiences during their lifetime. Geographers attempt to understand why those differences exist through data utilization and comparative analysis. For example, as has been previously mentioned, population density, infant mortality rates, and literacy rates are all useful tools in analyzing human characteristics of a place; however, those are not the only tools geographers utilize. In fact, organizations such as the *Population Reference Bureau* and the *Central Intelligence Agency* both provide an incredible amount of *demographic* data useful to researchers, students, or really anyone curious about the world in which they live.

The *CIA World Factbook* is an indispensable resource for anyone interested in the field of human or physical geography. Providing information such as land area, literacy rates, birth rate, and economic data, this resource is one of the most comprehensive on the Internet. In addition to the CIA World Factbook, the *Population Reference Bureau* (*PRB*) also provides students of geography with an abundant supply of information. In contrast to the CIA source, the *PRB* provides a treasure trove of analyses related to human populations including HIV rates, immigration rates, poverty rates, etc.

In addition to the aforementioned sources, the *United States Census Bureau* provides similar information about the dynamics of the American population. Not only does this source focus on the data geographers need to understand the world, but it also provides information about upcoming classes, online workshops, and even includes an online library of resources for both students and teachers.

Websites for each source can be found below:

- Population Reference Bureau: www.prb.org
- United States Census Bureau: www.census.gov
- CIA World Factbook: https://www.cia.gov/library/publications/the-world-factbook/

Spatial Concepts

Location is the central theme in understanding spatial concepts. In geography, there are two primary types of locations that people utilize on a daily basis. The first type, *relative location*, is used frequently and involves locating objects by notating their proximity to another, better known object. For example, directions from person to person may relate directly to massive shopping centers, major highways, or well-known intersections. Although relative location is important, in the modern world, it's common to use digital satellite-based technologies, which rely on *GPS* (*Global Positioning System*). To determine *Absolute Location*, or the exact latitudinal and longitudinal position on the globe, GPS uses sensors that interact with satellites orbiting the Earth. *Coordinates* correspond with the positions on a manmade grid system using imaginary lines known as *latitude* (also known as *parallels*) and *longitude* (also known as *meridians*).

In order to understand latitude and longitude, one should think of a simple X and Y-axis. The *equator* serves as the X-axis at zero degrees, and measures distance from north to south. The Y-axis is at zero degrees and is represented by the *Prime Meridian*.

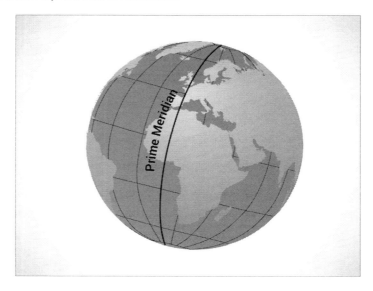

In addition to anchoring the grid system to create the basis for absolute location, these major lines of latitude and longitude also divide the Earth into *hemispheres*. The Equator divides the Earth into the northern and southern hemispheres, while the Prime Meridian establishes the eastern and western hemispheres. Coordinates are always expressed in the following format:

Degrees north or south, degrees east or west, or 40°N, 50°E. Since lines of latitude and longitude are great distance from one another, absolute locations are often found in between two lines. In those cases, degrees are broken down into *minutes* and *seconds*, which are expressed in this manner: (40° 53' 44" N, 50° 22' 65" E).

In addition to the Equator and the Prime Meridian, other major lines of latitude and longitude exist to divide the world into regions relative to the direct rays of the sun. These lines correspond with the Earth's *tilt*, and are responsible for the seasons. For example, the northern hemisphere is tilted directly toward the sun from June 22 to September 23, which creates the summer season in that part of the world. Conversely, the southern hemisphere is tilted away from the direct rays of the sun and experiences winter during those same months.

The transitions from season to season involve two factors: the 23 ½ degree tilt of the Earth and the movement of the direct rays of the sun relative to the Earth's revolution. To clarify, the area between the *Tropic of Cancer* (23 ½ degrees north) and the *Tropic of Capricorn* (23 ½ degrees south) can be envisioned as the playing field for the direct rays of the sun. These rays never leave the playing field, and, as a result, the area between those two lines of latitude—the *tropics*—tends to be warmer and experience fewer variations in seasonal temperatures. In contrast, the area between the Tropic of Cancer and the *Arctic Circle* (66 ½ degrees north) is in the *middle latitudes*—the region where most of the Earth's population resides. In the Southern Hemisphere, the middle latitudes exist between the Tropic of Capricorn and the *Antarctic Circle* (66 ½ degrees south). In both of these places, indirect rays of the sun strike the Earth, so seasons are more pronounced and milder temperatures generally prevail. The final region, known as the *high latitudes*, is found north of the Arctic Circle and south of the Antarctic Circle. These regions generally tend to be cold all year, and experience nearly twenty-four

hours of sunlight during their respective *summer solstice* and twenty-four hours of darkness during the *winter solstice*.

Regarding the seasons, it is important to understand that those in the Southern Hemispheres are opposite of those in the Northern Hemisphere, due to the position of the direct rays of the sun. When the sun's direct rays are over the Equator, it is known as an *equinox*, and day and night are almost of equal length throughout the world. Equinoxes occur twice a year; the fall, or autumnal equinox, occurs on September 22nd, while the spring equinox occurs on March 20th. Obviously, if seasons are opposite of one another depending on the hemisphere, the corresponding names flip-flop depending on one's location (i.e. when the Northern Hemisphere is experiencing summer, it is winter in the Southern Hemisphere).

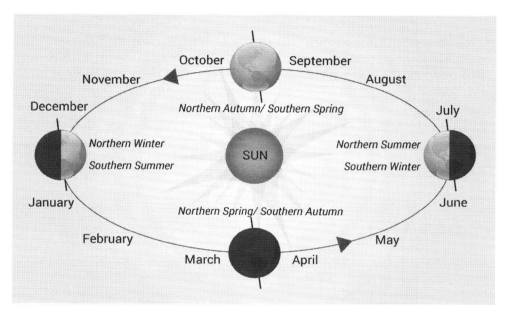

Place
Both absolute and relative location help humans understand their sense of place. Place is a simple concept that helps to define the characteristics of the world around us. For example, people may create *toponyms* to further define and orient themselves with their sense of place. Toponyms are simply names given to locations to help develop familiarity within a certain location. Although not always the case, toponyms generally utilize geographical features, important people in an area, or even wildlife commonly found in a general location. For example, many cities in the state of Texas are named in honor of military leaders who fought in the Texas Revolution of 1836 (such as Houston and Austin), while other places, such as Mississippi and Alabama, utilize Native American toponyms to define their sense of place.

Regions
In addition to location and place, geographers also divide the world into regions in order to more fully understand differences inherent with the world, its people, and its environment. As mentioned previously, lines of latitude such as the Equator, the Tropics, and the Arctic and Antarctic Circles already divide the Earth into solar regions relative to the amount of either direct or indirect sunlight that they receive. Although not the same throughout, the middle latitudes generally have a milder climate than areas found within the tropics. Furthermore, tropical locations are usually warmer than places in the middle latitudes, but that is not always the case. For example, the lowest place in the United States—

Death Valley, California—is also home to the nation's highest-ever recorded temperature. Likewise, the Andes Mountains in Peru and Ecuador, although found near the Equator, are also home to heavy snow, low temperatures, and dry conditions, due to their elevation.

Formal regions are spatially defined areas that have overarching similarities or some level of *homogeneity* or *uniformity*. Although not exactly alike, a formal region generally has at least one characteristic that is consistent throughout the entire area. For example, the United States could be broken down into one massive formal region due to the fact that in all fifty states, English is the primary language. Of course, English isn't the only language spoken in the United States, but throughout that nation, English is heavily used. As a result, geographers are able to classify the United States as a formal region; but, more specifically, the United States is a *linguistic region*—a place where everyone generally speaks the same language.

Functional regions are similar to formal regions in that they have similar characteristics, but they do not have clear boundaries. The best way to understand these sorts of regions is to consider large cities. Each large city encompasses a large *market area*, whereby people in its vicinity generally travel there to conduct business, go out to eat, or watch a professional sporting event. However, once anyone travels farther away from that *primate city*, they transition to a different, more accessible city for their needs. The functional region, or *area of influence*, for that city, town, or sports team transitions, depending upon the availability of other primate cities. For example, New York City has two primary professional baseball, basketball, and football teams. As a result, its citizens may have affinities for different teams even though they live in the same city. Conversely, a citizen in rural Idaho may cheer for the Seattle Seahawks, even though they live over 500 miles from Seattle, due to the lack of a closer primate city.

Effects of Physical Processes, Climate Patterns, and Natural Hazards on Human Societies

The Earth's surface, like many other things in the broader universe, does not remain the same for long; in fact, it changes from day to day. The Earth's surface is subject to a variety of physical processes that continue to shape its appearance. In each process, water, wind, temperature, or sunlight play a role in continually altering the Earth's surface.

Erosion can be caused by a variety of different stimuli including ice, snow, water, wind, and ocean waves. *Wind erosion* is a specific phenomenon that occurs in generally flat, dry areas with loose topsoil. Over time, the persistent winds can dislodge significant amounts of soil into the air, reshaping the land and wreaking havoc on those who depend on agriculture for their livelihoods. Erosion can also be caused by water and is responsible for changing landscapes as well. For example, the Grand Canyon was carved over thousands of years by the constant movement of the Colorado River. Over time, the river moved millions of tons of soil, cutting a huge gorge in the Earth along the way. In all cases, erosion involves the movement of soil from one place to another. In water erosion, material carried by the water is referred to as *sediment*. With time, some sediment can collect at the mouths of rivers, forming *deltas*, which become small islands of fertile soil. This process of detaching loose soils and transporting them to a different location where they remain for an extended period of time is referred to as *deposition*, and is the end result of the erosion process.

In contrast to erosion, *weathering* does not involve the movement of any outside stimuli. In this physical process, the surface of the Earth is either broken down physically or chemically. *Physical weathering* involves the effects of atmospheric conditions such as water, ice, heat, or pressure. Through the process of weathering over the course of centuries, large rocks can be broken down with the effects of icy

conditions. *Chemical weathering* generally occurs in warmer climates and involves organic material that breaks down rocks, minerals, or soil. This process is what scientists believe led to the creation of fossil fuels such as oil, coal, and natural gas.

Climate Patterns

Weather is defined as the condition of the Earth's atmosphere at a particular time. *Climate* is different; instead of focusing on one particular day, climate is the relative pattern of weather in a place for an extended period of time. For example, the city of Atlanta is in the American South and generally has a humid subtropical climate; however, Atlanta also occasionally experiences snowstorms in the winter months. Despite the occasional snow and sleet storm, over time, geographers, meteorologists, and other Earth scientists have determined the patterns that are indicative to north Georgia, where Atlanta is located. Almost all parts of the world have predictable climate patterns, which are influenced by the surrounding geography.

The Central Coast of California is an example of a place with a predictable climate pattern. Santa Barbara, California, one of the region's larger cities, has almost the same temperature for most of the summer, spring, and fall, with only minimal fluctuation during the winter months. The temperatures there, which average between 75 and 65 degrees Fahrenheit daily regardless of the time of year, are influenced by a variety of different climatological factors including elevation, location relative to the mountains and ocean, and ocean currents. In the case of Santa Barbara, the city's location on the Pacific Coast and its position near mountains heavily influences its climate. The cold California current, which sweeps down the west coast of the United States, causes the air near the city to be temperate, while the mountains trap cool air over the city and the surrounding area. This pattern, known as the *orographic effect*, or *rain shadow*, also affects temperatures on the leeward side of the mountains by blocking most of the cool air and causing dry conditions to dominate. Temperatures can fluctuate by more than 20 degrees Fahrenheit on opposite sides of the mountain.

Other factors affecting climate include elevation, prevailing winds, vegetation, and latitudinal position on the globe.

Like climate, *natural hazards* also affect human societies. In tropical and subtropical climates, hurricanes and typhoons form over warm water and can have devastating effects. Additionally, tornadoes, which are powerful cyclonic windstorms, also are responsible for widespread destruction in many parts of the United States and in other parts of the world. Like storms, earthquakes, usually caused by shifting plates along faults deep below the Earth's surface, also cause widespread devastation, particularly in nations with a poor or crumbling infrastructure. For example, San Francisco, which experiences earthquakes regularly due to its position near the San Andreas Fault, saw relatively little destruction and deaths (67 total) as a result of the last major earthquake to strike there. However, in 2010, an earthquake of similar magnitude reportedly killed over 200,000 people in the western hemisphere's poorest nation, Haiti. Although a variety of factors may be responsible for the disparity, modern engineering methods and better building materials most likely helped to minimize destruction in San Francisco. Other natural hazards, such as tsunamis, mudslides, avalanches, forest fires, dust storms, flooding, volcanic eruptions, and blizzards, also affect human societies throughout the world.

Characteristics and Spatial Distribution of Earth's Ecosystems

Earth is an incredibly large place filled with a variety of different land and water *ecosystems*. *Marine ecosystems* cover over 75 percent of the Earth's surface and contain over 95 percent of the Earth's water. Marine ecosystems can be broken down into two primary subgroups: *freshwater ecosystems*, which only encompass around 2 percent of the earth's surface; and *ocean ecosystems*, which make up over 70 percent. On land, *terrestrial ecosystems* vary depending on a variety of factors, including latitudinal distance from the equator, elevation, and proximity to mountains or bodies of water. For example, in the high latitudinal regions north of the Arctic Circle and south of the Antarctic Circle, frozen *tundra* dominates. Tundra, which is characterized by low temperatures, short growing seasons, and minimal vegetation, is only found in regions that are far away from the direct rays of the sun.

In contrast, *deserts* can be found throughout the globe and are created by different ecological factors. For example, the world's largest desert, the Sahara, is almost entirely within the tropics; however, other deserts like the Gobi in China, the Mojave in the United States, and the Atacama in Chile, are the result of the orographic effect and their close proximity to high mountain ranges such as the Himalayas, the Sierra Nevada, and the Andes, respectively. In the Middle Latitudes, greater varieties of climatological zones are more common due to fluctuations in temperatures relative to the sun's rays, coupled with the particular local topography. In the Continental United States, *temperate deciduous forest* dominates the southeastern portion of the country. However, the Midwestern states such as Nebraska, Kansas, and the Dakotas, are primarily *grasslands*. Additionally, the states of the Rocky Mountains can have decidedly different climates relative to elevation. In Colorado, Denver, also known as the "Mile High City," will often see snowfalls well into late April or early May due to colder temperatures, whereas towns and cities in the eastern part of the state, with much lower elevations, may see their last significant snowfall in March.

In the tropics, which are situated between the Tropics of Cancer and Capricorn, temperatures are generally warmer, due to the direct rays of the sun's persistence. However, like most of the world, the tropics also experience a variety of climatological regions. In Brazil, Southeast Asia, Central America, and even Northern Australia, tropical rainforests are common. These forests, which are known for abundant vegetation, daily rainfall, and a wide variety of animal life, are absolutely essential to the health of the world's ecosystems. For example, the *Amazon Rain Forest* is also referred to as "the lungs of the world," as its billions of trees produce substantial amounts of oxygen and absorb an equivalent amount of carbon dioxide—the substance that many climatologists assert is causing climate change or *global warming*. Unlike temperate deciduous forests whose trees lose their leaves during the fall and winter months, *tropical rain forests* are always lush, green, and warm. In fact, some rainforests are so dense with vegetation that a few indigenous tribes have managed to exist within them without being influenced by any sort of modern technology, virtually maintaining their ancient way of life in the modern era.

The world's largest ecosystem, the *taiga*, is found primarily in high latitudinal areas, which receive very little of the sun's indirect rays. These forests are generally made up of *coniferous* trees, which do not lose their leaves at any point during the year as *deciduous* trees do. Taigas are cold-climate regions that make up almost 30 percent of the world's land area. These forests dominate the northern regions of Canada, Scandinavia, and Russia, and provide the vast majority of the world's lumber.

Overall, it is important to remember that climates are influenced by five major factors: elevation, latitude, proximity to mountains, ocean currents, and wind patterns. For example, the cold currents off the coast of California provide the West Coast of the United States with pleasant year-round temperatures. Conversely, Western Europe, which is at the nearly the same latitude as most of Canada, is influenced by the warm waters of the *Gulf Stream*, an ocean current that acts as a conveyor belt, moving warm tropical waters to the icy north. In fact, the Gulf Stream's influence is so profound that it even keeps Iceland—an island nation in the far North Atlantic—relatively warm.

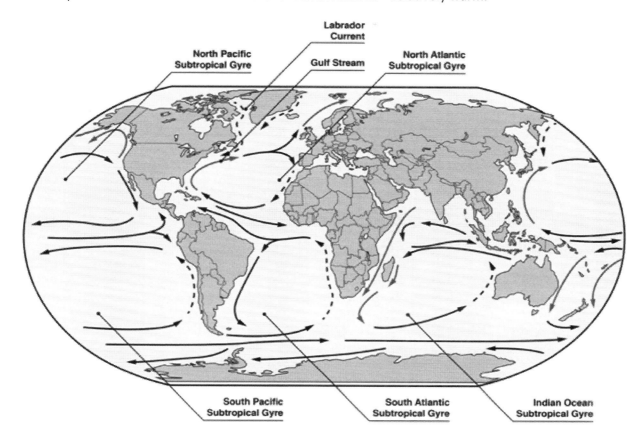

Interrelationships Between Humans and Their Environment

Like any other animal, humans adapt to their environment; but, unlike other animals, humans also adapt their environment to suit their needs. For example, human social systems are created around the goal of providing people with access to what they need to live more productive, fulfilling, and meaningful lives. Sometimes, humans create systems that are destructive, but generally speaking, humans tend to use their environment to make their lives easier. For example, in warmer climates, people tend to wear more comfortable clothing such as shorts, linen shirts, and hats. Additionally, in the excessively sun-drenched nations of the Middle East, both men and women wear flowing white clothing complete with both a head and neck covering, in order to prevent the blistering effects of sun exposure to the skin. Likewise, the native Inuit peoples of northern Canada and Alaska use the thick furs from the animals they kill to insulate their bodies against the bitter cold.

Humans also adapt to their environment to ensure that they have access to enough food and water for survival. Irrigation, or the process of moving water from its natural location to where it's needed, is an example of how humans change their environment in order to survive. For example, the city of Los Angeles, America's second most populous city, did not have adequate freshwater resources to sustain its population. However, city and state officials realized that abundant water resources existed approximately three hundred miles to the east. Rather than relocating some of its population to areas with more abundant water resources, the State of California undertook one of the largest construction projects in the history of the world, the Los Angeles Aqueduct, which is a massive concrete irrigation ditch connecting water-rich areas with the thirsty citizens of Los Angeles.

The Los Angeles Aqueduct is just one example of a human-environment interaction. In other cases, humans utilize what nature provides in close proximity. For example, the very first permanent British Colony in North America, Jamestown, VA, was heavily influenced by its environment. In contrast to the Pilgrims who settled in Plymouth, Massachusetts, Jamestown settlers found themselves in a hot, humid climate with fertile soil. Consequently, its inhabitants engaged in agriculture for both food and profit. Twelve years after Jamestown's foundation in 1607, it was producing millions of dollars of tobacco each year. In order to sustain this booming industry, over time, millions of African slaves and indentured servants from Europe were imported to provide labor. Conversely, the poor soils around Plymouth did not allow for widespread cash crop production, and the settlers in New England generally only grew enough food for themselves on small subsistence farms. Furthermore, slavery failed to take a strong foothold in the New England states, thus creating significantly different cultures within the same country, all due in part to human interaction with the environment.

Renewable and Nonrenewable Resources

When gas prices are high, prices on virtually everything increase. After all, there are very few products that humans can buy that are not transported by either a gasoline- or diesel-powered engine. As a result, an increase in fuel prices leads to an increase in the price of food, goods, or other cargo. Recently, there has been considerable debate regarding the reliance on *nonrenewable resources* like oil, natural gas, and coal. These resources, which are also known as *fossil fuels*, are quite common throughout the world and are generally abundant, and cheaper to use than *renewable resources* like solar, wind, and geothermal energy. While solar energy is everywhere, the actual means to convert the sun's rays into energy is not. Conversely, coal-fired power plants and gasoline-powered engines, which are older technologies in use during the industrial revolution, remain quite common throughout the world. In fact, reliance on non-renewable resources continues to grow, due to the availability coupled with the existing infrastructure. However, use of renewable energy is increasing, as it becomes more economically competitive with nonrenewable resources.

In addition to sources of energy, nonrenewable resources also include anything that can be exhausted. These can include precious metals like gold, silver, and platinum, freshwater underground aquifers, and precious stones such as diamonds, emeralds, and opals. Although abundant, most nonrenewable sources of energy are not sustainable because their creation takes a long time and cannot therefore be reproduced. Renewable resources are sustainable, but must be properly overseen so that they remain renewable. For example, the beautiful African island of Madagascar is home to some of the most amazing rainforest trees in the world. As a result, logging companies cut, milled, and sold thousands of them in order to make quick profits without planning how to ensure the continued health of the forests. As a result of severe deforestation on the island, mudslides became more and more common as the forests gradually shrank from widespread logging. In this case, renewable resources were mismanaged, and thus essentially became nonrenewable, due to the length of time for growth for the replacement of rainforest trees. In the United States, paper companies harvest pine trees to create paper; and because it can take almost twenty years for a pine tree to reach maturity, most of the companies utilize planning techniques to ensure that mature pine trees will always be available. In this manner, these resources remain renewable for human use in a sustainable fashion.

Renewable sources of energy are relatively new in the modern economy. Even though electric cars, wind turbines, and solar panels are becoming more common, they still do not provide enough energy to

power the world's economy. As a result, reliance on older, reliable forms of energy continues, which has a devastating effect on the environment. Beijing, China, which has seen a massive boom in industrial jobs, is also one of the most polluted places on Earth. Furthermore, developing nations with very little modern infrastructure also rely heavily on fossil fuels, due to the ease in which they are converted into usable energy. Even the United States, which has one of the most developed infrastructures in the world, still relies almost exclusively on fossil fuels, with only ten percent of the required energy coming from renewable sources.

Spatial Patterns of Cultural and Economic Activities

Spatial patterns refer to where things are in the world. Biomes, regions, and landforms all have spatial patterns regarding where they exist. Additionally, elements of *human geography*—the study of human culture and its effect on the world—also have certain patterns regarding where they appear on Earth.

Ethnicity

An ethnic group, or ethnicity, is essentially a group of people with a common language, society, culture, or ancestral heritage. Different ethnicities developed over centuries through historical forces, the impact of religious traditions, and other factors. Thousands of years ago, it was more common for ethnic groups to remain in one area with only the occasional interaction with outside groups. In the modern world, different ethnicities interact on a daily, if not hourly, basis, due to better transportation resources and the processes of globalization. For example, in the United States, it is not uncommon for a high school classroom to encompass people of Asian, African, Indian, European, or Native descent. That's not to suggest that all American classrooms have ethnic diversity, but, in general, due to a variety of pull-factors, the United States continues to attract people from all over the world. In less developed parts of the world, travel is limited due to the lack of infrastructure. Consequently, ethnic groups develop in small areas that can differ greatly from other people just a few miles away. For example, on the Balkan Peninsula in southeastern Europe, a variety of different ethnic groups live in close proximity to one another. Croats, Albanians, Serbs, Bosnians, and others all share the same land, but have very different worldviews, traditions, and religious influences. In the case of the Balkan Peninsula, such diversity has not always been a positive characteristic. For example, the First World War began there in 1914 related to a dispute regarding Serbia's national independence. Additionally, Bosnia was the scene of a horrible genocide against Albanians in an "ethnic cleansing" effort that continued throughout the late 20th century.

Linguistics

Linguistics, or the study of language, groups certain languages together according to their commonalities. For example, the Romance Languages—French, Spanish, Italian, Romansh, and Portuguese—all share language traits from Latin, the language of the former Roman Empire. These languages, also known as *vernaculars*, or more commonly spoken *dialects*, evolved over centuries of physical isolation on the European continent. The Spanish form of Latin emerged into today's Spanish language. In other parts of the world, the same pattern is true. The Bantu people of Africa travelled extensively and spread their language, now called *Swahili*, which is the first Pan-African language. When thinking of the world as a whole, it is important to understand that thousands of languages exist; however, to interconnect the world, it is important to have a means of communication with which everyone is at least somewhat familiar. A *lingua franca* is essentially the language of business. In other words, when executives from multinational corporations need to communicate regarding business, they often communicate in English, which is considered to be the world's lingua franca, due to the economic dominance of the United States.

Religion

Religion has played a tremendous role in creating the world's cultures. Devout Christians crossed the Atlantic in hopes of finding religious freedom in New England, Muslim missionaries and traders travelled to the Spice Islands of the East Indies to teach about the Koran, and Buddhist monks traversed the Himalayan Mountains into Tibet to spread their faith. In some countries, religion helps to shape legal systems. These nations, termed *theocracies*, have no separation of church and state and are more common in Islamic nations such as Saudi Arabia, Iran, and Qatar. In contrast, even though religion has played a tremendous role in the history of the United States, its government remains *secular*, or nonreligious, due to the influence of European Enlightenment philosophy at the time of its inception. Like ethnicity and language, religion is also a primary way that people self-identify. As a result, religious influences can shape a region's laws, architecture, literature, and music. For example, when the Ottoman Turks, who are Muslim, conquered Constantinople, which was once the home of the Eastern Orthodox Christian Church, they replaced Christian places of worship with mosques. Additionally, different forms of Roman architecture were replaced with those influenced by Arabic traditions.

Economics

Economic activity also has a spatial component. For example, nations with few natural resources generally tend to import what they need from nations willing to export raw materials to them. Furthermore, areas that are home to certain raw materials generally tend to alter their environment in order to maintain production of those materials. In the San Joaquin Valley of California, an area known for extreme heat and desert-like conditions, local residents have engineered elaborate drip irrigation systems to adequately water lemon, lime, olive, and orange trees, utilizing the warm temperatures to constantly produce citrus fruits. Additionally, other nations with abundant petroleum reserves build elaborate infrastructures in order to pump, house, refine, and transport their materials to nations who require gasoline, diesel, or natural gas. Essentially, different spatial regions on Earth create jobs, infrastructure, and transportation systems that seek to ensure the continued flow of goods, raw materials, and resources out of their location, so long as financial resources keep flowing into the area.

Patterns of Migration and Settlement

Migration is governed by two primary causes: *push factors*, which are reasons causing someone to leave an area, and *pull factors*, which are factors luring someone to a particular place. These two factors often work in concert with one another. For example, the United States of America has experienced significant *internal migration* from the industrial states in the Northeast (such as New York, New Jersey, Connecticut) to the Southern and Western states. This massive migration, which continues into the present-day, is due to high rents in the northeast, dreadfully cold winters, and lack of adequate retirement housing, all of which are push factors. These push factors lead to migration to the *Sunbelt*, a term geographers use to describe states with warm climates and less intense winters.

In addition to internal migrations within nations or regions, international migration also takes place between countries, continents, and other regions. The United States has long been the world's leading nation in regard to *immigration*, the process of having people come into a nation's boundaries. Conversely, developing nations that suffer from high levels of poverty, pollution, warfare, and other violence all have significant push factors, which cause people to leave and move elsewhere. This process, known as *emigration*, is when people in a particular area leave in order to seek a better life in a different—usually better—location.

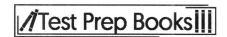

The Development and Changing Nature of Agriculture

Agriculture is essential to human existence. The *Neolithic Revolution*, or the use of farming to produce food, had a profound effect on human societies. Rather than foraging and hunting for food, human societies became more stable and were able to grow due to more consistent food supplies. In modern times, farming has changed drastically in order to keep up with the increasing world population.

Until the twentieth century, the vast majority of people on Earth engaged in *subsistence farming*, or the practice of growing only enough food to feed one's self, or one's family. Over time, due to inventions such as the steel plow, the mechanical reaper, and the seed drill, farmers were able to produce more crops on the same amount of land. As food became cheaper and easier to obtain, populations grew, but rather than leading to an increase in farmers, fewer people actually farmed. After the advent of mechanized farming in developed nations, small farms became less common, and many were either abandoned or absorbed by massive commercial farms producing both foodstuffs, staple crops, and cash crops.

In recent years, agricultural practices have undergone further changes in order to keep up with the rapidly growing population. Due in part to the *Green Revolution*, which introduced the widespread use of fertilizers to produce massive amounts of crops, farming techniques and practices continue to evolve. For example, *genetically modified organisms*, or *GMOs*, are plants or animals whose genetic makeup has been modified using different strands of DNA in hopes of producing more resilient strains of staple crops, livestock, and other foodstuffs. This process, which is a form of *biotechnology*, attempts to solve the world's food production problems through the use of genetic engineering. Although these crops are abundant and resistant to pests, drought, or frost, they are also the subject of intense scrutiny. For example, the international food company, Monsanto, has faced an incredible amount of criticism regarding its use of GMOs in its products. Many activists assert that "rewiring" mother nature is inherently problematic and that foods produced through such methods are dangerous to human health. Despite the controversy, GMOs and biotechnologies continue to change the agricultural landscape by changing the world's food supply.

Like Monsanto, other agribusinesses exist throughout the world. Not only do these companies produce food for human consumption, but they also provide farming equipment, fertilizers, agrichemicals, and breeding and slaughtering services for livestock. While these companies are found all over the world, they are generally headquartered near the product they produce. For example, General Mills, a cereal manufacturer, is headquartered in the Midwestern United States, near its supply of wheat and corn—the primary ingredients in its cereals.

Contemporary Patterns and Impacts of Development, Industrialization, and Globalization

As mentioned previously, *developing nations* are nations that are struggling to modernize their economy, infrastructure, and government systems. Many of these nations may struggle to provide basic services to their citizens like clean water, adequate roads, or even police protection. Furthermore, government corruption makes life even more difficult for these countries' citizens. In contrast, *developed nations* are those who have relatively high *Gross Domestic Products (GDP)*, or the total value of all goods and services produced in the nation in a given year. To elucidate, the United States, which is one of the wealthiest nations on Earth when ranked by overall GDP, has nearly a 19 trillion dollar GDP; while Haiti, one of the poorest nations in the Western Hemisphere, has nearly a nine billion dollar GDP. This is a difference of almost seventeen trillion dollars. This is not to disparage Haiti or other developing nations; the comparison is simply used to show that extreme inequities exist in very close proximity to one another, and it may be difficult for developing nations to meet the needs of their citizens and move their economic infrastructure forward toward modernization.

In the modern world, industrialization is the initial key to modernization and development. For developed nations, the process of industrialization took place centuries ago. England, where the *Industrial Revolution* began, actually began to produce products in factories in the early 1700s. Later, the United States and some nations of Western Europe followed suit, using raw materials brought in from their colonies abroad to make finished products. For example, cotton was spun into fabric on elaborate weaving machines that mass-produced textiles. As a result, nations that perfected the textile process were able to sell their products around the world, which produced enormous profits. Over time, those nations were able to accumulate wealth, improve their nation's infrastructure, and provide more services for their citizens. Similar to the events of the eighteenth and nineteenth centuries, nations throughout the world are undergoing the same process in today's world. China exemplifies this concept. In China, agriculture was once the predominant occupation, and although it is true that agriculture is still a dominant sector of the Chinese economy, millions of Chinese citizens are flocking to major cities like Beijing, Shanghai, and Hangzhou, due to the availability of factory jobs that allow its workers a certain element of *social mobility*, or the ability to rise up out of one's socioeconomic situation.

Due to improvements in transportation and communication, the world has become figuratively smaller. For example, university students on the Indian Subcontinent now compete directly with students all over the world to obtain the skills employers desire to move their companies forward. Additionally, many corporations in developed nations have begun to *outsource* labor to nations with high levels of educational achievement but lower wage expectations. The process of opening the marketplace to all nations throughout the world, or *globalization*, has only just started to take hold in the modern economy. As industrial sites shift to the developing world, so does the relative level of opportunity for those nation's citizens. However, due to the massive amounts of pollution produced by factories, the process of globalization also has had significant ecological impacts. The most widely known impact, *climate change*, which most climatologists assert is caused by an increase of carbon dioxide in the atmosphere, remains a serious problem that has posed challenges for developing nations, who need industries in order to raise their standard of living, and developed nations, whose citizens use a tremendous amount of fossil fuels to run their cars, heat their homes, and maintain their ways of life.

Demographic Patterns and Demographic Change

Demography, or the study of human populations, involves a variety of closely related stimuli. First, as has been previously addressed, economic factors play a significant role in the movement of people, as do climate, natural disasters, or internal unrest. For example, in recent years, millions of immigrants from the war-torn country of Syria have moved as far as possible from danger. Although people are constantly moving, some consistencies remain throughout the world. First, people tend to live near reliable sources of food and water, which is why the first human civilizations sprung up in river valleys like the Indus River Valley in India, the Nile River Valley in Egypt, and the Yellow River Valley in Asia. Second, extreme temperatures tend to push people away, which is why the high latitudinal regions near the North and South Poles have such few inhabitants. Third, the vast majority of people tend to live in the Northern Hemisphere, due to the simple fact that more land lies in that part of the Earth. In keeping with these factors, human populations tend to be greater where human necessities are easily accessible, or at least more readily available. In other words, such areas have a greater chance of having a higher population density than places without such characteristics.

Demographic patterns on Earth are not always stagnant. In contrast, people move and will continue to move as both push and pull factors fluctuate along with the flow of time. For example, in the 1940s, thousands of Europeans fled their homelands due to the impact of the Second World War. Today, thousands of migrants arrive on European shores each month due to conflicts in the Levant and difficult economic conditions in Northern Africa. Furthermore, as previously discussed, people tend to migrate to places with a greater economic benefit for themselves and their families. As a result, developed nations such as the United States, Germany, Canada, and Australia have a net gain of migrants, while developing nations such as Somalia, Zambia, and Cambodia generally tend to see thousands of their citizens seek better lives elsewhere.

It is important to understand the key variables in changes regarding human population and its composition worldwide. Religion and religious conflict play a role in where people choose to live. For example, the Nation of Israel won its independence in 1948 and has since attracted thousands of people of Jewish descent from all over the world. Additionally, the United States has long attracted people from all over the world, due to its promise of religious freedom inherent within its own Constitution. In contrast, nations like Saudi Arabia and Iran do not typically tolerate different religions, resulting in a decidedly uniform religious—and oftentimes ethnic—composition. Other factors such as economic opportunity, social unrest, and cost of living also play a vital role in demographic composition.

Basic Concepts of Political Geography

Nations, states, and nation-states are all terms with very similar meanings, but knowing the differences aids in a better understanding of geography. A nation is a people group with similar cultural, linguistic, and historical experiences. A state is a political unit with sovereignty, or the ability to make its own decisions within defined borders; and a nation-state is an entity that combines states into one, singular government system. For example, in the United States, the state of Texas is not an independent state. Instead, it is part of the United States and thus, is subject to its laws. In a similar fashion, the United Kingdom encompasses four member states: England, Wales, Northern Ireland, and Scotland. Although people in those states may consider themselves to be *sovereign*, or self-governing, the reality is that those states cannot make decisions regarding international trade, declarations of war, or other important decisions regarding the rest of the world. Instead, they are *semi-autonomous*, meaning that they can make some decisions regarding how their own state is run, but must yield more major powers to a centralized authority. In the United States, this sort of system is called *Federalism*, or the sharing of power among Local, State, and Federal entities, each of whom is assigned different roles in the overall system of government.

Nation-states, and the boundaries that define where they are, are not always permanent. For example, after the fall of the Soviet Union in 1991, new nations emerged that had once been a part of the larger entity called the Union of Soviet Socialists Republics. These formerly sovereign nations were no longer forced to be a part of a unifying communist government, and as a result, they regained their autonomy and became newly independent nations that were no longer *satellite nations* of the Soviet Union. In a historical sense, the United States can be seen as a prime example of how national boundaries change. After the conclusion of the American Revolution in 1781, the Treaty of Paris defined the United States' western boundary as the Mississippi River; today, after a series of conflicts with Native American groups, the Mexican government, Hawaiian leadership, the Spanish, and the purchase of Alaska from the Russians, the boundaries of the United States have changed drastically. In a similar fashion, nations in Europe, Africa, and Asia have all shifted their boundaries due to warfare, cultural movements, and language barriers.

In the modern world, boundaries continue to change. For example, the Kurds, an ethnic minority and an excellent example of a nation, are still fighting for the right to control their people's' right to *self-determination*, but have not yet been successful in establishing a state for themselves. In contrast, the oil-rich region of South Sudan, which has significant cultural, ethnic, and religious differences from Northern Sudan, successfully won its independence in a bloody civil war, which established the nation's newest independent state. In recent years, Russia has made the world nervous by aggressively annexing the Crimean Peninsula, an area that has been part of the Ukraine since the end of the Cold War. Even the United Kingdom and Canada have seen their own people nearly vote for their own rights to self-determination. In 1995, the French-speaking Canadian province of Quebec narrowly avoided becoming a sovereign nation through a tightly contested referendum. In a similar fashion, Scotland, which is part of the UK, also voted to remain a part of the Crown, even though many people in that state see themselves as inherently different from those in other regions within the nation.

Political geography is constantly changing. Boundaries on maps from ten years ago are not consistent with those of 2016 and beyond. *Decolonization*, or the removal of dependency on colonizers, has altered the political landscape of Africa, allowed more autonomy for the African people, and has forever redefined the boundaries of the entire continent.

Economics

Fundamental Economic Concepts

Economics is the study of human behavior in response to the production, consumption, and distribution of assets or wealth. Economics can help individuals or societies make decisions or plans for themselves or communities, dependent upon their needs, wants, and resources. Economics is divided into two subgroups: microeconomics and macroeconomics.

Microeconomics is the study of individual or small group behaviors and patterns in relationship to such things as earning and spending money. It focuses on particular markets within the economy, and looks at single factors that could potentially affect individuals or small groups. For example, the use of coupons in a grocery store can affect an individual's product choice, quantity purchased, and overall savings that a person may later roll into a different purchase. Microeconomics is the study of scarcity, choice, opportunity costs, economics systems, factors of production, supply and demand, market efficiency, the role of government, distribution of income, and product markets.

Macroeconomics examines a much larger scale of the economy. It focuses on how a society or nation's goods, services, spending habits, and other factors affect the people of that entity. It focuses on aggregate factors such as demand and output. For example, if a national company moves its production overseas to save on costs, how will production, labor, and capital be affected? Macroeconomics analyzes all aggregate indicators and the microeconomic factors that influence the economy. Government and corporations use macroeconomic models to help formulate economic policies and strategies.

Microeconomics

Scarcity
People have different needs and wants, and the question arises, are the resources available to supply those needs and wants? Limited resources and high demand create scarcity. When a product is scarce, there is a short supply of it. For example, when the newest version of a cellphone is released, people line

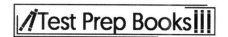

up to buy the phone or put their name on a wait list if the phone is not immediately available. The product, the new cellphone, may become a scarce commodity. In turn, because of the scarcity, companies may raise the cost of the commodity, knowing that if it is immediately available, people may pay more for the instant gratification—and vice versa. If a competing company lowers the cost of the phone but has contingencies, such as extended contracts or hidden fees, the buyer will still have the opportunity to purchase the scarce product. Limited resources and extremely high demand create scarcity and, in turn, cause companies to acquire opportunity costs.

Factors of Production

There are four factors of production:

- Land: both renewable and nonrenewable resources
- Labor: effort put forth by people to produce goods and services
- Capital: the tools used to create goods and services
- Entrepreneurship: persons who combine land, labor, and capital to create new goods and services

The four factors of production are used to create goods and services to make economic profit. All four factors strongly impact one another.

Supply and Demand

Supply and demand is the most important concept of economics in a market economy. Supply is the amount of a product that a market can offer. Demand is the quantity of a product needed or desired by buyers. The price of a product is directly related to supply and demand. The correlation between the price of a product and the demand necessary to distribute resources to the market go hand in hand in a market economy. For example, when there are a variety of treats at a bakery, certain treats are in higher demand than others. The bakery can raise the cost of the more demanded items as supplies get limited. Conversely, the bakery can sell the less desirable treats by lowering the cost of those items as an incentive for buyers to purchase them.

Product Markets

Product markets are marketplaces where goods and services are bought and sold. Product markets provide sellers a place to offer goods and services to consumers, and for consumers to purchase those goods and services. The annual value of goods and services exchanged throughout the year is measured by the Gross Domestic Product (GDP), a monetary measure of goods and services made either quarterly or annually. Department stores, gas stations, grocery stores, and other retail stores are all examples of product markets. However, product markets do not include any raw, scarce, or trade materials.

Theory of the Firm

The behavior of firms is composed of several theories varying between short- and long-term goals. There are four basic firm behaviors: perfect competition, profit maximization, short run, and long run. Each firm follows a pattern, depending on its desired outcome. Theory of the Firm posits that firms, after conducting market research, make decisions that will maximize their profits since they are for-profit entities.

- Perfect competition:
- In perfect competition, several businesses are selling the same product at the same time.
- There are so many businesses and consumers that none will directly impact the market.

- Each business and consumer is aware of the competing businesses and markets.
- Profit maximization:
- Firms decide the quantity of a product that needs to be produced in order to receive maximum profit gains. Profit is the total amount of revenue made after subtracting costs.
- Short run:
- A short amount of time where fixed prices cannot be adjusted
- The quantity of the product depends on the varying amount of labor. Less labor means less product.
- Long run:
- An amount of time where fixed prices can be adjusted
- Firms try to maximize production while minimizing labor costs.

Overall, microeconomics operates on a small scale, focusing on how individuals or small groups use and assign resources.

Macroeconomics

Macroeconomics analyzes the economy as a whole. It studies unemployment, interest rates, price levels, and national income, which are all factors that can affect the nation as a whole, and not just individual households. Macroeconomics studies all large factors to determine how, or if, they will affect future trend patterns of production, consumption, and economic growth.

Measures of Economic Performance

It is important to measure economic performance to determine if an economy is growing, stagnant, or deteriorating. To measure the growth and sustainability of an economy, several indicators can be used. Economic indicators provide data that economists can use to determine if there are faulty processes or if some form of intervention is needed.

One of the main indicators to measure economic performance is the growth of the country's Gross Domestic Product (GDP). GDP growth provides important information that can be used to determine fiscal or financial policies. The GDP does not measure income distribution, quality of life, or losses due to natural disasters. For example, if a community lost everything to a hurricane, it would take a long time to rebuild the community and stabilize its economy. That is why there is a need to take into account more balanced performance measures when factoring overall economic performance.

Other indicators used to measure economic performance are unemployment or employment rates, inflation, savings, investments, surpluses and deficits, debt, labor, trade terms, the HDI (Human Development Index), and the HPI (Human Poverty Index).

Unemployment

Unemployment occurs when an individual does not have a job, is actively trying to find employment, and is not getting paid. Official unemployment rates do not factor in the number of people who have stopped looking for work, unlike true unemployment rates that do, causing them to be higher.

There are three types of unemployment: cyclical, frictional, and structural.

Cyclical
The product of a business cycle. This usually occurs during a recession.
Frictional
The difficulty of matching qualified workers for specific jobs. An example would be a person changing careers.
Structural
When a person no longer qualifies for a specific job, or failing out of a retraining course for a job.

Given the nature of a market economy and the fluctuations of the labor market, a 100 percent employment rate is impossible to reach.

Inflation

Inflation is when the cost of goods and services rises over time. Supply, demand, and money reserves all affect inflation. Generally, inflation is measured by the Consumer Price Index (CPI), a tool that tracks price changes of goods and services over time. The CPI measures goods and services such as gasoline, cars, clothing, and food. When the cost of goods and services increase, the quantity of the product may decrease due to lower demand. This decreases the purchasing power of the consumer. Basically, as more money is printed, it holds less and less value in purchasing power. For example, when inflation occurs, consumers in the United States are spending and saving less because the U.S. dollar is worth less, and therefore the consumer cannot buy or save as much money. However, if inflation occurs steadily over time, the people can better plan and prepare for future necessities.

Inflation can vary from year to year, usually never fluctuating more than 2 percent. Central banks try to prevent drastic increases or decreases of inflation to prohibit prices from rising or falling far from the minimum. Inflation can also vary based on different monetary currencies. Although rare, any country's economy may experience hyperinflation (when inflation rates increase to over 50 percent), while other economies may experience deflation (when the cost of goods and services decrease over time). Deflation occurs when the inflation rate drops below zero percent.

Business Cycle

A business cycle is when the Gross Domestic Product (GDP) moves downward and upward over a long-term growth trend. These cycles help determine where the economy currently stands, as well as where the economy could be heading. Business cycles usually occur almost every six years, and have four phases: expansion, peak, contraction, and trough. Here are some characteristics of each phase:

- Expansion:
- Increased employment rates and economic growth
- Production and sales increase
- On a graph, expansion is where the lines climb.
- Peak:
- Employment rates are at or above full employment and the economy is at maximum productivity.
- On a graph, the peak is the top of the hill, where expansion has reached its maximum.
- Contraction:
- When growth starts slowing
- Unemployment is on the rise.

- On a graph, contraction is where the graph begins to slide back down or contract.
- Trough:
- The cycle has hit bottom and is waiting for the next cycle to start again.
- On a graph, the trough is the bottom of the contraction prior to when it starts to climb back up.

When the economy is expanding or "booming," the business cycle is going from a trough to a peak. When the economy is headed down and toward a recession, the business cycle is going from a peak to a trough.

Four phases of a business cycle:

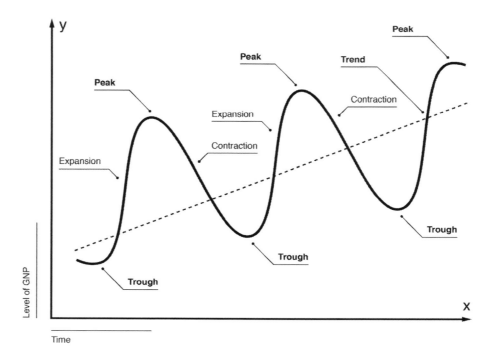

Economic Growth

Economic growth is measured by the increase in the Gross National Product (GNP) or Gross Domestic Product (GDP). The increase of goods and services over time indicates positive movement in economic growth. Keep in mind that the quantity of goods and services produced is not necessarily an indicator of economic growth. The value of the goods and services produced matters more than the quantity.

There are many causes of economic growth, which can be short- or long-term. In the short term, if aggregate demand (the total demand for goods and services produced at a given time) increases, then the overall Gross Domestic Product (GDP) increases as well. Not only will the GDP increase, interest rates may decrease. With reduced interest rates, spending and investing will increase. Consumer and government spending will also increase because there will be more disposable income. Real estate prices will rise, and there will be lower income taxes. All of these short-term factors can stimulate economic growth.

In the long term, if aggregate supply (the total supply of goods or services in a given time period) increases, then there is potential for an increase in capital as well. With more working capital, more infrastructure and jobs can be created. With more jobs, there is an increased employment rate, and education and training for jobs will improve. New technologies will be developed, and new raw materials may be discovered. All of these long-term factors can also stimulate economic growth.

Outside of the short- and long-term causes for economic growth, other factors include low inflation and stability. Lower inflation rates encourage more investing versus higher inflation rates that cause instability in the market. Stability encourages businesses to continue investing. If the market is unstable, investors may question the volatility of the market.

Potential Costs of Economic Growth:

- Inflation: When economic growth occurs, inflation tends to be high. If supply cannot keep up with demand, then the inflation rate may be unmanageable.

- Economic booms and recessions: The economy goes through cycles of booms and recessions. This causes inflation to increase and decrease over time, which puts the economy into a continuous cycle of rising and falling.
- Account inefficiencies: When the economy grows, consumers and businesses increase their import spending. The increase of import spending affects the current account and causes a shortage.
- Environmental costs: When the economy is growing, there is an abundance of output, which may result in more pollutants and a reduction in quality of life.
- Inequalities: Growth occurs differently among members of society. While the wealthy may be getting richer, those living in poverty may just be getting on their feet. So, while economic growth is happening, it may happen at two very different rates.

While these potential costs could affect economic growth, if the growth is consistent and stable, then growth can occur without severe inflation swings. Also, as technology improves, new ways of production can reduce negative environmental factors as well.

Government Involvement in the Economy

Governments have considerable influence over the flow of economies, which makes it important to understand the relationships between them. When a government has full control over the economic decisions of a nation, it is called a command system. This was the case in many absolute monarchies such as eighteenth-century France; King Louis XIV built his economy on the concept of mercantilism, which believed that the state should manage all resources, particularly by accumulating gold and silver. This system of economics discouraged exports and thereby limited trade.

In contrast, the market system is guided by the concept of capitalism, in which individuals and businesses have the freedom to manage their economic decisions. This allows for private property and increases the opportunities for entrepreneurship and trade. Early proponents of capitalism emphasized *laissez-faire* policies, which means "let it be," and argued that the government should not be involved with the economy at all. They believe the market is guided by the concept of self-interest and that individuals will optimally work for their personal success. However, individuals' interests do not necessarily correlate with the needs of the overall economy. For example, during a financial recession, consumers may decide to save up their money rather than make purchases; doing so helps them in the short run but further reduces demand in a slumping economy. Therefore, most capitalist governments still assert a degree of control over their economies while still allowing for private business.

Likewise, many command system economies, such as monarchical France, still relied heavily on private businesses maintained by wealthy businessmen. With the end of most absolute monarchies, communism has been the primary form of command system economies in the modern era. Communism is a form of socialism that emphasizes communal ownership of property and government control over production. The high degree of government control gives more stability to the economy, but it also creates considerable flaws. The monopolization of the economy by the government limits its ability to respond to local economic conditions because certain regions often have unique resources and needs. With the collapse of the Soviet Union and other communist states, command systems have been largely replaced with market systems.

The U.S. government helps to manage the nation's economy through a market system in several ways. First and foremost, the federal government is responsible for the production of money for use within the economy; depending on how the government manages the monetary flow, it may lead to a stable economy, deflation, or inflation. Second, state and federal governments impose taxes on individuals, corporations, and goods. For example, a tariff might be imposed on imports in order to stimulate demand for local goods in the economy. Third, the government can pass laws that require additional regulation or inspections. In addition, the government has passed antitrust laws to inhibit the growth of private monopolies, which could limit free growth in the market system. Debates continue over whether the government should take further action to manage private industries or reduce its control over the private sector.

Just as governments can affect the direction of the economy, the state of the economy can have significant implications on government policies. Financial stability is critical in maintaining a prosperous state. A healthy economy will allow for new developments that contribute to the nation's growth and create jobs. On the other hand, an economic crisis, such as a recession or depression, can gravely damage a government's stability. Without a stable economy, business opportunities plummet, and people begin to lose income and employment. This, in turn, leads to frustration and discontent in the population, which can lead to criticism of the government. This could very well lead to demands for new leadership to resolve the economic crisis.

The dangers of a destabilized economy can be seen with the downfall of the French monarchy. The mercantilist approach to economics stifled French trade. Furthermore, regional aristocracies remained exempt from government taxes, which limited the government's revenues. This was compounded by expensive wars and poor harvests that led to criticism of King Louis XIV's government. The problems persisted for decades, and Louis XIV was forced to convene the Estates-General, a legislative body of representatives from across France, to address the crisis. The economic crises at the end of the eighteenth century were critical in the beginning of the French Revolution. Those financial issues, in turn, at least partially stemmed from both the government's control of the economy through mercantilism and its inability to impose economic authority over local regions.

Economic Systems

Economic systems determine what is being produced, who is producing it, who receives the product, and the money generated by the sale of the product. There are two basic types of economic systems: market economies (including free and competitive markets), and planned or command economies.

- Market Economies are characterized by:

- Privately owned businesses, groups, or individuals providing goods or services based on demand.

- The types of goods and services provided (supply) are based on that demand.

- Two types: competitive market and free market.

Competitive Market	Free Market
Due to the large number of both buyers and sellers, there is no way any one seller or buyer can control the market or price.	Voluntary private trades between buyers and sellers determine markets and prices without government intervention or monopolies.

- Planned or Command Economies:

- In planned or command economies, the government or central authority determines market prices of goods and services.

- The government or central authority determines what is being produced as well as the quantity of production.

- Some advantages to command economies include a large number of shared goods such as public services (transportation, schools, or hospitals).

- Disadvantages of command economies include wastefulness of resources.

Market Efficiency and the Role of Government (Taxes, Subsidies, and Price Controls)

Market efficiency is directly affected by supply and demand. The government can help the market stay efficient by either stepping in when the market is inefficient and/or providing the means necessary for markets to run properly. For example, society needs two types of infrastructure: physical (bridges, roads, etc.) and institutional (courts, laws, etc.). The government may impose taxes, subsidies, and price controls to increase revenue, lower prices of goods and services, ensure product availability for the government, and maintain fair prices for goods and services.

The Purpose of Taxes, Subsidies, and Price Controls

Taxes	Subsidies	Price Controls
-Generate government revenue -Discourage purchase or use of "bad" products such as alcohol or cigarettes	-Lower the price of goods and services -Reassure the supply of goods and services -Allow opportunities to compete with overseas vendors	-Act as emergency measures when government intervention is necessary -Set a minimum or maximum price for goods and services

Money and Banking

Money is the universal form of currency used throughout goods and services exchanges that holds its value over time. Money provides a convenient way for sellers and consumers to understand the value of their goods and services. As opposed to bartering (when sellers and consumers exchange goods or services as equal trades), money is quick and easy for both buyers and sellers.

There are three main forms of money: commodity, fiat, and bank. Here are characteristics of each form:

- Commodity money: Money as a valuable good, such as precious metals
- Fiat money: The value of the good set by supply and demand rather than the actual value it represents, such as paper money
- Bank money: Money that is credited by a bank to those who deposit it into bank accounts, such as checking and savings accounts or credit

While price levels within the economy set the demand for money, most countries have central banks that supply the actual money. Essentially, banks buy and sell money. Borrowers can take loans and pay back the bank, with interest, providing the bank with extra capital.

A central bank has control over the printing and distribution of money. Central banks serve three main purposes: manage monetary growth to help steer the direction of the economy, be a backup to commercial banks that are suffering, and provide options and alternatives to government taxation.

The Federal Reserve is the central bank of the United States. The Federal Reserve controls banking systems and determines the value of money in the United States. Basically, the Federal Reserve is the bank for banks.

All Western economies have to keep a minimum amount of protected cash called *required reserve*. Once banks meet those minimums, they can then lend or loan the excess to consumers. The required reserves are used within a fractional reserve banking system (fractional because a small portion is kept separate and safe). Not only do banks reserve, manage, and loan money, but they also help form monetary policies.

Monetary Policy

The central bank and other government committees control the amount of money that is made and distributed. The money supply determines monetary policy. Three main features sustain monetary policy:

- Assuring the minimum amount held within banks (bank reserves). When banks are required to hold more money in reserve funds, banks are less willing to lend money to help control inflation.

- Adjusting interest rates. For example, if the value of money is low, companies and consumers buy more (products, employees, stocks) because prices are cheap. Just like an investment, it is risky, but can pay off in the long term.

- The purchase and sales of bonds (otherwise known as open market operations). When buying bonds to increase money and selling bonds to reduce the supply, the central bank helps control the money supply.

In the United States, the Federal Reserve maintains monetary policy. There are two main types of monetary policy: expansionary monetary policy and contractionary monetary policy.

- Expansionary monetary policy:
- Increases the money supply
- Lowers unemployment
- Increases consumer spending
- Increases private sector borrowing
- Possibly decreases interest rates to very low levels, even near zero
- Decreases reserve requirements and federal funds
- Contractionary monetary policy:
- Decreases the money supply
- Helps control inflation
- Possibly increases unemployment due to slowdowns in economic growth
- Decreases consumer spending
- Decreases loans and/or borrowing

The Federal Reserve uses monetary policy to try to achieve maximum employment and secure inflation rates. Because the Federal Reserve is the "bank of banks," it truly strives to be the last-resort option for distressed banks. This is because once these kinds of institutions begin to rely on the Federal Reserve for help, all parts of the banking industry—such as those dealing with loans, bonds, interest rates, and mortgages—are affected.

International Trade and Exchange Rates

International trade is when countries import and export goods and services. Countries often want to deal in terms of their own currency. Therefore, when importing or exporting goods or services, consumers and businesses need to enter the market using the same form of currency. For example, if the United States would like to trade with China, the U.S. may have to trade in China's form of currency, the *Yuan*, versus the dollar, depending on the business.

The exchange rate is what one country's currency will exchange for another. The government and the market (supply and demand) determine the exchange rate. There are two forms of exchange rates: fixed and floating. Fixed exchange rates involve government interventions (like central banks) to help keep

the exchange rates stable. Floating or "flexible" exchange rates constantly change because they rely on supply and demand needs. While each type of exchange rate has advantages and disadvantages, the rate truly depends on the current state of each country's economy. Therefore, each exchange rate may differ from country to country.

Advantages and Disadvantages of Fixed Versus Floating Exchange Rates			
Fixed Exchange Rate: government intervention to help keep exchange rates stable		Floating or "Flexible" Exchange Rate: Supply and demand determines the exchange rate	
Advantages	*Disadvantages*	*Advantages*	*Disadvantages*
-Stable prices -Stable foreign exchange rates -Exports are more competitive and in turn more profitable	-Requires a large amount of reserve funds -Possibly mispricing currency values -Inflation increases	-Central bank involvement is not needed. -Facilitates free trade	-Currency speculation -Exchange rate risks -Inflation increases

While each country may have differing economic statuses and exchange rates, countries rely on one another for goods and services. Prices of imports and exports are affected by the strength of another country's currency. For example, if the United States dollar is at a higher value than another country's currency, imports will be less expensive because the dollar will have more value than that of the country selling its good or service. On the other hand, if the dollar is at a low value compared to the currency of another country, importers will tend to defer away from buying international items from that country. However, U.S. exporters to that country could benefit from the low value of the dollar.

Fiscal Policy

A fiscal policy is when the government is involved in adjusting spending and tax rates to assist the way in which an economy financially functions. Fiscal policies can either increase or decrease tax rates and spending. These policies represent a tricky balancing act, because if the government increases taxes too much, consumer spending and monetary value will decrease. Conversely, if the government lowers taxes, consumers will have more money in their pockets to buy more goods and services, which increases demand and the need for companies to supply those goods and services. Due to the higher demand, suppliers can add jobs to fulfill that demand. While the increase of supply, demand, and jobs are positive for the overall economy, they may result in a devaluation of the dollar and less purchasing power.

Consumer Economics

Economics are closely linked with the flow of resources, technology, and population in societies. The use of natural resources, such as water and fossil fuels, has always depended in part on the pressures of the economy. A supply of a specific good may be limited in the market, but with sufficient demand the sellers are incentivized to increase the available quantity. Unfortunately, the demand for certain objects can often be unlimited, and a high price or limited supply may prevent consumers from obtaining the product or service. If the sellers succumb to the consumers' demand and continue to exploit a scarce resource, supply could potentially be exhausted.

The resources for most products, both renewable and nonrenewable, are finite. This is a particularly difficult issue with nonrenewable resources, but even renewable resources often have limits: organic products such as trees and animals require stable populations and sufficient habitats to support those populations. Furthermore, the costs of certain decisions can have detrimental effects on other

resources. For example, industrialization provides economic benefits in many countries but also has had the negative effect of polluting surrounding environments; the pollution, in turn, often eliminates or harms fish, plants, and other potential resources.

The control of resources within an economy is particularly important in determining how resources are used. While the demand may change with the choices of consumers, the range of supply depends on the objectives of the people producing the goods. They determine how much of their supply they allot for sale, and in the case of monopolies, they might have sole access to the resource. They might choose to limit their use of the resources or instead gather more to meet the demand. As they pay for the products, consumers can choose which sellers they rely on for the supply. In the case of a monopoly, though, consumers have little influence over the company's decision because there is no alternative supplier. Therefore, the function of supply within an economy can drastically influence how the resources are exploited.

The availability of resources, in turn, affects the human population. Humans require basic resources such as food and water for survival, as well as additional resources for healthy lifestyles. Therefore, access to these resources helps determine the survival rate of humans. For much of human existence, economies have had limited ability to extract resources from the natural world, which restricted the growth rate of populations. However, the development of new technologies, combined with increasing demand for certain products, has pushed resource use to a new level. On the one hand, this led to higher living standards that ensured that fewer people would die. However, this has also brought mass population growth. Admittedly, countries with higher standards of living often have lower birthrates. Even so, the increasing exploitation of resources has sharply increased the world's population as a whole to unsustainable levels. The rising population leads, in turn, to more demand for resources that cannot be met. This creates poverty, reduced living conditions, and higher death rates. As a result, economics can significantly influence local and world population levels.

Technology is also intricately related to population, resources, and economics. The role of demand within economies has incentivized people to innovate new technologies that enable societies to have a higher quality of life and greater access to resources. Entrepreneurs expand technologies by finding ways to create new products for the market. The Industrial Revolution, in particular, illustrates the relationship between economics and technology because the ambitions of businessmen led to new infrastructure that enabled more efficient and sophisticated use of resources. Many of these inventions reduced the amount of work necessary for individuals and allowed the development of leisure activities, which in turn created new economic markets. However, economic systems can also limit the growth of technology. In the case of monopolies, the lack of alternative suppliers reduces the incentive to meet and exceed consumer expectations. Moreover, as demonstrated by the effects of economics on resources, technology's increasing ability to extract resources can lead to their depletion and create significant issues that need to be addressed.

Distribution of Income

Distribution of income refers to how wages are distributed across a society or segments of a society. If everyone made the same amount of money, the distribution of income would be equal. That is not the case in most societies. The wealth of people and companies varies. Income inequality gaps are present in America and many other nations. Taxes provide an option to redistribute income or wealth because they provide revenue to build new infrastructure and provide cash benefits to some of the poorest members in society.

Choice and Opportunity Costs

When an individual decides between possibilities, that individual is making a choice. Choices allow people to compare opportunity costs. **Opportunity cost**s are benefits that a person could have received, but gave up, in choosing another course of action. What is an individual willing to trade or give up for a different choice? For example, if an individual pays someone to mow the lawn because he or she would rather spend that time doing something else, then the opportunity cost of paying someone to mow the lawn is worth the time gained from not doing the job himself or herself.

On a larger scale, governments and communities have to assess different opportunity costs when it comes to using taxpayers' money. Should the government or community build a new school, repair roads, or allocate funds to local hospitals are all examples of choices taxpayers may have to review at some point in time. How do they decide which choice is the best, since each one has a trade off? By comparing the opportunity cost of each choice, they may decide what they are willing to live without for the sake of gaining something else.

Practice Questions

1. What's the term for the ability of a ruling body to influence the actions, behavior, and attitude of a person or group of people?
    a. Politics
    b. Power
    c. Authority
    d. Legitimacy

2. Which of the following is NOT a shared characteristic sufficient to form a nation?
    a. Culture and traditions
    b. History
    c. Sovereignty
    d. Beliefs and religion

Question 3 is based on the following passage:

> Upon this, one has to remark that men ought either to be well treated or crushed, because they can avenge themselves of lighter injuries, of more serious ones they cannot; therefore the injury that is to be done to a man ought to be of such a kind that one does not stand in fear of revenge.
>
> From Niccolo Machiavelli's *The Prince*, 1513

3. What advice is Machiavelli giving to the prince?
    a. Lightly injured enemies will overthrow the prince.
    b. Seek to injure everyone you meet.
    c. Hurting people is always the correct course of action.
    d. If you are going to cause an enemy some injury, ensure the injury is fatal.

Question 4 is based on the following passage:

> The creed which accepts as the foundation of morals, Utility, or the Greatest-Happiness Principle, holds that actions are right in proportion as they tend to promote happiness, wrong as they tend to produce the reverse of happiness. By happiness is intended pleasure, and the absence of pain; by unhappiness, pain, and the privation of pleasure.
>
> The utilitarian morality does recognise in human beings the power of sacrificing their own greatest good for the good of others. It only refuses to admit that the sacrifice is itself a good. A sacrifice which does not increase, or tend to increase, the sum total of happiness, it considers as wasted.
>
> From John Stuart Mill's *Utilitarianism*, 1861

4. What is the meaning of the "Utility"?
    a. Actions should be judged based on the net total of pleasure.
    b. Actions requiring sacrifice can never be valuable.
    c. Actions promoting sacrifice that increase happiness are more valuable than actions that only increase happiness.
    d. Actions can be valuable even if the pain outweighs the pleasure.

Question 5 is based on the following passage:

> The history of all hitherto existing society is the history of class struggles.
>
> Freeman and slave, patrician and plebeian, lord and serf, guildmaster and journeyman, in a word, oppressor and oppressed, stood in constant opposition to one another, carried on an uninterrupted, now hidden, now open fight, that each time ended, either in the revolutionary reconstitution of society at large, or in the common ruin of the contending classes.
>
> Let the ruling classes tremble at a Communistic revolution. The proletarians have nothing to lose but their chains. They have a world to win.
>
> Workingmen of all countries unite!
>
> Karl Marx and Friedrich Engels, *The Communist Manifesto,* 1848

5. What's the main idea presented in the excerpt?
    a. Working men are morally superior to the ruling class.
    b. Every society will come to an end at some point.
    c. History is defined by class struggle, and working men must now unite and fight the ruling class to gain freedom.
    d. Working men are in the same position as the slave, plebeian, serf, and journeyman.

Question 6 is based on the following image:

Spectrum of Political Ideologies

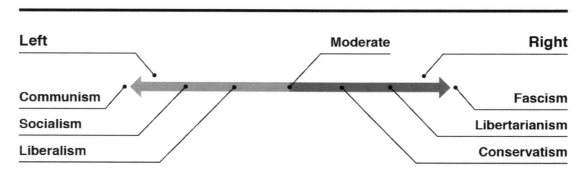

6. Of the following ideologies, which one advocates for the most radical government intervention to achieve social and economic equality?
    a. Socialism
    b. Liberalism
    c. Libertarianism
    d. Fascism

7. Of the following ideologies, which one prioritizes stability and traditional institutions within a culture?
    a. Socialism
    b. Liberalism
    c. Conservatism
    d. Libertarianism

8. The central government established under the Articles of Confederation held which of the following powers?
    a. The power to impose taxes
    b. The power to declare war
    c. The power to regulate trade
    d. The power to enforce laws enacted by Congress

Question 9 is based on the following diagram:

Separation of Powers

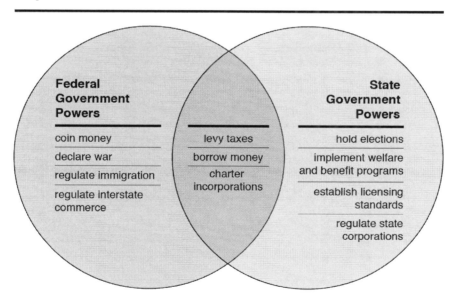

9. Which of the following terms best describes the missing title?
    a. Reserved powers
    b. Implied powers
    c. Delegated powers
    d. Concurrent powers

Question 10 is based on the following passage:

> Ambition must be made to counteract ambition. The interest of the man must be connected with the constitutional rights of the place. It may be a reflection on human nature, that such devices should be necessary to control the abuses of government. But what is government itself, but the greatest of all reflections on human nature?
>
> If men were angels, no government would be necessary. If angels were to govern men, neither external nor internal controls on government would be necessary. In framing a government which is to be administered by men over men, the great difficulty lies in this: you must first enable the government to control the governed; and in the next place oblige it to control itself.
>
> Alexander Hamilton or James Madison, aka *Publius*, "Federalist No. 50," 1788

10. What is the main idea presented in the excerpt?
    a. Men are inherently immoral and abusive.
    b. The best form of government is the type that angels would construct.
    c. Government reflects human nature.
    d. An effective government requires a separation of powers to regulate itself.

Question 11 is based on the following diagram:

Checks and Balances

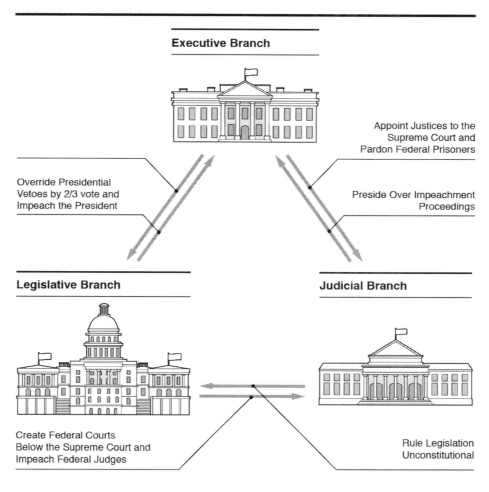

Executive Branch

Appoint Justices to the
Supreme Court and
Pardon Federal Prisoners

Override Presidential
Vetoes by 2/3 vote and
Impeach the President

Preside Over Impeachment
Proceedings

Legislative Branch

Judicial Branch

Create Federal Courts
Below the Supreme Court and
Impeach Federal Judges

Rule Legislation
Unconstitutional

11. Which of the following answer choices best completes the blank line at the top left section of the diagram by the Executive Branch?
   a. Impeach congressmen and veto legislation
   b. Call special sessions of Congress and refuse to enforce laws
   c. Call special sessions of Congress and veto legislation
   d. Impeach congressmen and refuse to enforce laws

12. What amendment guarantees American citizens the right to keep and bear arms?
   a. First Amendment
   b. Second Amendment
   c. Third Amendment
   d. Fourth Amendment

Question 13 is based on the following table:

Presidential Election of 1824			
Candidate	Electoral Votes	Popular Votes	State Votes in the House of Representatives
Andrew Jackson	99	153,544	7
John Quincy Adams	84	108,740	13
William H. Crawford	41	46,618	4
Henry Clay	37	47,136	0

13. Who won the presidential election of 1824?
    a. Andrew Jackson
    b. John Quincy Adams
    c. William H. Crawford
    d. Henry Clay

14. What electoral system can result in a second round of voting commonly referred to as a runoff?
    a. Majority systems
    b. Plurality systems
    c. Single transferable systems
    d. Party list systems

15. In international relations, which of the following is NOT a basic tenet of realism?
    a. States are the central actors.
    b. States act rationally to advance their self-interest.
    c. States should seek to form international organizations to increase global cooperation and respond to international issues.
    d. All states are interested in maintaining or expanding their power as a means of self-preservation.

16. How did the outcome of the French and Indian War impact the life of American colonists?
    a. The colonies expanded west of the Allegheny Mountains.
    b. Great Britain imposed taxes on the colonies to pay off the British war debt.
    c. A lasting peace developed between the colonists and Native Americans.
    d. The power of self-government increased in the colonies.

Question 17 is based on the following passage:

We hold these Truths to be self-evident: that all Men are created equal; that they are endowed by their creator with certain inalienable rights; that among these are life, liberty, and the pursuit of happiness: that to secure these rights, governments are instituted among men, deriving their just powers from the consent of the governed; that whenever any form of government becomes destructive of these ends, it is the right of the people to alter or abolish it, and to institute new government, laying its foundation on such principles, and organizing its powers in such form, as to them shall seem most likely to affect their safety and happiness.

Prudence indeed will dictate that governments long established should not be changed for light and transient causes; and accordingly all experience hath shown that mankind are more disposed to suffer while evils are sufferable, than to right themselves by abolishing the forms to which they are accustomed. But when a long train of abuses and usurpations begun at a

distinguished period and pursuing invariably the same object, evinces a design to reduce them under absolute despotism, it is their right, it is their duty to throw off such government, and to provide new guards for their future security

Declaration of Independence, adopted July 4, 1776

17. What is the main purpose of the excerpt?
    a. Provide a justification for revolution when the government infringes on "certain inalienable rights"
    b. Provide specific evidence of the "train of abuses"
    c. Provide an argument why "all Men are created equal"
    d. Provide an analysis of the importance of "life, liberty, and the pursuit of happiness"

Question 18 is based on the following map:

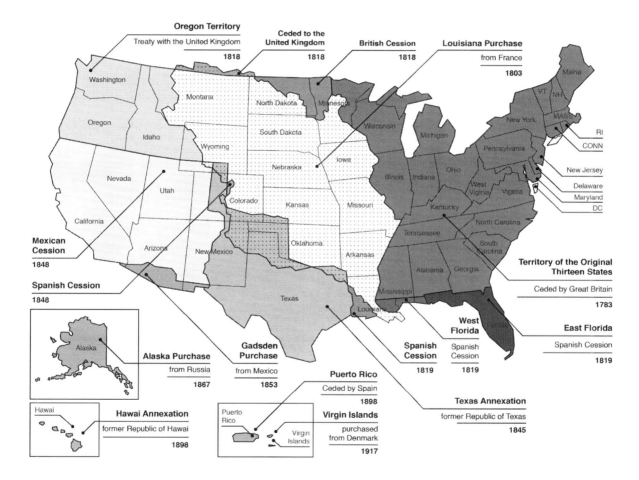

18. What current state did the United States gain through military force with a non-native nation-state?
    a. Nebraska
    b. Missouri
    c. Alaska
    d. Nevada

Question 19 is based on the following passage:

> Now, therefore I, Abraham Lincoln, President of the United States, by virtue of the power in me vested as Commander-in-Chief, of the Army and Navy of the United States in time of actual armed rebellion against the authority and government of the United States, and as a fit and necessary war measure for suppressing said rebellion...
>
> And by virtue of the power, and for the purpose aforesaid, I do order and declare that all persons held as slaves within said designated States, and parts of States, are, and henceforward shall be free; and that the Executive government of the United States, including the military and naval authorities thereof, will recognize and maintain the freedom of said persons.
>
> President Abraham Lincoln, Emancipation Proclamation, January 1, 1863

19. How does President Lincoln justify freeing the slaves in designated areas of the South?
    a. Emancipation is necessary since slavery is evil.
    b. Emancipation is necessary to boost the morale of the North.
    c. Emancipation is necessary to punish for the South seceding from the Union.
    d. Emancipation is necessary to strengthen the war effort of the North.

20. What was a consequence of the industrialization that followed the Civil War?
    a. Decreased immigration
    b. Increased urbanization
    c. Decreased socioeconomic inequality
    d. Increased rights for workers

21. Which of the following best describes how the Treaty of Versailles contributed to the outbreak of World War II?
    a. Forced Germany to assume responsibility for all damage incurred during the war and pay billions of dollars in reparations.
    b. Failed to adequately end the violence of World War I.
    c. Left large tracts of territory unclaimed by any nation-state.
    d. Created the League of Nations.

Question 22 is based on the following passage:

Hand in hand with this we must frankly recognize the overbalance of population in our industrial centers and, by engaging on a national scale in a redistribution, endeavor to provide a better use of the land for those best fitted for the land. The task can be helped by definite efforts to raise the values of agricultural products and with this the power to purchase the output of our cities. It can be helped by preventing realistically the tragedy of the growing loss through foreclosure of our small homes and our farms. It can be helped by insistence that the Federal, State, and local governments act forthwith on the demand that their cost be drastically reduced. It can be helped by the unifying of relief activities which today are often scattered, uneconomical, and unequal. It can be helped by national planning for and supervision of all forms of transportation and of communications and other utilities which have a definitely public character. There are many ways in which it can be helped, but it can never be helped merely by talking about it. We must act and act quickly.

Finally, in our progress toward a resumption of work we require two safeguards against a return of the evils of the old order: there must be a strict supervision of all banking and credits and investments, so that there will be an end to speculation with other people's money; and there must be provision for an adequate but sound currency.

President Franklin D. Roosevelt, Inaugural Address, March 4, 1933

22. Which of the following best describes President Roosevelt's underlying approach to government?
   a. Government must be focused on redistribution of land.
   b. Government must "act and act quickly" to intervene and regulate the economy.
   c. Government must exercise "strict supervision of all banking."
   d. Government must prevent the "growing loss through foreclosure."

Question 23 is based on the following passage:

What, to the American slave, is your 4th of July? I answer: a day that reveals to him, more than all other days in the year, the gross injustice and cruelty to which he is the constant victim. To him, your celebration is a sham; your boasted liberty, an unholy license; your national greatness, swelling vanity; your sounds of rejoicing are empty and heartless; your denunciations of tyrants, brass fronted impudence; your shouts of liberty and equality, hollow mockery; your prayers and hymns, your sermons and thanksgivings, with all your religious parade, and solemnity, are, to him, mere bombast, fraud, deception, impiety, and hypocrisy—a thin veil to cover up crimes which would disgrace a nation of savages. There is not a nation on the earth guilty of practices, more shocking and bloody, than are the people of these United States, at this very hour.

Frederick Douglass, "What to the Slave is the 4th of July?" July 5, 1852

23. What is the specific hypocrisy that Douglass repudiates?
   a. The Declaration of Independence declared that all men are created equal, but Thomas Jefferson owned slaves.
   b. Americans are free, but they do not value their freedom.
   c. The Fourth of July is a celebration about freedom, and slavery remained legal in the United States.
   d. The United States is a Christian nation, but American traditions contradict their faith.

Question 24 is based on the following passage:

> May it please your honor, I shall never pay a dollar of your unjust penalty. All the stock in trade I possess is a $10,000 debt, incurred by publishing my paper—The Revolution—four years ago, the sole object of which was to educate all women to do precisely as I have done, rebel against your man-made, unjust, unconstitutional forms of law, that tax, fine, imprison and hang women, while they deny them the right of representation in the government; and I shall work on with might and main to pay every dollar of that honest debt, but not a penny shall go to this unjust claim. And I shall earnestly and persistently continue to urge all women to the practical recognition of the old revolutionary maxim, that "Resistance to tyranny is obedience to God."

> *An Account of the Proceedings on the Trial of Susan B. Anthony on the Charge of Illegal Voting,* 1874.

24. What is the main idea presented in the excerpt?
    a. Taxation without representation is tyranny.
    b. Domestic abuse and violence against women is the cause of tyranny.
    c. Anthony cannot pay her fine due to debt accumulated from fighting for women's rights.
    d. Denying women the right to vote is tyranny and must be resisted.

25. Which of the following most accurately describes the platform of Ronald Reagan?
    a. Christianity, optimism, and preserving social safety nets
    b. Increased defense spending, deregulation, and tax cuts
    c. Moral majority, international cooperation, and compromise
    d. Conservatism, opposition to abortion, and organized labor

26. How do market economies differ from planned economies?
    a. Unlike market economies, planned economies have a larger number of both buyers and sellers.
    b. Unlike market economies, planned economies distribute resources more efficiently.
    c. Unlike planned economies, market economies allow demand to set prices.
    d. Unlike planned economies, market economies prioritize public services.

Question 27 is based on the following passage:

> Those who are opposed to this proposition tell us that the issue of paper money is a function of the bank and that the government ought to go out of the banking business. I stand with Jefferson rather than with them, and tell them, as he did, that the issue of money is a function of the government and that the banks should go out of the governing business.

> If they dare to come out in the open field and defend the gold standard as a good thing, we shall fight them to the uttermost, having behind us the producing masses of the nation and the world. Having behind us the commercial interests and the laboring interests and all the toiling masses, we shall answer their demands for a gold standard by saying to them, you shall not

press down upon the brow of labor this crown of thorns. You shall not crucify mankind upon a cross of gold.

William Jennings Bryan, "Cross of Gold" speech, 1896

27. What is the main idea presented in the excerpt?
    a. Banks prefer the gold standard.
    b. Most Americans dislike the gold standard.
    c. Violence is justified when the government oppresses the masses.
    d. The government should set the monetary policy based on the will of the people.

Question 28 is based on the following graph:

History of Unemployment in the United States

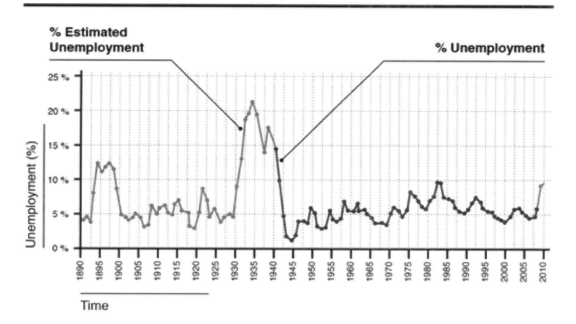

28. Which event caused the second largest increase in unemployment in American history?
    a. Panic of 1893
    b. Depression of 1920
    c. Depression of 1929
    d. Great Recession of 2007

Question 29 is based on the following diagram:

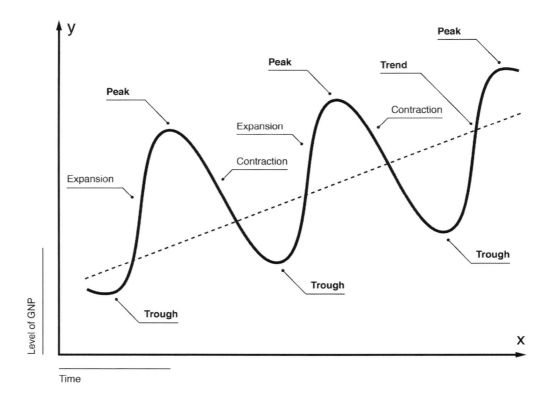

29. Which of the following phases of a business cycle occurs when there is continual growth?
    a. Expansion
    b. Peak
    c. Contraction
    d. Trough

30. Which of the following types of government intervention lowers prices, reassures the supply, and creates opportunity to compete with foreign vendors?
    a. Income redistribution
    b. Price controls
    c. Taxes
    d. Subsidies

31. What type of map would be the most useful for calculating data and differentiating between the characteristics of two places?
    a. Topographic maps
    b. Dot-density maps
    c. Isoline maps
    d. Flow-line maps

Question 32 is based on the following map:

Map of the United States by population density

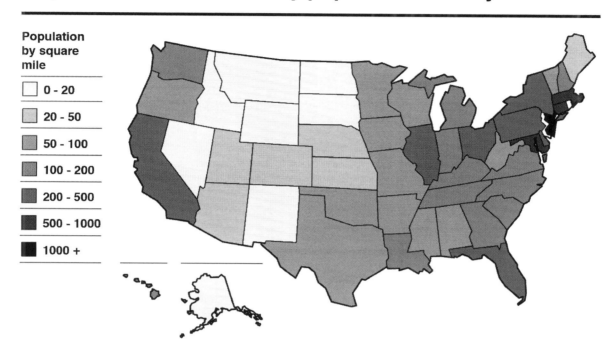

Population
by square
mile

☐ 0 - 20

20 - 50

50 - 100

100 - 200

200 - 500

500 - 1000

1000 +

32. According to the map, what area of the United States has the highest population density?
   a. Northwest
   b. Northeast
   c. Southwest
   d. Southeast

33. What accounts for different parts of the Earth experiencing different seasons at the same time?
   a. Differences in the rate of Earth's rotation
   b. Ocean currents
   c. Tilt of the Earth's rotational axis
   d. Elevation

34. Which of the following is NOT a reason why nonrenewable energy sources are used more often than renewables?
   a. Nonrenewable energy is currently cheaper.
   b. Infrastructure was built specifically for nonrenewable sources.
   c. Renewable energy is more difficult and expensive to store for long periods.
   d. Renewable energy cannot be converted into a power source.

Question 35 is based on the following map:

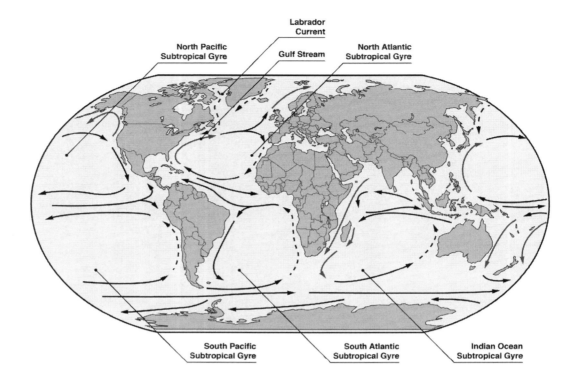

35. Which pair of ocean currents or gyres has the most impact on the weather of the United States?
    a. North Atlantic Subtropical Gyre and South Atlantic Subtropical Gyre
    b. Gulf Stream and North Pacific Subtropical Gyre
    c. North Pacific Subtropical Gyre and South Pacific Subtropical Gyre
    d. Alaskan Current and Indian Ocean Subtropical Gyre

36. Which of the following countries was a U.S. ally during World War II?
    a. The Soviet Union
    b. Italy
    c. Germany
    d. Japan

37. The North Atlantic Treaty Organization (NATO) was formed between which countries or regions?
    a. Canada, the U.S., and South America
    b. Western Europe, the U.S., and Canada
    c. The U.S., Western Europe, Canada, and the Soviet Union
    d. Asia, the U.S., and Western Europe

38. Which of these events was not a driving force for the passage of the Civil Rights Act in 1964?
    a. *Brown vs. the Board of Education*
    b. Freedom rides
    c. The G.I. Bill
    d. The Montgomery bus boycott

39. What program launched by the U.S. government under President Ronald Reagan was designed to shield the U.S. from nuclear attack by the Soviet Union?
    a. The Strategic Arms Limitation Talks (SALT I and II)
    b. The Strategic Defense Initiative (SDI)
    c. The Iran-Contra Affair
    d. *Glasnost*

40. After the terrorist attacks initiated by Islamic fundamentalist Osama bin Laden on September 11, 2001, President George W. Bush ordered bombing raids on various locations in what country in an attempt to bring down bin Laden and his al-Qaeda network?
    a. Afghanistan
    b. Iraq
    c. Kuwait
    d. Pakistan

41. What are the two largest rivers in the U.S. called?
    a. The Mississippi and the Colorado
    b. The Mississippi and the Missouri
    c. The Missouri and the Ohio
    d. The Mississippi and the Ohio

42. What is used to pinpoint location on a map?
    a. Scale and longitude
    b. Contour lines and scale
    c. Latitude and longitude
    d. Latitude and contour lines

43. Under America's democratic form of government, which of the following are citizens not obligated to do?
    a. Obey the law
    b. Pay taxes
    c. Serve on a jury if asked to do so
    d. Vote in elections

44. How many times has the U.S. Constitution been amended in order to accommodate changes and updates?
    a. Fourteen
    b. Eighteen
    c. Twenty-one
    d. Twenty-seven

45. What is interaction of consumers, households, and companies within individual markets and the relationships between them called?
    a. Macroeconomics
    b. Microeconomics
    c. Boom and bust
    d. Economic output

46. What are the types of productive resources used to create products, such as machinery, tools, buildings, and equipment called?
    a. Natural resources
    b. Human resources
    c. Capital resources
    d. Entrepreneurship

47. What is the business sector of the economy that provides consumer or business services, including industries such as entertainment, travel and tourism, and banking called?
    a. Primary
    b. Secondary
    c. Tertiary
    d. Quaternary

48. Consumers must make choices regarding the goods and services to buy with their limited income. By purchasing one good or service, they are giving up the chance to purchase another. This is referred to as which of the following?
    a. The circular flow model
    b. Opportunity cost
    c. Savings account
    d. Assets

Answer Explanations

1. B: Choice *B* is correct, as power is the ability of a ruling body to influence the actions, behavior, and attitude of a person or group of people. Choice *A* is incorrect, as politics is the process of governance typically exercised through the enactment and enforcement of laws over a community, most commonly a state. Although closely related to power, Choice *C* is incorrect, because authority refers to a political entity's justification to exercise power. Legitimacy is synonymous with authority, so Choice *D* is also incorrect.

2. C: Choice *C* is correct. There are no definitive requirements to be a nation. Rather, the nation only needs a group bound by some shared characteristic. Examples include language, culture, religion, homeland, ethnicity, and history. Choice *C* isn't a requirement to be a nation, though it is required to be a state.

3. D: Choice *D* is correct. Machiavelli was an Italian diplomat, politician, and historian, and *The Prince* is his best-known political treatise. The excerpt instructs the Prince that if he injures a man, then he must ensure the injury is "of such a kind that one does not stand in fear of revenge." Choices *B* and *C* contradict the first sentence of the excerpt, which says that men "ought either to be well treated or crushed." Choice *A* is close, but the selection goes too far, assuming revenge will result in overthrowing the Prince.

4. A: Choice *A* is correct. John Stuart Mill was an English philosopher and political economist who advocated for utilitarianism and women's rights. In the excerpt, "utility" is defined as actions that are "right in proportion as they tend to promote happiness, wrong as they tend to produce the reverse of happiness." The excerpt then explains that happiness is measured by pleasure, and the reverse is pain. Therefore, Mill calls for actions to be evaluated based on the net total of pleasure. Choice *D* contradicts the definition provided in the excerpt. The excerpt doesn't support Choice *C*, as there's no evidence that pleasure-generating sacrifices merit special status. Choice *B* is incorrect because sacrifice can still be valuable if it leads to more pleasure than pain.

5. C: Choice *C* is correct. Karl Marx, a philosopher, social scientist, historian, and revolutionary, is considered the father of communism. All the answer choices contain true statements or reasonable assumptions from the passage; however, Choice *C* best articulates the main idea—society is the history of class struggle, and working men must unite and fight a revolutionary battle like their historical ancestors.

6. A: Choice *A* is correct. On the political spectrum, ideologies on the left side of the axis emphasize socioeconomic equality and advocate for government intervention, while ideologies on the right axis seek to preserve society's existing institutions and oppose government intervention. Therefore, the answer will be the farthest left on the axis, making Choice *A* correct.

7. C: Choice *C* is correct, as it most closely corresponds to the provided definition. Conservatism prioritizes traditional institutions. In general, conservatives oppose modern developments and value stability. Socialism and liberalism both feature the desire to change the government to increase equality. Libertarianism is more concerned with establishing a limited government to maximize personal autonomy.

8. B: Choice *B* is correct. The Articles of Confederation were the first form of government adopted in the American colonies. Under the Articles of Confederation, the central government (the Continental Congress) was granted very limited powers, rendering it largely ineffective. Although the choices describe what would appear to be basic functions of government, the central government could only declare war.

9. D: Choice *D* is correct. The missing title is in the overlap between federal and state government powers. Concurrent powers are shared between federal and state governments. Reserved powers are the unspecified powers of the states not expressly granted to the federal government or denied to the state by the Constitution, and left to the states by the Tenth Amendment. Implied powers are the unstated powers that can be reasonably inferred from the Constitution. Delegated powers are the specific powers granted to the federal government by the Constitution.

10. D: Choice *D* is correct. The Federalists supported the expansion of the federal government, and the anti-Federalists feared that a stronger central government would weaken the states. *The Federalist Papers* argued for the ratification of the Constitution to establish a more powerful central government. The main idea of this excerpt is to argue that the Constitution establishes a central government powerful enough to rule, while also providing checks and balances to ensure the government doesn't abuse its power. Separation of powers is the concept behind checks and balances, so Choice *D* is the correct answer. Choices *A* and *C* are true statements, but they don't identify the main idea. Choice *B* references a theoretical assertion from the excerpt, but it's not the main idea.

11. C: Choice *C* is the correct answer. Checks and balances refer to the powers granted to ensure other branches don't overstep their authority. The other arrows in the diagram identify checks and balances, so the correct answer is the executive branch's checks and balances on the legislative branch. The executive branch can call special sessions of Congress and veto legislation, so Choice *C* is correct. Unlike the judicial and executive branches, members of the legislative branch cannot be impeached by another branch, though the legislative branch can expel its own members. The executive branch cannot refuse to enforce laws.

12. B: Choice *B* is correct. The Second Amendment states, "A well regulated Militia, being necessary to the security of a free State, the right of the people to keep and bear Arms, shall not be infringed." The First Amendment provides freedom of religion, speech, and the press, the right to assemble, and the right to petition the government. The Third Amendment establishes the right to refuse to house soldiers in times of war. The Fourth Amendment establishes a series of protections for citizens accused and charged with crimes.

13. B: Choice *B* is correct. The Electoral College determines the winner of presidential races, but if a candidate doesn't win a majority of electoral votes, the Twelfth Amendment requires the House of Representatives to decide the presidency, with each state delegation voting as a single bloc. The candidate with the most votes in the House wins the election. The table shows that Andrew Jackson won a plurality of electoral and popular votes, but he didn't receive a majority. John Quincy Adams received the most votes in the House of Representatives, so he won the presidency.

14. A: Choice *A* is correct. Electoral systems dictate how the members of the ruling body are selected, how votes translate into positions, and how seats are filled in the political offices at each level of government. In a majority system, a candidate must receive a majority of votes in order to be awarded a seat, but if none of the candidates reach a majority, a second round of voting occurs, commonly referred to as a runoff.

15. C: Choice *C* is correct. The two major theories of international relations are Realism and Liberalism. Realism analyzes international relations through the interactions of states under the assumption that states act rationally to maintain or expand power a means of self-preservation, which inevitably leads to conflict in an anarchical system. The question asks for the choice that doesn't adhere to Realism, and the other choices state three of the four basic tenets of Realism. In contrast, Choice *C* states a principle of Liberalism. Realists don't value international organizations or prioritize global cooperation.

16. B: Choice *B* is correct. Following the French and Indian War, the British government amassed an enormous war debt, and Great Britain imposed taxes on the colonists to generate more revenue. King George III argued that British resources defended the colonists from French and Native American forces, so the colonists should share in the expenses. The other choices are factually incorrect. The Royal Proclamation of 1763 prevented the colonies from expanding west of the Allegheny Mountains. No lasting peace ever occurred between the colonists and Native Americans. Self-government decreased in the colonies after the French and Indian War.

17. A: Choice *A* is correct. Heavily influenced by the Enlightenment, the Declaration of Independence repudiated the colonies' allegiance to Great Britain. The main purpose of the excerpt is to justify the colonists' revolutionary ambitions due to Great Britain's tyranny and the role of consent in government to protect the natural rights of citizens. Although the excerpt alludes to abuses, the purpose isn't to list specific evidence. This occurs later in the Declaration of Independence. Choices *C* and *D* are supporting evidence for the main purpose.

18. D: Choice *D* is correct. The question asks for a territory that the United States didn't gain from war. Any territory gained via purchase is incorrect. Missouri and Nebraska became American territories through the Louisiana Purchase, and the United States purchased Alaska from Russia. In contrast, Mexico ceded Nevada as part of the peace agreement ending the Mexican-American War.

19. D: Choice *D* is correct. President Lincoln issued the Emancipation Proclamation to free the slaves in the Confederacy, allowing the institution to continue in states and territories that didn't secede. The excerpt justifies the decision as a "fit and necessary war measure for suppressing said rebellion." Therefore, per the excerpt, emancipation was necessary to strengthen the war effort for the North. Choice *C* is the second-best answer, but the excerpt supports the contention that emancipation was part of an active war effort, rather than merely a punishment. Nothing in the excerpt describes the evil of slavery or the effect of emancipation on morale in the North.

20. B: Choice *B* is correct. Industrialization directly caused an increase in urbanization. Factories were located near cities to draw upon a large pool of potential employees. Between 1860 and 1890, the urbanization rate increased from about 20 percent to 35 percent. The other three choices are factually incorrect. Immigration increased during industrialization, as immigrants flooded into America to search for work. Socioeconomic problems plagued the period due to the unequal distribution of wealth and the social ills caused by rapid urbanization. Labor unrest was common as unions advocated for workers' rights and organized national strikes.

21. A: Choice *A* is correct. The Treaty of Versailles contained a clause that required Germany to assume responsibility for damages incurred during the conflict. Thus, the Treaty ordered Germany to pay $31.4 billion, the equivalent of $442 billion in 2017. World War I ravaged the German economy, and the country couldn't afford the war debt. The resulting poverty contributed to the rise of the Nazi Party, leading to World War II.

22. B: Choice *B* is correct. President Franklin D. Roosevelt introduced the New Deal, a series of executive orders and laws passed by Congress in response to the Great Depression. The excerpt describes how President Roosevelt intended to fight poverty by using the government's power to intervene and regulate the economy. Although the other answer choices correctly identify specific activities referenced in the excerpt, they are examples of the underlying philosophy in action. The underlying philosophy is an active role for government in the nation's economic affairs.

23. C: Choice *C* is correct. Frederick Douglass escaped from slavery and worked as an abolitionist for the rest of his life. The excerpt references the hypocrisy of the Fourth of July, as the holiday celebrates freedom in a country with millions of slaves. The other answer choices identify hypocritical aspects surrounding the slavery debate, but Choice *C* directly states the specific hypocrisy attacked in the excerpt.

24. D: Choice *D* is correct. Along with Lucy Stone and Elizabeth Cady Stanton, Susan B. Anthony was one of the most outspoken advocates for women's suffrage. Women couldn't vote in the United States until Congress passed the Nineteenth Amendment in 1920. Choice *D* accurately expresses the main idea of the excerpt. Denying women the right to vote is tyranny, so Anthony will not pay a fine for voting illegally. Choice *A* is the second-best answer, but it's too general to be the main idea of an excerpt specifically about women's suffrage.

25. B: Choice *B* is correct. Ronald Reagan won the presidential election of 1980 and promised to restore America's military power through defense spending, cutting government regulations, and reducing taxes. Evangelical Christians and the Moral Majority fiercely supported President Reagan's agenda, particularly his opposition to abortion and his conservative approach to social issues. The other answer choices include at least one mischaracterization. Choice *A* is incorrect because President Reagan generally opposed social programs. Choice *C* is incorrect because President Reagan valued American leadership more than international cooperation. In addition, his platform was far more radically conservative than compromising. Choice *D* is incorrect because President Reagan fought labor unions on several fronts, most notably when he broke a strike organized by an air traffic controllers' union.

26. C: Choice *C* is correct. In a market economy, privately owned businesses, groups, or individuals price goods or services and set prices based on demand. In contrast, the government or central authority determines prices and quantity of production in a command economy. Of the two economic systems, the market economy more efficiently uses resources. Choice *C* is the only accurate statement.

27. D: Choice *D* is correct. William Jennings Bryan's "Cross of Gold" is one of the most famous speeches in American history, launching his candidacy in the 1896 presidential election. The speech advocates for abolishing the gold standard and adopting a bimetallic system to provide more government control over monetary policy. The excerpt condemns the influence of banks in monetary policy, and without some reform, the masses should act to remove the gold standard. Although the other answer choices accurately state assertions from the excerpt, they aren't the main idea.

28. C: Choice *A* is correct. The Depression of 1929, commonly referred to as the Great Depression, is the largest increase to unemployment, but the question stem asks for the second-largest increase. According to the graph, the Panic of 1893 increased unemployment by approximately ten percent; the Depression of 1920 increased unemployment by approximately six percent; the Depression of 1929 increased unemployment by approximately fifteen percent; and the Great Recession of 2007 increased unemployment by approximately four percent. Thus, the Panic of 1893 marks the second-largest increase to unemployment.

29. A: Choice *A* is correct. A business cycle is when the gross domestic product (GDP) moves downward and upward over a long-term growth trend, and the four phases are expansion, peak, contraction, and trough. An expansion is the only phase where employment rates and economic growth continually grow. Contraction is the opposite of expansion. The peak and trough are the extreme points on the graph.

30. D: Choice *D* is correct. The government can intervene in the economy by imposing taxes, subsidies, and price controls to increase revenue, lower prices of goods and services, ensure product availability for the government, and maintain fair prices for goods and services. Subsidies lower prices, reassure the supply, and create opportunity to compete with foreign vendors, so Choice *D* is correct.

31. C: Choice *C* is correct. Isoline maps are used to calculate data and differentiate between the characteristics of two places. In an isoline map, symbols represent values, and lines can be drawn between two points to determine differences. The other answer choices are maps with different purposes. Topographic maps display contour lines, which represent the relative elevation of a particular place. Dot-density maps and flow-line maps are types of thematic maps. Dot-density maps illustrate the volume and density of a characteristic of an area. Flow-line maps use lines to illustrate the movement of goods, people, or even animals between two places.

32. B: Choice *B* is correct. The map is a density map illustrating population density by state in the United States. Accordingly, the darker areas have higher population density. The darkest area of the map is the Northeast, so Choice *B* is correct.

33. C: Choice *C* is correct. The tilt of the Earth's rotation causes the seasons due to the difference in direct exposure to the Sun. For example, the northern hemisphere is tilted directly toward the Sun from June 22 to September 23, which creates the summer in that part of the world. Conversely, the southern hemisphere is tilted away from the Sun and experiences winter during those months. Choice *A* is factually incorrect—the rate of Earth's rotation is constant. Choice *B* and *D* are factors in determining climate, but differences in climate don't cause the seasons.

34. D: Choice *D* is correct. Nonrenewable energy resources are oil, natural gas, and coal, collectively referred to as fossil fuels. Nonrenewable energy is more widely used due to its abundance and relatively cheap price. In addition, countries have tailored their existing infrastructure to nonrenewable energy. Currently, the technology to store renewable energies for long periods is either nonexistent or expensive. Choice *D* is correct because it's inaccurate. Renewable energy can be converted into a power source, but the issue is scale of use. For example, the United States converts renewable resources to derive ten percent of the country's energy.

35. B: Choice *B* is correct. Ocean currents dramatically impact the climate by storing heat from the Sun and transporting the warmth around the globe. The evaporation of ocean water increases the temperature and humidity in the nearby landmasses. A gyre is a system of circulating currents. Countries are most impacted by the currents and gyres closest to their shores. The question stem asks what currents have the most impact on the United States. According to the map, the North Atlantic Gyre, North Pacific Gyre, California Current, Alaskan Current and Gulf Stream impact the United States. Choice *B* is the only answer with a pair of those currents or gyres.

36. A: The Soviet Union was invaded by Germany in 1941 and allied with Britain and subsequently the U.S. President Roosevelt, British Prime Minister Winston Churchill, and Soviet director Joseph Stalin met in 1945 to plan their final assault on Germany and discuss postwar strategies. Germany aligned with Italy and Japan in 1940 to form the Axis Alliance. Their goal was to establish a German empire in Europe and place Japan in control over Asia. Thus, Choices *B*, *C*, and *D* are incorrect.

37. B: The North Atlantic Treaty Organization (NATO) was formed between Western Europe, Canada, and the U.S. in defense of Soviet hostility after the Soviet Union introduced Communism into Eastern Europe. The Soviet Union countered by creating the Warsaw Pact.

38. C: The G.I. Bill was a government program started in the 1950s that gave military veterans a free education. In the revolutionary 1954 case, *Brown vs. the Board of Education,* the Supreme Court ruled that school segregation was illegal, thereby setting the Civil Rights Movement in motion, making Choice *A* incorrect. *Freedom Rides,* Choice *B,* and the Montgomery bus boycott, Choice *D,* were among the non-violent protests against segregation that took place in the U.S. in the 1960s.

39. B: President Reagan advocated *peace through strength*, building up the U.S. military and launching the Strategic Defense Initiative (SDI), also called *Star Wars*. Choice *A,* the Strategic Arms Limitation Talks (SALT I and II), negotiated between 1972 and 1979, resulted in limits on nuclear weapons for both the U.S. and Russia. Choice *C,* the Iran-Contra Affair, was a scandal involving the secret sale of weapons to Iran in exchange for American hostages. Choice *D, Glasnost,* was a policy of political openness launched by Soviet leader Mikhail Gorbachev.

40. A: Afghanistan was the site of the bombing raids. Bush invaded Iraq, Choice *B,* in 2003 when Iraqi dictator Saddam Hussein defied the terms of the truce agreed upon in 1991 after the Gulf War. Kuwait, Choice *C,* was invaded by Iraq in 1990, sparking the Gulf War. Pakistan, Choice *D,* is where Osama bin Laden was killed by a group of Navy SEALs under orders from President Obama.

41. B: The Mississippi and the Missouri are the two largest rivers in the U.S., winding through the Great Plains in the center of the country. The Colorado and Ohio Rivers are about half the length of the Mississippi and Missouri.

42. C: Latitude – imaginary lines covering the globe from east to west – and longitude – imaginary lines running north to south – are used to pinpoint location on a map. Scale is used to show the relationship between the map measurements and the equivalent distance on the world's surface. Contour lines are used to show detailed elevation on a map.

43. D: Under America's democratic form of government, voting is a *right*, but it is not an *obligation*. U.S. citizens are *obliged* to obey the law, pay taxes, and serve on a jury if asked to do so, making Choices *A, B,* and *C* incorrect.

44. D: There are twenty-seven amendments to the U.S. Constitution. The 14th Amendment was adopted in 1868 to abolish slavery. The 18th Amendment was passed in 1919 and prohibited the production and sale of alcoholic beverages, but the 21st Amendment repealed it in 1933.

45. **B:** Microeconomics looks at the interplay of consumers, households, and companies within individual markets and the relationships between them. Macroeconomics, Choice *A,* is the study of entire economies. Booms and busts, Choice *C,* are terms used to describe the cyclical nature of economic activity, typically prompted by extreme changes in the economy. Economic output, Choice *D,* is the total amount of goods and services produced by an ***economy.***

46. C: Capital resources are the man-made physical resources used to create products, such as machinery, tools, buildings, and equipment. Natural resources, Choice *A*, are raw materials taken from the land, such as corn, beef, lumber, water, oil, and iron. Human resources, Choice *B*, refer to the human labor—both mental and physical—required to produce goods. Entrepreneurship, Choice *D*, is the capability and motivation to cultivate, organize, and oversee the other three resources into a business venture.

47. C: The tertiary sector provides consumer or business services, including industries such as entertainment, retail sales, and restaurants. The primary sector, Choice *A*, takes raw materials from the Earth, such as coal, timber, copper, and wheat. The secondary sector, Choice *B*, converts raw materials into goods, such as textile manufacturing, food processing, and car manufacturing. The quaternary sector, Choice *D*, provides informational and knowledge services, such as education, business consulting, and financial services.

48. B: Opportunity cost is the term used to describe the choices that determine how consumers spend or save their money. Choice *A*, the circular flow model, is used by economists to describe the movement of supply, demand, and payment between businesses and consumers. A savings account, Choice *C*, is considered low-risk because the bank will pay the saver a low interest rate to keep it safe. Assets, Choice *D*, are valuable items purchased by investors in the hopes that they will increase in worth over time and yield returns or profits.

Science

Earth and Space Sciences

Earth's Place in the Universe

Structure of the Solar System

The *solar system* is an elliptical planetary system with a large sun in the center that provides gravitational pull on the planets.

Laws of Motion

Planetary motion is governed by three scientific laws called Kepler's laws:

1. The orbit of a planet is elliptical in shape, with the Sun as one focus.

2. An imaginary line joining the center of a planet and the center of the Sun sweeps out equal areas during equal intervals of time.

3. For all planets, the ratio of the square of the orbital period is the same as the cube of the average distance from the Sun.

The most relevant of these laws is the first. Planets move in elliptical paths because of gravity; when a planet is closer to the Sun, it moves faster because it has built up gravitational speed. As illustrated in the diagram below, the second law states that it takes planet 1 the same time to travel along the A1 segment as the A2 segment, even though the A2 segment is shorter.

Kepler's Laws of Planetary Motion

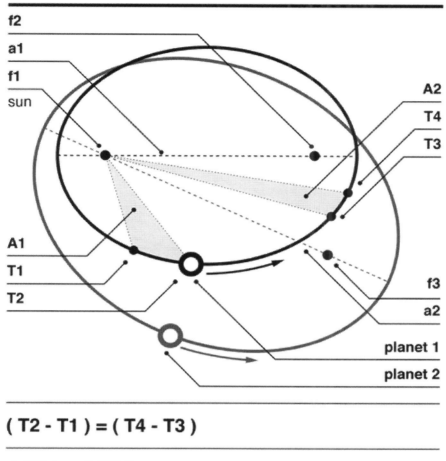

$$(T2 - T1) = (T4 - T3)$$

$$\Delta\ TA1 = \Delta\ TA2$$

Characteristics of the Sun, Moon, and Planets

The Sun is comprised mainly of hydrogen and helium. Metals make up only about 2% of its total mass. The Sun is 1.3 million kilometers wide, weighs 1.989×10^{30} kilograms, and has temperatures of 5,800 Kelvin (9980 °F) on the surface and 15,600,000 Kelvin (28 million °F) at the core. The Sun's enormous size and gravity give it the ability to provide sunlight. The gravity of the Sun compresses hydrogen and helium atoms together through nuclear fusion and releases energy and light.

The Moon has a distinct core, mantle, and crust. It has elevations and craters created by impacts with large objects in the solar system. The Moon makes a complete orbit around the Earth every 27.3 days. It's relatively large compared to other moons in the Solar System, with a diameter one-quarter of the Earth and a mass 1/81 of the Earth.

The eight planets of the Solar System are divided into four inner (or terrestrial) planets and four outer (or Jovian) planets. In general, terrestrial planets are small, and Jovian planets are large and gaseous. The planets in the Solar System are listed below from nearest to farthest from the Sun:

- Mercury: the smallest planet in the Solar System; it only takes about 88 days to completely orbit the Sun

- Venus: around the same size, composition, and gravity as Earth and orbits the Sun every 225 days

- Earth: the only known planet with life

- Mars: called the Red Planet due to iron oxide on the surface; takes around 687 days to complete its orbit

- Jupiter: the largest planet in the system; made up of mainly hydrogen and helium

- Saturn: mainly composed of hydrogen and helium along with other trace elements; has 61 moons; has beautiful rings, which may be remnants of destroyed moons

- Uranus: the coldest planet in the system, with temperatures as low as -224.2 °C (-371.56 °F)

- Neptune: the last and third-largest planet; also, the second-coldest planet

Asteroids, Meteoroids, Comets, and Dwarf/Minor Planets

Several other bodies travel through the universe. *Asteroids* are orbiting bodies composed of minerals and rock. They're also known as *minor planets*—a term given to any astronomical object in orbit around the Sun that doesn't resemble a planet or a comet. *Meteoroids* are mini-asteroids with no specific orbiting pattern. *Meteors* are meteoroids that have entered the Earth's atmosphere and started melting from contact with greenhouse gases. *Meteorites* are meteors that have landed on Earth. *Comets* are composed of dust and ice and look like a comma with a tail from the melting ice as they streak across the sky.

Theories of Origin of the Solar System

One theory of the origins of the Solar System is the *nebular hypothesis*, which posits that the Solar System was formed by clouds of extremely hot gas called a *nebula*. As the nebula gases cooled, they became smaller and started rotating. Rings of the nebula left behind during rotation eventually condensed into planets and their satellites. The remaining nebula formed the Sun.

Another theory of the Solar System's development is the *planetesimal hypothesis*. This theory proposes that planets formed from cosmic dust grains that collided and stuck together to form larger and larger bodies. The larger bodies attracted each other, growing into moon-sized protoplanets and eventually planets.

The Earth's Rotation and Orbital Revolution Around the Sun

Besides revolving around the Sun, the Earth also spins like a top. It takes one day for the Earth to complete a full spin, or rotation. The same is true for other planets, except that their "days" may be shorter or longer. One Earth day is about 24 hours, while one Jupiter day is only about nine Earth hours, and a Venus day is about 241 Earth days. Night occurs in areas that face away from the Sun, so one side of the planet experiences daylight and the other experiences night. This phenomenon is the reason that the Earth is divided into time zones. The concept of time zones was created to provide people around the world with a uniform standard time, so the Sun would rise around 7:00 AM, regardless of location.

The Earth's tilted axis creates the seasons. When Earth is tilted toward the Sun, the Northern Hemisphere experiences summer while the Southern Hemisphere has winter—and vice versa. As the Earth rotates, the distribution of direct sunlight slowly changes, explaining how the seasons gradually change.

Phases of the Moon

The Moon goes through two phases as it revolves around Earth: waxing and waning. Each phase lasts about two weeks:

- Waxing: the right side of the Moon is illuminated
- New moon (dark): the Moon rises and sets with the Sun
- Crescent: a tiny sliver of illumination on the right
- First quarter: the right half of the Moon is illuminated
- Gibbous: more than half of the Moon is illuminated
- Full moon: the Moon rises at sunset and sets at sunrise
- Waning: the left side of the Moon is illuminated
- Gibbous: more than half is illuminated, only here it is the left side that is illuminated
- Last quarter: the left half of the Moon is illuminated
- Crescent: a tiny sliver of illumination on the left
- New moon (dark): the Moon rises and sets with the Sun

Although the Earth is much larger, the Moon still has a significant gravitational force that pulls on Earth's oceans. At its closest to Earth, the Moon's gravitation pull is greatest and creates high tide. The opposite is true when the Moon is farthest from the Earth: less pull creates low tide.

Solar and Lunar Eclipses

Eclipses occur when the Earth, the Sun, and the Moon are all in line. If the three bodies are perfectly aligned, a total eclipse occurs; otherwise, it's only a partial eclipse. A *solar eclipse* occurs when the Moon is between the Earth and the Sun, blocking sunlight from reaching the Earth. A *lunar eclipse* occurs when the Earth interferes with the Sun's light reflecting off the full moon. The Earth casts a shadow on the

Moon, but the particles of the Earth's atmosphere refract the light, so some light reaches the Moon, causing it to look yellow, brown, or red.

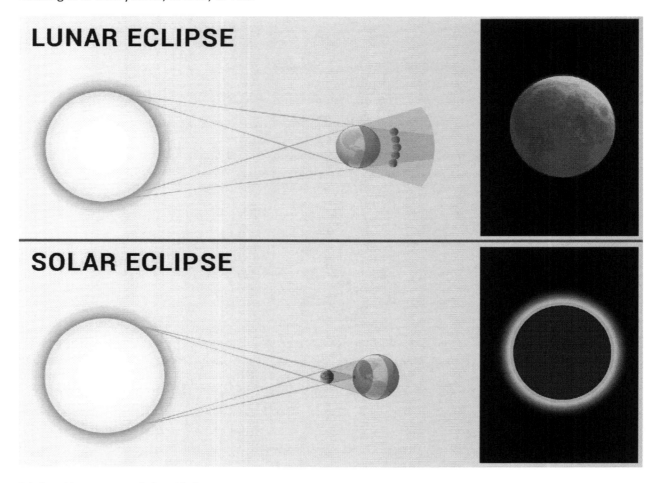

Major Features of the Universe

Galaxies

Galaxies are clusters of stars, rocks, ice, and space dust. Like everything else in space, the exact number of galaxies is unknown, but there could be as many as a hundred billion. There are three types of galaxies: spiral, elliptical, and irregular. Most galaxies are *spiral galaxies*; they have a large, central galactic bulge made up of a cluster of older stars. They look like a disk with spinning arms. *Elliptical*

galaxies are groups of stars with no pattern of rotation. They can be spherical or extremely elongated, and they don't have arms. *Irregular galaxies* vary significantly in size and shape.

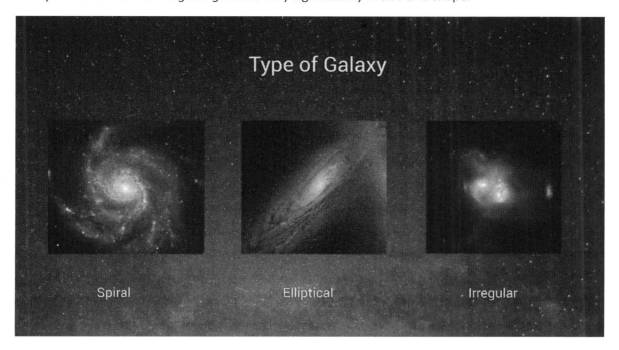

To say that galaxies are large is an understatement. Most galaxies are 1,000 to 100,000 parsecs in diameter, with one *parsec* equal to about 19 trillion miles. The Milky Way is the galaxy that contains Earth's Solar System. It's one of the smaller galaxies that has been studied. The diameter of the Milky Way is estimated to be between 31,000 to 55,000 parsecs.

Life Cycle of Stars

All stars are formed from nebulae. Depending on their mass, stars take different pathways during their evolution. Low- and medium-mass stars start as nebulae and then become red giants and white dwarfs. High-mass stars become red supergiants, supernovas, and then either neutron stars or black holes. Official stars are born as red dwarfs because they have plentiful amounts of gas—mainly hydrogen—to undergo nuclear fusion. Red dwarfs mature into white dwarfs before expending their hydrogen fuel source. When the fuel is spent, it creates a burst of energy that expands the star into a red giant. Red giants eventually condense to form white dwarfs, which is the final stage of a star's life.

Stars that undergo nuclear fusion and energy expenditure extremely quickly can burst in violent explosions called *supernovas*. These bursts can release as much energy in a few seconds as the Sun can release in its entire lifetime. The particles from the explosion then condense into the smallest type of star—a neutron star—and eventually form a *black hole*, which has such a high amount of gravity that not even light energy can escape. The Sun is currently a red dwarf, early in its life cycle.

Color, Temperature, Apparent Brightness, Absolute Brightness, and Luminosity

The color of a star depends on its surface temperature. Stars with cooler surfaces emit red light, while the hottest stars give off blue light. Stars with temperatures between these extremes, such as the Sun, emit white light. The *apparent brightness* of a star is a measure of how bright a star appears to an observer on the Earth. The *absolute brightness* is a measure of the intrinsic brightness of a star and is

measured at a distance of exactly 10 parsecs away. The *luminosity* of a star is the amount of light emitted from its surface.

Hertzsprung-Russell Diagrams

Hertzsprung-Russell diagrams are scatterplots that show the relationship of a star's brightness and temperature, or color. The general layout shows stars of greater luminosity toward the top of the diagram. Stars with higher surface temperatures appear toward the left side of the diagram. The diagonal area from the top-left of the diagram to the bottom-right is called the *main sequence*. Stars may or may not follow the main sequence during their evolutionary period.

Earth's Systems

Earth's Layers

Earth has three major layers: a thin solid outer surface or *crust*, a dense *core,* and a *mantle* between them that contains most of the Earth's matter. This layout resembles an egg, where the eggshell is the crust, the mantle is the egg white, and the core is the yolk. The outer crust of the Earth consists of igneous or sedimentary rocks over metamorphic rocks. Together with the upper portion of the mantle, it forms the *lithosphere*, which is broken into tectonic plates.

Major plates of the lithosphere

![Map showing the major tectonic plates: North American Plate, Euroasian Plate, Pacific Plate, African Plate, South American Plate, Nazca Plate, Indo-Australian Plate, and Antarctic Plate.]

The mantle can be divided into three zones. The *upper mantle* is adjacent to the crust and composed of solid rock. Below the upper mantle is the *transition zone*. The *lower mantle* below the transition zone is a layer of completely solid rock. Underneath the mantle is the molten *outer core* followed by the

compact, solid *inner core*. The inner and outer cores contain the densest elements, consisting of mostly iron and nickel.

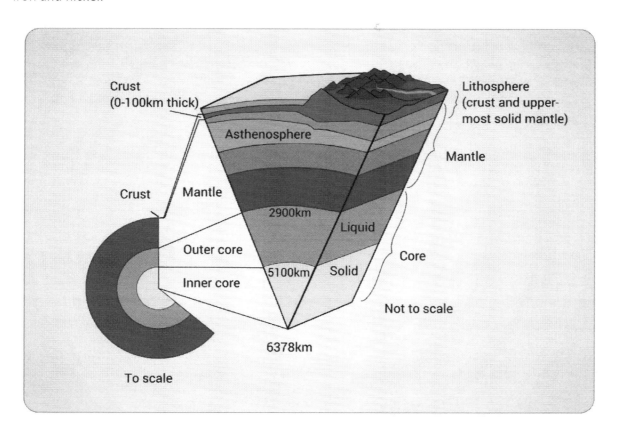

Shape and Size of the Earth

The Earth isn't a perfect sphere; it's slightly elliptical. From center to surface, its radius is almost 4,000 miles, and its circumference around the equator is about 24,902 miles. In comparison, the Sun's radius is 432,288 miles—over 1,000 times larger than the Earth's—and the Moon's radius is about 1,000 miles.

Geographical Features

The Earth's surface is dynamic and consists of various landforms. As tectonic plates are pushed together, *mountains* are formed. *Canyons* are deep trenches that are usually created by plates moving apart, but can also be created by constant weathering and erosion from rivers and runoff. *Deltas* are flat, triangular stretches of land formed by rivers that deposit sediment and water into the ocean. *Sand dunes* are mountains of sand located in desert areas or the bottom of the ocean. They are formed by wind and water movement when there's an absence of plants or other features that would otherwise hold the sand in place.

The Earth's Magnetic Field

The Earth's magnetic field is created by the magnetic forces that extend from the Earth's interior to outer space. It can be modeled as a magnetic dipole tilted about 10 degrees from the Earth's rotational axis, as if a bar magnet was placed at an angle inside the Earth's core. The geomagnetic pole located near Greenland in the northern hemisphere is actually the south pole of the Earth's magnetic field, and vice versa for the southern geomagnetic pole. The *magnetosphere* is the Earth's magnetic field, which extends tens of thousands of kilometers into space and protects the Earth and the atmosphere from damaging solar wind and cosmic rays.

Plate Tectonics Theory and Evidence

The theory of *plate tectonics* hypothesizes that the continents weren't always separated like they are today, but were once joined and slowly drifted apart. Evidence for this theory is based upon evolution and the fossil record. Fossils of one species were found in regions of the world now separated by an ocean. It's unlikely that a single species could have travelled across the ocean or that two separate species evolved into a single species.

Folding and Faulting

The exact number of tectonic plates is debatable, but scientists estimate there are around nine to fifteen major plates and almost 40 minor plates. The line where two plates meet is called a *fault*. The San Andreas Fault is where the Pacific and North American plates meet. Faults or boundaries are classified depending on the interaction between plates. Two plates collide at *convergent boundaries*. *Divergent boundaries* occur when two plates move away from each other. Tectonic plates can move vertically and horizontally.

Continental Drift, Seafloor Spreading, Magnetic Reversals

The movement of tectonic plates is similar to pieces of wood floating in a pool of water. They can bob up and down as well as bump, slide, and move away from each other. These different interactions create the Earth's landscape. The collision of plates can create mountain ranges, while their separation can create canyons or underwater chasms. One plate can also slide atop another and push it down into the Earth's hot mantle, creating magma and volcanoes, in a process called *subduction*.

Unlike a regular magnet, the Earth's magnetic field changes over time because it's generated by the motion of molten iron alloys in the outer core. Although the magnetic poles can wander geographically, they do so at such a slow rate that they don't affect the use of compasses in navigation. However, at irregular intervals that are several hundred thousand years long, the fields can reverse, with the north and south magnetic poles switching places.

Characteristics of Volcanoes

Volcanoes are mountainous structures that act as vents to release pressure and magma from the Earth's crust. During an *eruption*, the pressure and magma are released, and volcanoes smoke, rumble, and throw ash and *lava*, or molten rock, into the air. *Hot spots* are volcanic regions of the mantle that are hotter than surrounding regions.

Characteristics of Earthquakes

Earthquakes occur when tectonic plates slide or collide as a result of the crust suddenly releasing energy. Stress in the Earth's outer layer pushes together two faults. The motion of the planes of the fault continues until something makes them stop. The *epicenter* of an earthquake is the point on the surface directly above where the fault is slipping. If the epicenter is located under a body of water, the earthquake may cause a *tsunami*, a series of large, forceful waves.

Seismic waves and Triangulation

Earthquakes cause *seismic waves*, which travel through the Earth's layers and give out low-frequency acoustic energy. Triangulation of seismic waves helps scientists determine the origin of an earthquake.

The Water Cycle

Evaporation and Condensation

The *water cycle* is the cycling of water between its three physical states: solid, liquid, and gas. The Sun's thermal energy heats surface water so it evaporates. As water vapor collects in the atmosphere from evaporation, it eventually reaches a saturation level where it condenses and forms clouds heavy with water droplets.

Precipitation

When the droplets condense as clouds get heavy, they fall as different forms of precipitation, such as rain, snow, hail, fog, and sleet. *Advection* is the process of evaporated water moving from the ocean and falling over land as precipitation.

Runoff and Infiltration

Runoff and *infiltration* are important parts of the water cycle because they provide water on the surface available for evaporation. Runoff can add water to oceans and aid in the advection process. Infiltration provides water to plants and aids in the transpiration process.

Transpiration

Transpiration is an evaporation-like process that occurs in plants and soil. Water from the stomata of plants and from pores in soil evaporates into water vapor and enters the atmosphere.

Basic Structure and Composition of the Earth's Atmosphere

Layers

The Earth's atmospheric layers are determined by their temperatures but are reported by their distance above sea level. Listed from closest to sea level on upward, the levels are:

- Troposphere: sea level to 11 miles above sea level
- Stratosphere: 11 miles to 31 miles above sea level
- Mesosphere: 31 miles to 50 miles above sea level
- Ionosphere: 50 miles to 400 miles above sea level
- Exosphere: 400 miles to 800 miles above sea level

The ionosphere and exosphere are together considered the thermosphere. The ozone layer is in the stratosphere and weather experienced on Earth's surface is a product of factors in the troposphere.

Composition of the Atmosphere

The Earth's atmosphere is composed of gas particles: 78% nitrogen, 21% oxygen, 1% other gases such as argon, and 0.039% carbon dioxide. The atmospheric layers are created by the number of particles in the air and gravity's pull upon them.

Atmospheric Pressure and Temperature

The lower atmospheric levels have higher atmospheric pressures due to the mass of the gas particles located above. The air is less dense (it contains fewer particles per given volume) at higher altitudes. The temperature changes from the bottom to top of each atmospheric layer. The tops of the troposphere and mesosphere are colder than their bottoms, but the reverse is true for the stratosphere and thermosphere. Some of the warmest temperatures are actually found in the thermosphere because of a type of radiation that enters that layer.

Earth and Human Activity

Advances in science and technology can have both positive and negative impacts on the environment and society. Often, new discoveries and technologies make daily living more convenient for the average person. However, these advancements initially require extra natural resources and the long-term effects on society take time to be fully understood. It's also important to note that many scientific and technological advancements focus on mitigating, or even eliminating, longstanding environmental and societal concerns.

Air and Water Pollution

Air and water are vital to human existence. Clean air and potable water greatly impact human health outcomes. As countries develop and become more industrialized, pollution is inevitable. In the United States, the Industrial Revolution, which shifted the economy's focus from agricultural practices to manufacturing, greatly increased air, water, and soil pollution. This occurred because factories burned more coal to operate, leading to increased levels of *smog* (a type of hazy air resulting from the presence of smoke, sulfur oxides, nitrogen oxides, and/or additional hazardous organic compounds) and the presence of *acid rain* (rain that is acidic as a result of air pollution and consequently harms trees, bodies of water, and animals when it falls). Additionally, factories often disposed of waste in the most convenient manner possible—usually by dumping it into bodies of water often used for drinking water. By the middle of the 20th century, both the United States and England had experienced deadly smog events that had resulted in the deaths and sickness of thousands of citizens. These events spurred environmental movements in the 1960s, and the United States passed the *Clean Air Act* in 1970 to combat environmental hazards resulting from air pollution. Currently, many companies focus on the development and implementation of "clean" technologies to manage these issues. The introduction of battery-operated vehicles intended to reduce the country's production of automotive emissions is just one example.

Climate Change and Greenhouse Gases

Greenhouse gases in the Earth's atmosphere include water vapor, carbon dioxide, methane, nitrous oxide, and *chlorofluorocarbons (CFCs)*, which trap heat between the surface of the Earth and the Earth's lowest atmospheric layer, the troposphere. The increase of these gases leads to warming or cooling trends that cause unpredictable or unprecedented meteorological shifts. These shifts can cause natural disasters, affect plant and animal life, and dramatically impact human health. *Water vapor* is a naturally found gas, but as the Earth's temperature rises, the presence of water vapor increases; as water vapor increases, the Earth's temperature rises. This creates a somewhat undesirable loop. *Carbon dioxide* is produced through natural causes, such as volcanic eruptions, but also is greatly affected by human activities, such as burning fossil fuels. A significant increase in the presence of atmospheric carbon dioxide has been noted since the Industrial Revolution; this is important as carbon dioxide is considered the most significant influencer of climate change on Earth. *Methane* is produced primarily from animal and agriculture waste and landfill waste. *Nitrous oxide* is primarily produced from the use of fertilizers and fossil fuels. CFCs are completely synthetic and were previously commonly found in aerosol and other high pressure containers; however, after being linked to ozone layer depletion, they have been stringently regulated internationally and are now in limited use. Scientists have stated that the climate

shifts recorded since the Industrial Revolution cannot be attributed to natural causes alone, as the patterns do not follow those of climate shifts that took place prior to the Industrial Revolution.

Natural Greenhouse Effect vs. Human Influence

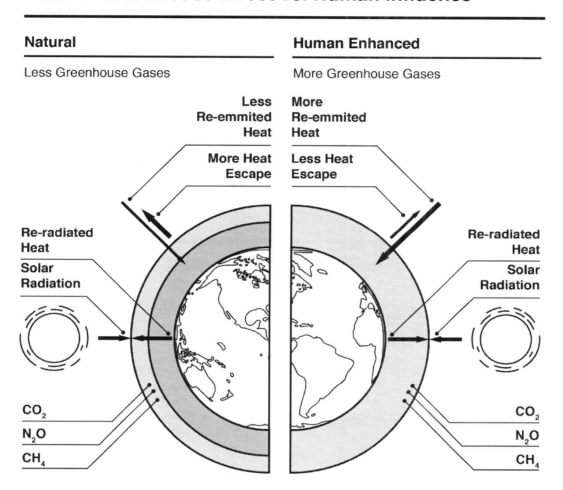

Irrigation

Irrigation refers to a systematic watering method, most pertinent to the agriculture and landscaping industries. It is not the only method of watering used in agricultural practices (some farmers and landscapers do utilize natural rainfall alone), but it is the primary method in which humans can control how to best utilize their most reliable water source for agricultural purposes. Irrigation systems can also be used in conjunction with a community's sewage and drainage system. There are multiple methods of irrigation, although they all have the same goal of supplying crops with a sufficient amount of water. Surface irrigation is the most commonly utilized method of irrigation, functioning by allowing water to freely flow across the desired area and naturally seeping into the soil below. Localized irrigation distributes water through piping and sprays directly onto a plant. An example is a residential sprinkler system that waters an entire lawn through in-ground nozzles, or a drip irrigation system that distributes water using low-flow pressure systems. Subsurface irrigation systems are underground and concentrate water at the root of a plant. Some methods of localized irrigation also combine fertilizer into the watering system.

Reservoirs and Levees

Reservoirs and *levees* are typically utilized as a means of directing and storing water from naturally occurring sources toward and within areas of need. They may be crucial in residential areas that are not close to a naturally existing source of water, such as in desert climates. Resourcing water in this way allows nearby communities to have drinking water, irrigation systems, recreation that centers on water activities, and may also serve as a power-generating source. Levees are used to control flooding of reservoirs, or may be built and used independently in flood-prone areas. Over the decades, both concepts have faced scrutiny from environmental and political experts. Creating a reservoir often utilizes dams, which create a barrier between the natural source of water and the created source of water. Damming can result in creating unnatural barriers for ecosystems that exist in the naturally occurring water source; sediment build-up in the reservoir, which affects its storage abilities; erosion that lowers water table levels and consequently affects crop output; and human errors such as poor construction. Poorly constructed dams and levees can break and lead to catastrophic flooding for nearby communities, especially those that are downstream from the water source. For example, Hurricane Katrina in 2005 caused enormous destruction to the city of New Orleans, Louisiana, and its surrounding areas as a result of inadequate levee systems that allowed for catastrophic flooding. Additionally, some studies indicate that reservoirs, acting as relatively stagnant water sources, can breed disease. Other studies show that warm-climate reservoirs contribute to excess greenhouse gas production due to the biomass production that occurs over time at the bottom of the reservoirs; this leads to the production and release of methane.

Depletion of Aquifers

Aquifers are naturally occurring sources of extractable freshwater water, normally found in permeable rock. These rocks can be drilled and pumped for water. The availability of water, and if it flows autonomously (without the use of a manmade pump), depends on the type of rock in which it is stored—more porous rocks may allow for more water flow—and seasonal precipitation. The rock also serves as a filter for the water; for example, clay and coal particles can often filter pesticide residue and other hazardous run-off that might taint fresh groundwater. Aquifers are becoming non-potable or depleted primarily from human activity. Residential and commercial use of septic tanks, overuse of fertilizers and pesticides on crops, sustained pumping along ocean coasts, and mining either degrade the quality of the water by allowing hazardous contaminants (such as waste or saltwater) to enter it, or by exposing the water to air and allowing it to evaporate. Depletion occurs when pumping occurs faster than the rocks can replenish their water stores. With exponential population growth, aquifer water is being used at a rate at which it cannot be replenished quickly enough.

Ozone Layer Depletion

Located in the stratosphere, the *ozone layer* protects the Earth from excessive *ultraviolet B (UVB)* ray exposure. The last century has shown signification depletion of the ozone layer, especially over Antarctica; this region is known as the *ozone hole*, missing almost 70% of its ozone layer. Chlorine molecules are especially harmful to ozone molecules. CFCs have been a major contributor to the ozone layer's depletion due to their high concentration of chlorine molecules. Almost all CFC production was a result of industrialization and human activity. In 1996, most CFC production was banned; however, it is expected that atmospheric chlorine levels will remain high for the next couple of decades. Additionally, other effects of climate change may prevent the stratosphere from ever reaching the gas composition that existed before CFCs were utilized. While ozone depletion does not contribute to global warming

directly, its impact on human health and disease is significant. The consequent increase in UVB exposure is linked to skin cancer in people, and ecosystem and food source disruption in animals. The effect on plants can lead to plant loss, which can indirectly impact the greenhouse effect, global warming, climate change, and human health.

Loss of Biodiversity

Biodiversity refers to the varied number of species on Earth—ranging from humans, to fish, to plants—and the way ecosystems are built within them. The biodiversity of an area strongly influences its air and soil quality, its energy availability, and how well its community thrives. Natural resources are currently being expended faster rate than they can be replaced, which is resulting in the extinction of species. As all species are interconnected in some way, the loss of an entire species can detrimentally impact the interactions and existence of other species. For example, if a particular animal feeds primarily on a plant species that becomes extinct, the animal species will have to radically change its feeding behaviors or it becomes prone to extinction. Decreasing supplies of water can impact the existence of plant and animal species as well, which, in turn, may affect how and what humans eat and grow. Plants are crucial to providing oxygen and reducing carbon dioxide—a greenhouse gas—on Earth. Additionally, many plants serve as ingredient sources in medicines; loss of plant life affects not only potential food sources but also medicinal sources. Overpopulation is likely the biggest threat to biodiversity, due to the inherent competing needs for land, water, and food production, as well as the risks of excess waste and pollution.

Space Exploration

Space, found beyond the Earth's atmosphere, has been a point of fascination for millennia. The development of rockets finally allowed for physical space exploration in the 1950s, with the intention of expanding knowledge in the fields of astronomy and physics. The 1950s were a time of competition between countries, as a number of countries attempted to pioneer the first space exploration. The Soviet Union launched the first orbiting satellite in 1957, and the United States succeeded in the first manned moon landing in 1969. China also sent manned missions to space in the 2000s. Additionally, satellites have been launched with the purpose of observing the Earth, serving as communication and navigation beacons, optimizing radio and television function, and relaying weather currently orbit the Earth. Present day has shifted some focus to exploring Mars and the potential for creating livable environments outside of the Earth. Some believe that limiting human existence to Earth only will ultimately lead to extinction as resources dwindle. Currently, the International Space Station is the only inhabitable, permanent structure outside of Earth and has been in use for almost two decades. The National Aeronautics and Space Administration (NASA) announced its intention to build a Moon station by 2024. There is also a focus on artificial intelligence and using automated, fully-functional, yet unmanned missions into deep space to conduct further exploration and research. However, much remains to be discovered about living in space. Conditions in space are harsh to human physiology, and no feasible long-term means of mitigating such effects yet exist.

Waste Disposal and Landfills

Waste disposal is a serious human concern. Waste production has almost doubled in the United States in the last 50 years, with the average household producing over 6,000 pounds of trash per year. Over half of that waste is disposed of in man-made sites in the Earth's ground. Piling (and even burying) trash is an ancient tradition. *Dumps* are open pits of trash, susceptible to rot, stench, and animal infestations. *Landfills* are designed structures intended to create a distinct boundary between the trash and the Earth. This boundary may be made of plastic, or with clay and soil. These structures try to prevent contamination of aquifers and crop soil. As waste breaks down in a landfill, methane is released into the air. Environmental groups and government regulations are pushing lifestyle changes and new technologies to reduce human trash generation. These include repurposing waste, extracting valuable materials from waste, turning waste into a renewable energy source, and advocating green behaviors such as using reusable grocery bags, using fewer plastic goods, and having *compost bins* at home. Many items in landfills could be disposed of in compost bins. These are composed of organic materials that decay quickly, such as food rinds and plant detritus. Once decayed, this material can be used to enrich soil and plant life, limit erosion, and even retain extra groundwater. The average household throws away between 20 to 50 percent of items to landfills that are compostable.

Finally, businesses and landfills in many countries have experienced new regulations aimed to limit waste production. Some governments offer tax breaks to companies that utilize green behaviors and focus on waste reduction.

Application

Using Information in a Concrete Situation

Finding Evidence that Supports a Finding

Science is one of the most objective, straightforward fields of study. Thus, it is no surprise that scientists and science articles are focused on **evidence**. When reading science passages, test takers are sometimes asked to find supporting evidence that reinforces a particular finding. A **finding** in science is a result of the investigation; it is what scientists find out. The majority of science passages tend to avoid opinions; instead, they focus on facts. Although no results are infallible just because the texts are scientific, most results are quantified. Quantified results mean they are expressed in numbers or measurements. Thus, when in doubt, go straight to the data, or numbers, that are offered. Sometimes data is embedded in the text; other times it appears in charts, tables, or graphs. These tools use numbers to demonstrate the patterns discussed in scientific texts, and they help readers to visualize concrete patterns. In order to find evidence to support a finding in scientific passage, all test takers should try collecting and analyzing the relevant data offered. Regardless of whether the data is coming from the text or a graph, it is helpful when making conclusions.

The following steps are helpful for identifying evidence that supports a finding in a science passage:

- Apply critical analysis and critical thinking by asking the right questions.
- Determine the weight of the information by figuring out its relevance.
- Identify trends in the numbers.
- Make inferences.
- Determine the most appropriate methods for either quantifying or communicating inferences.

Science Theories and Processes

Theories, models, and laws have one thing in common: *they develop on the basis of scientific evidence that has been tested and verified by multiple researchers on many different occasions*. Listed below are their exact definitions:

- **Theory:** An explanation of natural patterns or occurrences—i.e., the theory of relativity, the kinetic theory of gases, etc.

- **Model:** A representation of a natural pattern or occurrence that's difficult or impossible to experience directly, usually in the form of a picture or 3-D representation—i.e., Bohr's atomic model, the double-helix model of DNA, etc.

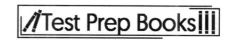

- **Law:** A mathematical or concise description of a pattern or occurrence in the observable universe—i.e., Newton's law of gravity, the laws of thermodynamics, etc.

The terms *theory, model,* and *law* are often used interchangeably in the sciences, although there's an essential difference: theories and models are used to explain *how* and *why* something happens, while laws describe exactly *what* happens. A common misconception is that theories develop into laws. But theories and models never become laws because they inherently describe different things.

Type	Function	Examples
Theory	To explain how and why something happens	Einstein's Theory of Special Relativity The Big Bang Theory
Model	To represent how and why something happens	A graphical model or drawing of an atom
Laws	To describe exactly what happens	$E = mc^2$ $F = ma$ $PV = nRT$

In order to ensure that scientific theories are consistent, scientists continually gather information and evidence on existing theories to improve their accuracy.

Analysis

Exploring Relationships Among Ideas

Designing a Science Investigation

Human beings are, by nature, very curious. Long before the scientific method was established, people have been making observations and predicting outcomes, manipulating the physical world to create extraordinary things—from the first man-made fire in 6000 B.C.E. to the satellite that orbited Pluto in 2016. Although the history of the scientific method is sporadic and attributed to many different people, it remains the most reliable way to obtain and utilize knowledge about the observable universe. Designing a science investigation is based on the scientific method, which consists of the following steps:

- Make an observation
- Create a question
- Form a hypothesis
- Conduct an experiment
- Collect and analyze data
- Form a conclusion

The first step is to identify a problem based on an observation—the who, what, when, where, why, and how. An **observation** is the analysis of information using basic human senses: sight, sound, touch, taste, and smell. Observations can be two different types—qualitative or quantitative. A **qualitative observation** describes what is being observed, such as the color of a house or the smell of a flower. **Quantitative observations** measure what is being observed, such as the number of windows on a house or the intensity of a flower's smell on a scale of 1–5.

Observations lead to the identification of a problem, also called an **inference**. For example, if a fire truck is barreling down a busy street, the inferences could be:

- There's a fire.
- Someone is hurt.
- Some kid pulled the fire alarm at a local school.

Inferences are logical predictions based on experience or education that lead to the formation of a hypothesis.

Forming and Testing a Hypothesis

A hypothesis is a testable explanation of an observed scenario and is presented in the form of a statement. It's an attempt to answer a question based on an observation, and it allows a scientist to predict an outcome. A hypothesis makes assumptions on the relationship between two different variables, and answers the question: "If I do this, what happens to that?"

In order to form a hypothesis, there must be an independent variable and a dependent variable that can be measured. The **independent variable** is the variable that is manipulated, and the **dependent variable** is the result of the change.

For example, suppose a student wants to know how light affects plant growth. Based upon what he or she already knows, the student proposes (hypothesizes) that the more light to which a plant is exposed, the faster it will grow.

- Observation: Plants exposed to lots of light seem to grow taller.
- Question: Will plants grow faster if there's more light available?
- Hypothesis: The more light the plant has, the faster it will grow.
- Independent variable: The amount of time exposed to light (able to be manipulated)
- Dependent variable: Plant growth (the result of the manipulation)

Once a hypothesis has been formed, it must be tested to determine whether it's true or false. (How to test a hypothesis is described in a subsequent section.) After it has been tested and validated as true over and over, then a hypothesis can develop into a theory, model, or law.

Experimental Design

To test a hypothesis, one must conduct a carefully designed experiment. There are four basic requirements that must be present for an experiment to be valid:

- A control
- Variables
- A constant
- Repeated and collected data

The control is a standard to which the resultant findings are compared. It's the baseline measurement that allows for scientists to determine whether the results are positive or negative. For the example of light affecting plant growth, the control may be a plant that receives no light at all.

The independent variable is manipulated (a good way to remember this is I manipulate the Independent variable), and the dependent variable is the result of changes to the independent variable. In the plant example, the independent variable is the amount of time exposed to light, and the dependent variable is the resulting growth (or lack thereof) of the plant. For this experiment, there may be three plants—one that receives a minimal amount of light, the control, and one that receives a lot of light.

Finally, there must be constants in an experiment. A constant is an element of the experiment that remains unchanged. Constants are extremely important in minimizing inconsistencies within the experiment that may lead to results outside the parameters of the hypothesis. For example, some constants in the above case are that all plants receive the same amount of water, all plants are potted in the same kind of soil, the species of the plant used in each condition is the same, and the plants are stored at the same temperature. If, for instance, the plants received different amounts of water as well as light, it would be impossible to tell whether the plants responded to changes in water or light.

Once the experiment begins, a disciplined scientist must always record the observations in meticulous detail, usually in a journal. A good journal includes dates, times, and exact values of both variables and constants. Upon reading this journal, a different scientist should be able to clearly understand the experiment and recreate it exactly. The journal includes all collected data, or any observed changes. In this case, the data is rates of plant growth, as well as any other phenomena that occurred as a result of the experiment. A well-designed experiment also includes repetition in order to get the most accurate possible readings and to account for any errors, so several trials may be conducted.

Even in the presence of diligent constants, there are an infinite number of reasons that an experiment can (and will) go wrong, known as sources of error. All experimental results are inherently accepted as imperfect, if ever so slightly, because experiments are conducted by human beings, and no instrument can measure anything perfectly. The goal of scientists is to minimize those errors to the best of their ability.

Identifying and Explaining Independent and Dependent Variables

In an experiment, variables are the key to analyzing data, especially when data is in a graph or table. Variables can represent anything, including objects, conditions, events, and amounts of time.

Covariance is a general term referring to how two variables move in relation to each other. Take for example an employee that gets paid by the hour. For them, hours worked and total pay have a positive covariance. As hours worked increases, so does pay.

Constant variables remain unchanged by the scientist across all trials. Because they are held constant for all groups in an experiment, they aren't being measured in the experiment, and they are usually ignored. Constants can either be controlled by the scientist directly like the nutrition, water, and sunlight given to plants, or they can be selected by the scientist specifically for an experiment like using a certain animal species or choosing to investigate only people of a certain age group.

Independent variables are also controlled by the scientist, but they are the same only for each group or trial in the experiment. Each group might be composed of students that all have the same color of car or each trial may be run on different soda brands. The independent variable of an experiment is what is being indirectly tested because it causes change in the dependent variables.

Dependent variables experience change caused by the independent variable and are what is being measured or observed. For example, college acceptance rates could be a dependent variable of an experiment that sorted a large sample of high school students by an independent variable such as test scores. In this experiment, the scientist groups the high school students by the independent variable (test scores) to see how it affects the dependent variable (their college acceptance rates).

Note that most variables can be held constant in one experiment, but also serve as the independent variable or a dependent variable in another. For example, when testing how well a fertilizer aids plant growth, its amount of sunlight should be held constant for each group of plants, but if the experiment is

being done to determine the proper amount of sunlight a plant should have, the amount of sunlight is an independent variable because it is necessarily changed for each group of plants.

Evaluation

Judging the Soundness or Accuracy of Scientific Information or Methods

Deciding Whether Conclusions are Supported by Data

Drawing conclusions is the process of analyzing patterns in data and determining whether the relationship is **causal**, meaning that one variable is the cause of the change in the other. There are many correlations that aren't casual, such as a city where alcohol sales increase as crime increases. Although there's a positive correlation between the two, crime may not be the factor that causes an increase in alcohol sales. There could be other factors, such as an increase in unemployment, which increases both alcohol sales and crime rates. Although crime and alcohol sales are positively correlated, they aren't causally correlated.

For this reason, it's important for scientists to carefully design their experiments with all the appropriate constants to ensure that the relationships are causal. If a relationship is determined to be causal by isolating the variables from all other factors, only then can conclusions be drawn based on data. In the plant growth experiment, the conclusion is that light affects plant growth because the data shows they are causally correlated since the two variables were entirely isolated.

Making Conclusions Based on Data

The Science section of the TASC will contain one data-driven science passage that requires the test taker to examine evidence within a particular type of graphic. The test taker will then be required to interpret the data and answer questions demonstrating their ability to draw logical conclusions.

In general, there are two types of data: qualitative and quantitative. Science passages may contain both, but simply put, **quantitative** data is reflected numerically and qualitative is not. **Qualitative** data is based on its qualities. In other words, qualitative data tends to present information more in subjective generalities (for example, relating to size or appearance). Quantitative data is based on numerical findings such as percentages. Quantitative data will be described in numerical terms. While both types of data are valid, the test taker will more likely be faced with having to interpret quantitative data through one or more graphic(s), and then be required to answer questions regarding the numerical data. A test taker should take the time to learn the skills it takes to interpret quantitative data so that they can make sound conclusions.

An example of a line graph is as follows:

Cell Phone Use in Kiteville, 2000-2006

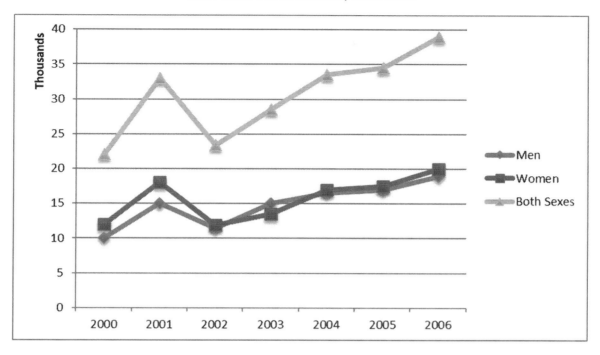

A **line graph** presents quantitative data on both horizontal (side to side) and vertical (up and down) axes. It requires the test taker to examine information across varying data points. When reading a line graph, a test taker should pay attention to any headings, as these indicate a title for the data it contains. In the above example, the test taker can anticipate the line graph contains numerical data regarding the use of cellphones during a certain time period. From there, a test taker should carefully read any outlying words or phrases that will help determine the meaning of data within the horizontal and vertical axes. In this example, the vertical axis displays the total number of people in increments of 5,000. Horizontally, the graph displays yearly markers, and the reader can assume the data presented accounts for a full calendar year. In addition, the line graph also uses different shapes to mark its data points. Some data points represent the number of men. Some data points represent the number of women, and a third type of data point represents the number of both sexes combined.

A test taker may be asked to read and interpret the graph's data, then answer questions about it. For example, the test may ask, *In which year did men seem to decrease cellphone use?* then require the test taker to select the correct answer. Similarly, the test taker may encounter a question such as *Which year yielded the highest number of cellphone users overall?* The test taker should be able to identify the correct answer as 2006.

A **bar graph** presents quantitative data through the use of lines or rectangles. The height and length of these lines or rectangles corresponds to numerical data. The data presented may represent information over time, showing shaded data over time or over other defined parameters. A bar graph will also utilize horizontal and vertical axes. An example of a bar graph is as follows:

Population Growth in Major U.S. Cities

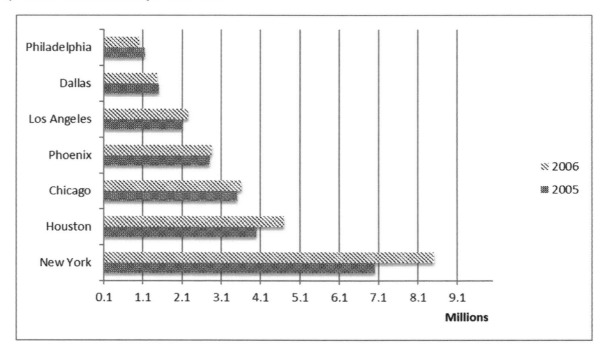

Reading the data in a bar graph is similar to the skills needed to read a line graph. The test taker should read and comprehend all heading information, as well as information provided along the horizontal and vertical axes. Note that the graph pertains to the population of some major U.S. cities. The "values" of these cities can be found along the left side of the graph, along the vertical axis. The population values can be found along the horizontal axes. Notice how the graph uses shaded bars to depict the change in population over time, as the heading indicates. Therefore, when the test taker is asked a question such as, *Which major U.S. city experienced the greatest amount of population growth during the depicted two year cycle,* the reader should be able to determine a correct answer of New York. It is important to pay particular attention to color, length, data points, and both axes, as well as any outlying header information in order to be able to answer graph-like test questions.

A **circle graph** (also sometimes referred to as a **pie chart**) presents quantitative data in the form of a circle. The same principles apply: the test taker should look for numerical data within the confines of the circle itself but also note any outlying information that may be included in a header, footer, or to the side of the circle. A circle graph will not depict horizontal or vertical axis information but will instead rely

on the reader's ability to visually take note of segmented circle pieces and apply information accordingly. An example of a circle graph is as follows:

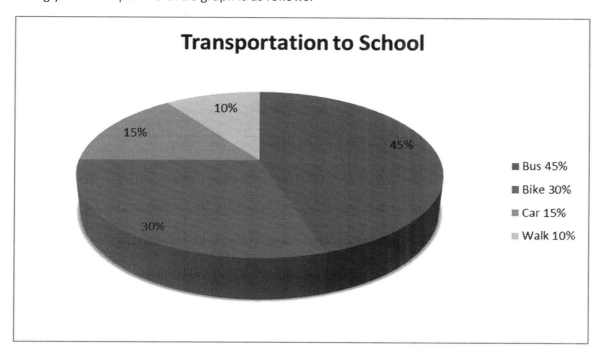

Notice the heading "Transportation to School." This should indicate to the test taker that the topic of the circle graph is how people traditionally get to school. To the right of the graph, the reader should comprehend that the data percentages contained within it directly correspond to the method of transportation. In this graph, the data is represented through the use shades and pattern. Each transportation method has its own shade. For example, if the test taker was then asked, *Which method of school transportation is most widely utilized,* the reader should be able to identify school bus as the correct answer.

Be wary of test questions that ask test takers to draw conclusions based on information that is not present. For example, it is not possible to determine, given the parameters of this circle graph, whether the population presented is of a particular gender or ethnic group. This graph does not represent data from a particular city or school district. It does not distinguish between student grade levels and, although the reader could infer that the typical student must be of driving age if cars are included, this is not necessarily the case. Elementary school students may rely on parents or others to drive them by personal methods. Therefore, do not read too much into data that is not presented. Only rely on the quantitative data that is presented in order to answer questions.

A **scatter plot** or **scatter diagram** is a graph that depicts quantitative data across plotted points. It will involve at least two sets of data. It will also involve horizontal and vertical axes.

An example of a scatter plot is as follows:

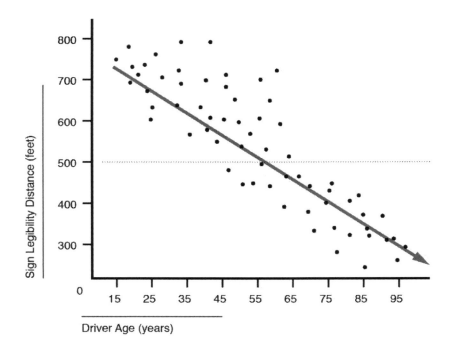

The skills needed to address a scatter plot are essentially the same as in other graph examples. Note any topic headings, as well as horizontal or vertical axis information. In the sample above, the reader can determine the data addresses a driver's ability to correctly and legibly read road signs as related to their age. Again, note the information that is absent. The test taker is not given the data to assess a time period, location, or driver gender. It simply requires the reader to note an approximate age to the ability to correctly identify road signs from a distance measured in feet. Notice that the overall graph also displays a trend. In this case, the data indicates a negative one and possibly supports the hypothesis that as a driver ages, their ability to correctly read a road sign at over 500 feet tends to decline over time. If the test taker were to be asked, *At what approximation in feet does a sixteen-year-old driver correctly see and read a street sign,* the answer would be the option closest to 700 feet.

Reading and examining scientific data in excerpts involves all of a reader's contextual reading, data interpretation, drawing logical conclusions based only on the information presented, and their application of critical thinking skills across a set of interpretive questions. Thorough comprehension and attention to detail is necessary to achieve test success.

Making Predictions Based on Data
Science is amazing in that it actually allows people to predict the future and see into the past with a certain degree of accuracy. Using numerical correlations created from quantitative data, one can see in a general way what will happen to *y* when something happens to *x*.

The best way to get a useful overview of quantitative data to facilitate predictions is to use a scatter plot, which plots each data point individually. As shown above, there may be slight fluctuations from the correlation line, so one may not be able to predict what happens with *every* change, but he or she will

be able to have a general idea of what is going to happen to *y* with a change in *x*. To demonstrate, the graph with a line of best fit created from the plant growth experiment is below.

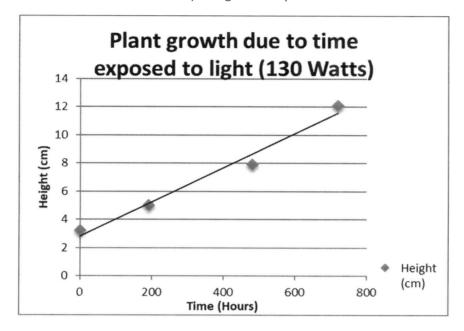

Using the trend line within the data, one can estimate what will happen to plant growth at a given length of time exposed to light. For example, it can be estimated that with 700 hours of time, the plant is expected to grow to a height of about 11 cm. The plant may not grow to exactly 11 cm, but it will likely grow to about that height based on previous data. This process allows scientists to draw conclusions based on data.

Identifying Possible Errors in a Science Investigation and Changing the Design to Correct Them

For a hypothesis to be proven true or false, all experiments are subject to multiple trials in order to verify accuracy and precision. A measurement is **accurate** if the observed value is close to the "true value." For example, if someone measured the pH of water at 6.9, this measurement would be considered accurate (the pH of water is 7). On the other hand, a measurement is **precise** if the measurements are consistent—that is, if they are reproducible. If someone had a series of values for a pH of water that were 6.9, 7.0, 7.2, and 7.3, their measurements would not be precise. However, if all measured values were 6.9, or the average of these values was 6.9 with a small range, then their measurements would be precise. Measurements can fall into the following categories:

- Both accurate and precise
- Accurate but not precise
- Precise but not accurate
- Neither accurate nor precise

The accuracy and precision of observed values most frequently correspond to the amount of error present in the experiment. Aside from general carelessness, there are two primary types of error: random and systematic. **Random errors** are unpredictable variations in the experiment that occur by chance. They can be difficult to detect, but they can often be nullified using a statistical analysis and minimized by taking repeated measurements and taking an average. **Systematic errors** occur when

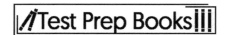

there are imperfections in the design of the experiment itself—usually errors that affect the accuracy of the measurements. These errors can be minimized by using the most accurate equipment available and by taking proper care of instruments and measuring techniques. Common examples of errors are listed below.

Random	Systematic
Environmental factors (random changes in vibration, temperature, humidity, etc.)	Poorly maintained instruments
	Old or out-of-date instruments
Differences in instrument use among scientists	Faulty calibration of instruments
Errors in judgment—can be affected by state of mind	Reading the instruments at an angle (parallax error) or other faulty reading errors
Incorrectly recorded observations	Not accounting for lag time

The most basic method to account for the possibility of errors is to take an average (also called a **mean**) of all observed values. To do so, one must divide the number of measurements taken from the sum of all measurements.

$$\frac{Sum\ of\ Measurements}{Total\ \#\ of\ Measurements}$$

For the above example of the pH values, the average is calculated by finding the sum of the pH values ascertained and dividing by the number of values recorded.

$$\frac{6.9 + 7.0 + 7.2 + 7.3}{4} = 7.1$$

The more observations recorded, the greater the precision. It's important to first assess the accuracy of measurements before proceeding to collect multiple trials of data. If a particular trial results in measurements that are vastly different from the average, it may indicate that a random or systematic error occurred during the trial. When this happens, a scientist might decide to "throw out" the trial and run the experiment again.

Identifying the Strengths and Weaknesses of Different Types of Science Investigations

In order to address the strengths and weaknesses of different types of scientific investigations, TASC test takers must first strengthen their capacity for scientific literacy and numeracy. It is important to familiarize oneself with methods for decoding highly specialized scientific terms, formulas, and symbols. Additionally, test takers can take the following suggestions to help identify unique weaknesses and strengths in different types of scientific investigations:

- Using critical analysis, test takers begin asking questions about the accuracy of the methods used to collect, analyze, and display data. They should carefully look at text and graphics that show scientific findings.

- Test takers should determine whether or not the words, data, and symbols provided by the author actually offer information that is relevant for testing a hypothesis or making an inference.

- When two or more passages on the same topic are offered, test takers should cross-analyze the findings to determine what data is accurate or relevant and which findings are most objective.

- Although scientific research strives for objectivity, test takers should highlight any subjective biases that may be embedded in a text. In particular, they should be aware of certain historical or ethical biases that might appear.

- Test takers should double check for any computational inaccuracies.

- Test takers should make suggestions for better ways to present the finding.

Physical Sciences

Matter and Its Interactions

Structure of Matter

Elements, Compounds, and Mixtures
Everything that takes up space and has mass is composed of **matter**. Understanding the basic characteristics and properties of matter helps with classification and identification.

An **element** is a substance that cannot be chemically decomposed to a simpler substance, while still retaining the properties of the element.

Compounds are composed of two or more elements that are chemically combined. The constituent elements in the compound are in constant proportions by mass.

When a material can be separated by physicals means (such as sifting it through a colander), it is called a **mixture.** Mixtures are categorized into two types: heterogeneous and homogeneous. **Heterogeneous mixtures** have physically distinct parts, which retain their different properties. A mix of salt and sugar is an example of a heterogeneous mixture. With heterogenous mixtures, it is possible that different samples from the same parent mixture may have different proportions of each component in the mixture. For example, in the sugar and salt mixture, there may be uneven mixing of the two, causing one random tablespoon sample to be mostly salt, while a different tablespoon sample may be mostly sugar.

A **homogeneous mixture**, also called a **solution,** has uniform properties throughout a given sample. An example of a homogeneous solution is salt fully dissolved in warm water. In this case, any number of samples taken from the parent solution would be identical.

Atoms, Molecules, and Ions
The basic building blocks of matter are **atoms**, which are extremely small particles that retain their identity during chemical reactions. Atoms can be singular or grouped to form **elements**. Elements are composed of one type of atom with the same properties.

Molecules are a group of atoms—either the same or different types—that are chemically bonded together by attractive forces. For example, hydrogen and oxygen are both atoms but, when bonded together, form water.

Ions are electrically-charged particles that are formed from an atom or a group of atoms via the loss or gain of electrons.

Basic Properties of Solids, Liquids, and Gases

Matter exists in certain **states,** or physical forms, under different conditions. These states are called solid, liquid, or gas.

A **solid** has a rigid, or set, form and occupies a fixed shape and volume. Solids generally maintain their shape when exposed to outside forces.

Liquids and **gases** are considered fluids, which have no set shape. Liquids are fluid, yet are distinguished from gases by their incompressibility (incapable of being compressed) and set volume. Liquids can be transferred from one container to another, but cannot be forced to fill containers of different volumes via compression without causing damage to the container. For example, if one attempts to force a given volume or number of particles of a liquid, such as water, into a fixed container, such as a small water bottle, the container would likely explode from the extra water.

A gas can easily be compressed into a confined space, such as a tire or an air mattress. Gases have no fixed shape or volume. They can also be subjected to outside forces, and the number of gas molecules that can fill a certain volume vary with changes in temperature and pressure.

Atomic Models

Theories of the atomic model have developed over the centuries. The most commonly referenced model of an atom was proposed by Niels Bohr. Bohr studied the models of J.J. Thomson and Ernest Rutherford and adapted his own theories from these existing models. Bohr compared the structure of the atom to that of the Solar System, where there is a center, or nucleus, with various sized orbitals circulating around this nucleus. This is a simplified version of what scientists have discovered about atoms, including the structures and placements of any orbitals. Modern science has made further adaptations to the model, including the fact that orbitals are actually made of electron "clouds."

Atomic Structure: Nucleus, Electrons, Protons, and Neutrons

Following the **Bohr model of the atom**, the **nucleus**, or core, is made up of positively-charged **protons** and neutrally-charged **neutrons**. The neutrons are theorized to be in the nucleus with the protons to provide greater "balance" at the center of the atom. The nucleus of the atom makes up the majority

(more than 99%) of the mass of an atom, while the orbitals surrounding the nucleus contain negatively-charged **electrons**. The entire structure of an atom is incredibly small.

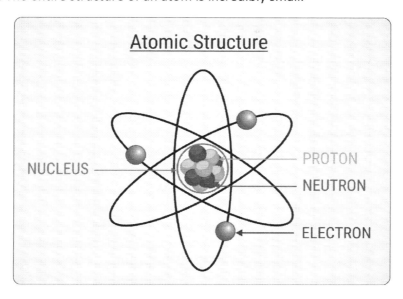

Atomic Number, Atomic Mass, and Isotopes

The **atomic number** of an atom is determined by the number of protons within the nucleus. When a substance is composed of atoms that all have the same atomic number, it is called an **element**. Elements are arranged by atomic number and grouped by properties in the **periodic table**.

An atom's **mass number** is determined by the sum of the total number of protons and neutrons in the atom. Most nuclei have a net neutral charge, and all atoms of one type have the same atomic number. However, there are some atoms of the same type that have a different mass number, due to an imbalance of neutrons. These are called **isotopes**. In isotopes, the atomic number, which is determined by the number of protons, is the same, but the mass number, which is determined by adding the protons and neutrons, is different due to the irregular number of neutrons.

Electron Arrangements

Electrons are most easily organized into distributions of subshells called **electron configurations**. Subshells fill from the inside (closest to the nucleus) to the outside. Therefore, once a subshell is filled, the next shell farther from the nucleus begins to fill, and so on. Atoms with electrons on the outside of a noble gas core (an atom with an electron inner shell that corresponds to the configuration of one of the noble gases, such as Neon) and pseudo-noble gas core (an atom with an electron inner shell that is similar to that of a noble gas core along with $(n-1)$ d^{10} electrons), are called **valence electrons**. Valence electrons are primarily the electrons involved in chemical reactions. The similarities in their configurations account for similarities in properties of groups of elements. Essentially, the **groups** (vertical columns) on the periodic table all have similar characteristics, such as solubility and reactivity, due to their similar electron configurations.

Describing Matter

Chemistry is the study of matter, including its properties and behavior. As mentioned, **matter** is the material that the universe is made of; it is any object that occupies space and has mass. Despite the diversity of items found in the universe, matter is comprised of only about 100 substances, known as elements. Elements cannot be broken down into simpler substances. Hydrogen and oxygen are two examples of elements. When different elements join together, they form compounds. Water is a compound made from hydrogen and oxygen. Atoms and molecules are among the smallest forms of matter.

Matter can be found in three different states: gas, liquid, or solid. **Gas** is a state that does not have a fixed volume or shape. It can expand to fill large containers or compress to fill smaller ones. Gas molecules are far apart from each other and float around at high speeds. They collide with each other and the container they are in, filling the container uniformly. When the gas is compressed, the space between the molecules decreases and the frequency of collisions between them increases. A **liquid** has an exact volume. It molds to the shape of the container that holds it. Liquid molecules are close together but still move rapidly. They cannot be compressed and slide over each other easily when liquids are poured. Solids have a definitive shape and volume. Similar to liquids, **solids** cannot be compressed. The molecules are packed together tightly in a specific arrangement that does not allow for much movement. When environmental changes occur, such as temperature or pressure changes, one state of matter can convert to another. For example, when water, a liquid, freezes, it becomes a solid, ice. When water is heated up to a certain temperature, it becomes water vapor, a gas.

The physical and chemical properties of matter can help distinguish different substances. Physical properties include color, odor, density, and hardness. These are properties that can be observed without changing the substance's identity or composition. When a substance undergoes a physical change, its physical appearance changes but its composition remains the same. Chemical properties are those that describe the way a substance might change to form another substance. Examples of chemical properties are flammability, toxicity, and ability to oxidize. These properties are observed by changing the environment of the substance and seeing how the substance reacts. A substance's composition is changed when it undergoes a chemical change.

Many properties of matter can be measured quantitatively, meaning the measurement is associated with a number. When a property of matter is represented by a number, it is important to include the unit of measure, otherwise the number is meaningless. For example, saying a pencil measures 10 is meaningless. It could be referring to 10 of something very short or 10 of something very long. The correct measurement notation would be 10 centimeters, because a centimeter has a designated length. Other examples of properties of matter that can be measured quantitatively are mass, time, and temperature, among others.

Chemical Equations and Reactions

Chemical reactions are represented by **chemical equations**. The equations help to explain how the molecules change during the reaction. For example, when hydrogen gas (H_2) combines with oxygen gas (O_2), two molecules of water are formed. The equation is written as follows, where the "+" sign means *reacts with* and the "$\rightarrow$" means *produces*:

$$2\ H_2 + O_2 \rightarrow 2\ H_2O$$

Two hydrogen molecules react with an oxygen molecule to produce two water molecules. In all chemical equations, the quantity of each element on the reactant side of the equation should equal the quantity of the same element on the product side of the equation due to the law of conservation of matter. If this is true, the equation is described as balanced. To figure out how many of each element there is on each side of the equation, the coefficient of the element should be multiplied by the subscript next to the element. Coefficients and subscripts are noted for quantities larger than one. The **coefficient** is the number located directly to the left of the element. The **subscript** is the small-sized number directly to

the right of the element. In the equation above, on the left side, the coefficient of the hydrogen is two and the subscript is also two, which makes a total of four hydrogen atoms. Using the same method, there are two oxygen atoms. On the right side, the coefficient two is multiplied by the subscript in each element of the water molecule, making four hydrogen atoms and two oxygen atoms. This equation is balanced because there are four hydrogen atoms and two oxygen atoms on each side. The states of the reactants and products can also be written in the equation: gas (g), liquid (l), solid (s), and dissolved in water (aq). If they are included, they are noted in parentheses on the right side of each molecule in the equation.

There are several types of chemical reactions: combination, decomposition, combustion single replacement, double replacement, and oxidation-reduction reactions. Reactions where two or more reactants combine to form one product, A + B → C, are called **combination** reactions. When one substance is broken down into two or more substances, C → B + A, it is known as a **decomposition** reaction. **Combustion** reactions involve oxygen gas as a reactant and generally involve the burning of a substance. One example is the combustion of propane:

$$C_3H_8(g) + 5O_2 \rightarrow 3CO_2(g) + 4H_2O$$

Oxygen is combined with propane gas to from carbon dioxide and water. **Double** and **single replacement** reactions involve ionic compounds and metals and the rearranging of cations and anions, and **oxidation-reduction** reactions involve the passing of electrons.

Periodic Table

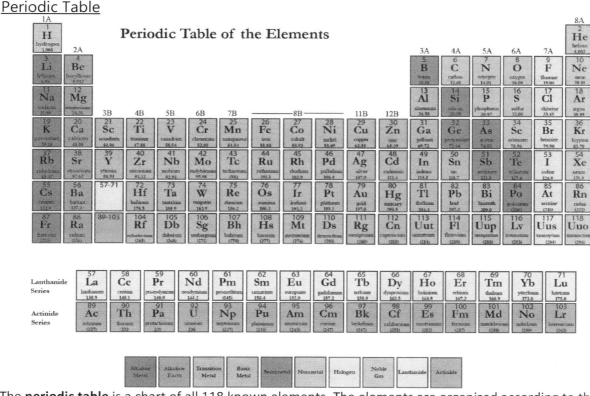

The **periodic table** is a chart of all 118 known elements. The elements are organized according to their quantity of protons, also known as their **atomic number**, their electron configurations, and their chemical properties. The rows are called **periods** and the columns are called **groups**. Groups have similar chemical behavior. For example, Group 8A is the noble gases, and because their outer electron shell is full, they are non-reactive. The closer the element is to having a full set of valence electrons it is, the more reactive it is. Group 1, the alkali metals, and Group 7, the halogens, are both highly reactive for that reason. The alkali metals form cations and lose their lone electron while halogens pick up an electron.

In each box in the period table, an element's symbol is the abbreviation in the center and its full name is located directly below that. The number in the top left corner is the atomic number and the atomic mass of the element is the number underneath. The atomic mass of the element is noted in atomic mass units, or amu, which represents the number of protons and neutrons combined. The number of protons defines the element, but the mass number can be different due to the existence of elements with a different number of neutrons, also called isotopes. The amu shown on the periodic table is a weighted mass (based on abundance) of all known isotopes of a particular element.

Atomic Radii

Atomic radius refers to the size of an atom. Going down a group, because each level represents an energy level, atomic radii increase. For example, when comparing lithium and potassium, potassium will have a greater atomic radius because it has two more energy levels.

Contrary to what might be predicted, going from left to right on the periodic table across a period, atomic radius actually decreases. This is due to proton pull. As each element gains a proton in its nucleus, it is like gaining a positive magnet in the middle. The more magnets in the middle, the stronger they will suck in any outer electrons. Comparing carbon and fluorine, fluorine has a smaller atomic radius due to more proton pull of the nucleus.

Ionic radius is a bit different. When sodium forms a cation, it loses an electron, losing an entire energy level. Thus, the ionic radii of cations are smaller than their elemental form. Anions, however, have a larger radius than their elemental form. This is because an extra electron is like adding a negative magnet to seven other negative magnets, resulting in a repulsive force that pushes them apart.

When atoms collide, in solutions, for example, they touch each other and then ricochet apart. The closest they come to each other is the combined length of each of their radii. This radius is known as **van der Waals radius** for each atom. When atoms bond with each other, however, they must come closer together than they would if they were just colliding. Their attraction also draws them closer together. The bonding atomic radius, or the **covalent radius**, is equal to one-half of the distance between the two atoms when they are bonded together. Within each group in the periodic table, the covalent radius increases from top to bottom. Within each period, the covalent radius decreases from left to right.

Ionization Energy and Electron Affinity

Energy is required to remove an electron from an atom or ion while it is in its ground state, usually as a gas. **Ionization energy** applies to all elements; it is just greater for those on the right side of the table. The ionization energy of an atom or ion is the minimum amount of energy required to do so. The greater the ionization energy, the harder it is to remove the electron. The energy required for the first electron is noted as I_1, the second as I_2, and then continues consecutively numbered for each successive electron. With each electron that is removed, the ionization energy of the element increases. I_1 increases in elements across the periods and moving down the columns of the period table. In contrast to ionization energy, energy is released from an atom when an electron is added. This negative change in energy is known as **electron affinity**. Ionization energies are always positive, because atoms require more energy to let go of an electron than to add one.

Looking at the periodic table, elements on the left, the alkali metals and alkali earth metals, have a low ionization energy. It takes little energy to peel off of their electron, because they have the least proton pull in their period. Elements on the right, like halogens (gases), have an extremely high ionization energy, because they have a very strong proton pull pulling electrons in, so it would require even more energy to peel off an electron. Noble gases have an extremely high ionization energy.

Going down a group, as energy levels increase, proton pull becomes less affected, so ionization energy decreases.

Molecular Weight
Elements can exist alone or combine to form compounds and molecules. Using the atomic mass, or atomic weights, that are noted on the periodic table, the **molecular weight** of a substance can be found by multiplying the subscript of each element by the atomic mass of that element and then adding together the weights of all of the elements in the molecule together. For example, to find the molecular weight of one glucose molecule, $C_6H_{12}O_6$, the atomic mass of carbon is multiplied by six, the atomic mass of hydrogen is multiplied by twelve, the atomic mass of oxygen is multiplied by six and then the three resulting numbers are added together. The answer is one glucose molecule has a molecular weight of 180 amu.

Nuclear Chemistry
Nuclear chemistry is the study of reactions in which the nuclei of atoms are transformed and their identities are changed. These reactions can involve large changes in energy—much larger than the energy changes that occur when chemical bonds between atoms are made or broken. Nuclear chemistry is also used to create electricity.

Nuclear reactions are described by nuclear equations, which have different notations than regular chemical equations. Nuclear equations are written as follows, with the superscript being the mass number and the subscript being the atomic number for each element:

$$^{238}_{92}U \rightarrow \, ^{234}_{90}Th + \, ^{4}_{2}He$$

The equation describes the spontaneous decomposition of uranium into thorium and helium via alpha decay. When this happens, the process is referred to as nuclear decay. Similar to chemical equations, nuclear equations must be balanced on each side; the sum of the mass numbers and the sum of the atomic numbers should be equal on both sides of the equation.

In some cases, the nucleus of an atom is unstable and constantly emits particles due to this instability. These atoms are described as radioactive and the isotopes are referred to as **radioisotopes.** There are three types of radioactive decay that occur most frequently: alpha (α), beta (β), and gamma (γ). **Alpha** radiation is emitted when a nucleus releases a stream of alpha particles, which are helium-4 nuclei. **Beta** radiation occurs when a stream of high-speed electrons is emitted by an unstable nucleus. The beta particles are often noted as β⁻. Gamma radiation occurs when the nucleus emits high-energy photons. In **gamma** radiation, the atomic number and the mass remain the same for the unstable nucleus. This type of radiation represents a rearrangement of an unstable nucleus into a more stable one and often accompanies other types of radioactive emission. Radioactive decay is often described in terms of its

half-life, which is the time that it takes for half of the radioactive substance to react. For example, the radioisotope strontium-90 has a half-life of 28.8 years. If there are 10 grams of strontium-90 to start with, after 28.8 years, there would be 5 grams left.

There are two distinct types of nuclear reactions: fission and fusion reactions. Both involve a large energy release. In **fission** reactions, a large atom is split into two or more smaller atoms. The nucleus absorbs slow-moving neutrons, resulting in a larger nucleus that is unstable. The unstable nucleus then undergoes fission. Nuclear power plants depend on nuclear fission reactions for energy. **Fusion** reactions involve the combination of two or more lighter atoms into a larger atom. Fusion reactions do not occur in Earth's nature due to the extreme temperature and pressure conditions required to make them happen. Fusion products are generally not radioactive. Fusion reactions are responsible for the energy that is created by the Sun.

Chemical Bonding

Chemical bonding occurs between two or more atoms that are joined together. There are three types of chemical bonds: ionic, covalent, and metallic. The characteristics of the different bonds are determined by how electrons behave in a compound. **Lewis structures** were developed to help visualize the electrons in molecules; they are a method of writing a compound structure formula and including its electron composition. A Lewis symbol for an element consists of the element symbol and a dot for each valence electron. The dots are located on all four sides of the symbol, with a maximum of two dots per side, and eight dots, or electrons, total. The octet rule states that atoms tend to gain, lose, or share electrons until they have a total of eight valence electrons.

Ionic Bonds

Ionic bonds are formed from the electrostatic attractions between oppositely charged atoms. They result from the transfer of electrons from a metal on the left side of the periodic table to a nonmetal on the right side. The metallic substance often has low ionization energy and will transfer an electron easily to the nonmetal, which has a high electron affinity. An example of this is the compound NaCl, which is sodium chloride or table salt, where the Na atom transfers an electron to the Cl atom. Due to strong bonding, ionic compounds have several distinct characteristics. They have high melting and boiling points, and are brittle and crystalline. They are arranged in rigid, well-defined structures, which allow them to break apart along smooth, flat surfaces. The formation of ionic bonds is a reaction that is exothermic. In the opposite scenario, the energy it takes to break up a one mole quantity of an ionic compound is referred to as **lattice energy**, which is generally endothermic. The Lewis structure for NaCl is written as follows:

$$Na\cdot \; + \; :\overset{..}{\underset{..}{Cl}}\cdot \; \longrightarrow \; Na^+ \; + :\overset{..}{\underset{..}{Cl}}:{}^-$$

Covalent Bonds

Covalent bonds are formed when two atoms share electrons, instead of transferring them as in ionic compounds. The atoms in covalent compounds have a balance of attraction and repulsion between their protons and electrons, which keeps them bonded together. Two atoms can be joined by single, double, or even triple covalent bonds. As the number of electrons that are shared increases, the length of the bond decreases. Covalent substances have low melting and boiling points, and are poor conductors of heat and electricity.

The Lewis structure for Cl_2 is written as follows:

Lewis structure Cl₂

$$:\ddot{C}l\cdot + \cdot\ddot{C}l: \longrightarrow :\ddot{C}l:\ddot{C}l:$$

Metallic Bonds

Metallic bonds are formed by electrons that move freely through metal. They are the product of the force of attraction between electrons and metal ions. The electrons are shared by many metal cations and act like glue that holds the metallic substance together, similar to the attraction between oppositely charged atoms in ionic substances, except the electrons are more fluid and float around the bonded metals and form a sea of electrons. Metallic compounds have characteristic properties that include strength, conduction of heat and electricity, and malleability. They can conduct electricity by passing energy through the freely moving electrons, creating a **current**. These compounds also have high melting and boiling points. Lewis structures are not common for metallic structures because of the free-roaming ability of the electrons.

Molar Relationships

One **mole** is the amount of matter contained in 6.02×10^{23} of any object, such as atoms, ions, or molecules. It is a useful unit of measure for items in large quantities. This number is also known as **Avogadro's number**. One mole of ^{12}C atoms is equivalent to 6.02×10^{23} ^{12}C atoms. Avogadro's number is often written as an inverse mole, or as $6.02 \times 10^{23}/mol$.

Molar Mass

A mole is always the same number, equivalent to Avogadro's number. The **molar mass** of a substance is the mass in grams of one mole of molecules of that substance. It is numerically equivalent to the molecular weight of the substance. The molecular weight of glucose ($C_6H_{12}O_6$) is 180 amu. Therefore, the molar mass of glucose is 180 grams per mole, written as 180 g/mol. In other words, one mole of glucose, or 6.02×10^{23} molecules of glucose, has a mass of 180 grams. Two substances with different molecular weights will have two different molar masses. Compared to glucose, O_2 has a molecular weight of 32 amu, which is less than that of glucose. So, the molar mass of O_2 is 32 g/mol. One mole of O_2 has an equivalent number of molecules as one mole of glucose, but it weighs less.

A simple calculation can determine how many moles are in a certain number of grams of a substance. The amount of substance in grams is divided by the molar mass of that substance and the result is the number of moles of the substance. Similarly, to convert the number of moles of a substance to the amount of grams, multiply the number of moles by the molar mass of the substance. The result is the amount of grams of the substance.

It is also possible to calculate the number of molecules in a certain number of grams of substance using Avogadro's number and the molar mass. The number of known grams is divided by the molar mass and then multiplied by Avogadro's number. The result is the number of molecules in the starting amount of grams.

Molarity

Molarity is the concentration of a solution. It is based on the number of moles of solute in one liter of solution and is written as the capital letter M. A 1.0 molar solution, or 1.0 M solution, has one mole of solute per liter of solution. The molarity of a solution can be determined by calculating the number of moles of the solute and dividing it by the volume of the solution in liters. The resulting number is the mol/L or M for molarity of the solution.

Ionic solutions can also be described by molarity values. Since ionic compounds dissolve in solution, the chemical formula of the compound can be used to determine the relative concentrations of the ions in the solution. For example, in a 1.0 M solution of $NaCl$, there is 1.0 M Na^+ ions and 1.0 M Cl^- ions. In a 1.0 M solution of Na_2SO_4, there are two Na^+ ions (2.0 M) for every one SO_4^{2-} ion (1.0 M).

Stoichiometry

Stoichiometry investigates the quantities of chemicals that are consumed and produced in chemical reactions. Chemical equations are made up of reactants and products; stoichiometry helps elucidate how the changes from reactants to products occur, as well as how to ensure the equation is balanced.

Limiting Reactants

Chemical reactions are limited by the amount of starting material, or reactants, available to drive the process forward. The reactant that has the smallest amount of substance is called the **limiting reactant**. The limiting reactant is completely consumed by the end of the reaction. The other reactants are called **excess reactants**. For example, gasoline is used in a combustion reaction to make a car move and is the limiting reactant of the reaction. If the gasoline runs out, the combustion reaction can no longer take place, and the car stops.

Reaction Yield

The quantity of product that should be produced after using up all of the limiting reactant can be calculated, and is called the **theoretical yield** of the reaction. Since the reactants do not always act as they should, the actual amount of resulting product is called **actual yield**. The actual yield is divided by the theoretical yield and then multiplied by 100 to find the **percent yield** for the reaction.

Solution Stoichiometry

Solution stoichiometry deals with quantities of solutes in chemical reactions that occur in solutions. The quantity of a solute in a solution can be calculated by multiplying the molarity of the solution by the volume. Similar to chemical equations involving simple elements, the number of moles of the elements that make up the solute should be equivalent on both sides of the equation.

When the concentration of a particular solute in a solution is unknown, a **titration** is used to determine that concentration. In a titration, the solution with the unknown solute is combined with a standard solution, which is a solution with a known solute concentration. The point at which the unknown solute has completely reacted with the known solute is called the **equivalence point**. Using the known information about the standard solution, including the concentration and volume, and the volume of the unknown solution, the concentration of the unknown solute is determined in a balanced equation. For example, in the case of combining acids and bases, the equivalence point is reached when the resulting solution is neutral. HCl, an acid, combines with NaOH, a base, to form water, which is neutral, and a solution of Cl^- ions and Na^+ ions. Before the equivalence point, there are an unequal number of cations and anions and the solution is not neutral.

Reaction Rates

The rate of a reaction is the measure of the change in concentration of the reactants or products over a certain period of time. Many factors affect how fast or slow a reaction occurs, such as concentration, pressure, or temperature. As the concentration of a reactant increases, the rate of the reaction also increases, because the frequency of collisions between elements increases. High-pressure situations for reactants that are gases cause the gas to compress and increase the frequency of gas molecule collisions, similar to solutions with higher concentrations. Reactions rates are then increased with the higher frequency of gas molecule collisions. Higher temperatures usually increase the rate of the reaction, adding more energy to the system with heat and increasing the frequency of molecular collisions.

Equilibrium

Equilibrium is described as the state of a system when no net changes occur. Chemical equilibrium occurs when opposing reactions occur at equal rates. In other words, the rate of reactants forming products is equal to the rate of the products breaking down into the reactants—the concentration of reactants and products in the system doesn't change. Although the concentrations are not changing, the forward and reverse reactions are likely still occurring. This type of equilibrium is called a **dynamic equilibrium**. In situations where all reactions have ceased, a **static equilibrium** is reached. Chemical equilibriums are also described as homogeneous or heterogeneous. **Homogeneous equilibrium** involves substances that are all in the same phase, while **heterogeneous equilibrium** means the substances are in different phases when equilibrium is reached.

When a reaction reaches equilibrium, the conditions of the equilibrium are described by the following equation, based on the chemical equation aA + bB ↔ cC + dD:

$$K_c = \frac{[C]^c[D]^d}{[A]^a[B]^b}$$

This equation describes the **law of mass action**. It explains how the reactants and products react during dynamic equilibrium. K_c is the equilibrium constant and is obtained when molarity values are put into the equation for the reactants and products. It is important to note that K_c is only dependent on the stoichiometry of the equation. If K_c is greater than 1, the equilibrium occurs when there are more products generated; the equilibrium lies to the right. If K_c is less than 1, the equilibrium occurs when there are more reactants generated and the equilibrium is to the left.

Similar to finding K_c, the quantity of reactants and products, as well as the direction of the reaction can be determined at any point of time by finding Q, the reaction quotient. Q_c is substituted for the K_c in the equation above. If Q is less than K, the concentration of the reactants is too large and the concentration of the products is too small, so the reaction must move from left to right to achieve equilibrium. If Q is equal to K, the system is at equilibrium. If Q is greater than K, the concentration of the products is too large and the concentration of the reactants is too small; the reaction must move from right to left to reach equilibrium.

Oxidation and Reduction

Oxidation and reduction reactions, also known as **redox** reactions, are those in which electrons are transferred from one element to another. Batteries and fuel cells are two energy-related technologies that utilize these reactions. When an atom, ion, or molecule loses its electrons and becomes more positively charged, it is described as being **oxidized**. When a substance gains electrons and becomes more negatively charged, it is **reduced**. In chemical reactions, if one element or molecule is oxidized, another must be reduced for the equation to be balanced. Although the transfer of electrons is obvious in some reactions where ions are formed, redox reactions also include those in which electrons are transferred but the products remain neutral.

Oxidation Numbers

Keep track of oxidation states or oxidation numbers to ensure the chemical equation is balanced. **Oxidation numbers** are assigned to each atom in a neutral substance or ion. For ions made up of a single atom, the oxidation number is equal to the charge of the ion. For atoms in their original elemental form, the oxidation number is always zero. Each hydrogen atom in an H_2 molecule, for example, has an oxidation number of zero. The sum of the oxidation numbers in a molecule should be equal to the overall charge of the molecule. If the molecule is a positively charged ion, the sum of the oxidation number should be equal to overall positive charge of the molecule. In ionic compounds that have a cation and anion joined, the sum of the oxidation numbers should equal zero.

Balancing Redox Reactions

All chemical equations must have the same number of elements on each side of the equation to be balanced. Redox reactions have an extra step of counting the electrons on both sides of the equation to be balanced. Separating redox reactions into oxidation reactions and reduction reactions is a simple way to account for all of the electrons involved. The individual equations are known as **half-reactions**. The number of electrons lost in the oxidation reaction must be equal to the number of electrons gained in the reduction reaction for the redox reaction to be balanced. The oxidation of tin (Sn) by iron (Fe) can be balanced by the following half-reactions:

Oxidation: $Sn^{2+} \rightarrow Sn^{4+} + 2e^-$

Reduction: $2Fe^{3+} + 2e^- \rightarrow 2Fe^{2+}$

Complete redox reaction: $Sn^{2+} + 2Fe^{3+} \rightarrow Sn^{4+} + 2Fe^{2+}$

Acids and Bases

Acids and bases are defined in many different ways. An **acid** can be described as a substance that increases the concentration of H^+ ions when it is dissolved in water, as a proton donor in a chemical equation, or as an electron-pair acceptor. A **base** can be a substance that increases the concentration of OH^- ions when it is dissolved in water, accepts a proton in a chemical reaction, or is an electron-pair donor.

Autoionization of Water

Water can act as either an acid or a base. When mixed with an acid, water can accept a proton and become an H_3O^+ ion. When mixed with a base, water can donate a proton and become an OH^- ion. Sometimes water molecules donate and accept protons from each other; this process is called **autoionization**. The chemical equation is written as follows: $H_2O + H_2O \rightarrow OH^- + H_3O^+$.

PH Scale

The **pH scale** is a numeric scale that determines whether a solution is acidic, basic, or neutral. The pH of a solution is equal to the inverse base 10 logarithm of its activity of H^+ ions ($pH = \log_{10}(1/(H^+ \text{ activity}))$). Solutions with a pH value greater than seven are considered **basic** and those with a pH value less than seven are considered **acidic**. Water is **neutral** and has a pH of 7.0.

Strength of Acids and Bases

Acids and bases are characterized as strong, weak, or somewhere in between. Strong acids and bases completely or almost completely ionize in aqueous solution. The chemical reaction is driven completely forward, to the right side of the equation, where the acidic or basic ions are formed. Weak acids and bases do not completely disassociate in aqueous solution. They only partially ionize and the solution becomes a mixture of the acid or base, water, and the acidic or basic ions. Strong acids are complemented by weak bases, and vice versa. A **conjugate acid** is an ion that forms when its base pair gains a proton. For example, the conjugate acid NH_4^+ is formed from the base NH_3. The **conjugate base** that pairs with an acid is the ion that is formed when an acid loses a proton. NO_2^- is the conjugate base of the acid HNO_2.

Motion and Stability: Forces and Interactions

Description of Motion in One and Two Dimensions

The description of motion is known as **kinetics**, and the causes of motion are known as **dynamics**. Motion in one dimension is known as a **scalar** quantity. It consists of one measurement such as length (length or distance is also known as displacement), speed, or time. Motion in two dimensions is known as a **vector** quantity. This would be speed with a direction, or velocity.

Velocity is the measure of the change in distance over the change in time. All vector quantities have a direction that can be relayed through the sign of an answer, such as -5.0 m/s or +5.0 m/s. The objects registering these velocities would be in opposite directions, where the change in distance is denoted by Δx and the change in time is denoted by Δt:

$$v = \frac{\Delta x}{\Delta t}$$

Acceleration is the measure of the change in an object's velocity over a change in time, where the change in velocity, $v_2 - v_1$, is denoted by Δv and the change in time, $t_1 - t_2$, is denoted by Δt:

$$a = \frac{\Delta v}{\Delta t}$$

The linear momentum, p, of an object is the result of the object's mass, m, multiplied by its velocity, v, and is described by the equation:

$$p = mv$$

This aspect becomes important when one object hits another object. For example, the linear momentum of a small sports car will be much smaller than the linear momentum of a large semi-truck. Thus, the semi-truck will cause more damage to the car than the car to the truck.

Newton's Three Laws of Motion

Sir Isaac Newton summarized his observations and calculations relating to motion into three concise laws.

First Law of Motion: Inertia

This law states that an object in motion tends to stay in motion or an object at rest tends to stay at rest, unless the object is acted upon by an outside force.

For example, a rock sitting on the ground will remain in the same place, unless it is pushed or lifted from its place.

The First Law also includes the relation of weight to gravity and force between objects relative to the distance separating them:

$$Weight = G\frac{Mm}{r^2}$$

In this equation, G is the gravitational constant, M and m are the masses of the two objects, and r is the distance separating the two objects.

Second Law of Motion: F = ma

This law states that the force on a given body is the result of the object's mass multiplied by any acceleration acting upon the object. For objects falling on Earth, an acceleration is caused by gravitational force ($9.8\ m/s^2$).

Third Law of Motion: Action-Reaction

This law states that for every action there is an equal and opposite reaction. For example, if a person punches a wall, the wall exerts a force back on the person's hand equal and opposite to his or her punching force. Since the wall has more mass, it absorbs the impact of the punch better than the person's hand.

Mass, Weight, and Gravity

Mass is a measure of how much of a substance exists, or how much inertia an object has. The mass of an object does not change based on the object's location, but the **weight** of an object does vary with its location.

For example, a 15-kg mass has a weight that is determined by acceleration from the force of gravity here on Earth. However, if that same 15-kg mass were to be weighed on the moon, it would weigh much less, since the acceleration force from the moon's gravity is approximately one-sixth of that on Earth.

Weight = mass × acceleration

W_{Earth} = 15 kg × 9.8 m/s^2	>	W_{Moon} = 15 kg × 1.62 m/s^2
W_{Earth} = 147N	>	24.3N

Analysis of Motion and Forces

Projectile motion describes the path of an object in the air. Generally, it is described by two-dimensional movement, such as a stone thrown through the air. This activity maps to a parabolic curve. However, the definition of projectile motion also applies to free fall, or the non-arced motion of an object in a path straight up and/or straight down. When an object is thrown horizontally, it is subject to the same influence of gravity as an object that is dropped straight down. The farther the projectile motion, the farther the distance of the object's flight.

Friction is a force that opposes motion. It can be caused by a number of materials; there is even friction caused by air. Whenever two differing materials touch, rub, or pass by each other, it will create friction, or an oppositional force, unless the interaction occurs in a true vacuum. To move an object across a floor, the force exerted on the object must overcome the frictional force keeping the object in place. Friction is also why people can walk on surfaces. Without the oppositional force of friction to a shoe pressing on the floor, a person would not be able to grip the floor to walk—similar to the challenge of walking on ice. Without friction, shoes slip and are unable to help people propel forward and walk.

When calculating the effects of objects hitting (or colliding with) each other, several things are important to remember. One of these is the definition of **momentum**: the mass of an object multiplied by the object's velocity. As mentioned, it is expressed by the following equation:

$$p = mv$$

Here, *p* is equal to an object's momentum, *m* is equal to the object's mass, and *v* is equal to the object's velocity.

Another important thing to remember is the principle of the **conservation of linear momentum**. The total momentum for objects in a situation will be the same before and after a collision. There are two primary types of collisions: elastic and inelastic. In an **elastic collision**, the objects collide and then travel in different directions. During an **inelastic collision**, the objects collide and then stick together in their final direction of travel. The total momentum in an elastic collision is calculated by using the following formula:

$$m_1 v_1 + m_2 v_2 = m_1 v_1 + m_2 v_2$$

Here, m_1 and m_2 are the masses of two separate objects, and v_1 and v_2 are the velocities, respectively, of the two separate objects.

The total momentum in an inelastic collision is calculated by using the following formula:

$$m_1 v_1 + m_2 v_2 = (m_1 + m_2) v_f$$

Here, v_f is the final velocity of the two masses after they stick together post-collision.

Example:

If two bumper cars are speeding toward each other, head-on, and collide, they are designed to bounce off of each other and head in different directions. This would be an elastic collision.

If real cars are speeding toward each other, head-on, and collide, there is a good chance their bumpers might get caught together and their direction of travel would be together in the same direction.

An **axis** is an invisible line on which an object can rotate. This is most easily observed with a toy top. There is actually a point (or rod) through the center of the top on which the top can be observed to be spinning. This is called the axis.

When objects move in a circle by spinning on their own axis, or because they are tethered around a central point (also an axis), they exhibit circular motion. Circular motion is similar in many ways to linear (straight line) motion; however, there are a few additional points to note. A spinning object is always accelerating because it is always changing direction. The force causing this constant acceleration on or around an axis is called **centripetal force** and is often associated with centripetal acceleration. Centripetal force always pulls toward the axis of rotation. An imaginary reactionary force, called **centrifugal force**, is the outward force felt when an object is undergoing circular motion. This reactionary force is not the real force; it just feels like it is there. For this reason, it has also been referred to as a "fictional force." The true force is the one pulling inward, or the centripetal force.

The terms *centripetal* and *centrifugal* are often mistakenly interchanged. If the centripetal force acting on an object moving with circular motion is removed, the object will continue moving in a straight line tangent to the point on the circle where the object last experienced the centripetal force. For example, when a traditional style washing machine spins a load of clothes to expunge the water from the load, it rapidly spins the machine barrel. A force is pulling in toward the center of the circle (centripetal force). At the same time, the wet clothes, which are attempting to move in a straight line, are colliding with the outer wall of the barrel that is moving in a circle. The interaction between the wet clothes and barrel wall cause a reactionary force to the centripetal force and this expels the water out of the small holes that line the outer wall of the barrel.

Conservation of Angular Momentum

An object moving in a circular motion also has momentum; for circular motion, it is called **angular momentum**. This is determined by rotational inertia, rotational velocity, and the distance of the mass from the axis or center of rotation. When objects exhibit circular motion, they also demonstrate the **conservation of angular momentum**, meaning that the angular momentum of a system is always constant, regardless of the placement of the mass. **Rotational inertia** can be affected by how far the mass of the object is placed with respect to the axis of rotation. The greater the distance between the mass and the axis of rotation, the slower the rotational velocity. Conversely, if the mass is closer to the axis of rotation, the rotational velocity is faster. A change in one affects the other, thus conserving the angular momentum. This holds true as long as no external forces act upon the system.

For example, ice skaters spinning in on one ice skate extends their arms out for a slower rotational velocity. When skaters bring their arms in close to their bodies (which lessens the distance between the mass and the axis of rotation), their rotational velocity increases and they spin much faster. Some skaters extend their arms straight up above their head, which causes an extension of the axis of rotation, thus removing any distance between the mass and the center of rotation, which maximizes their rotational velocity.

Another example is when a person selects a horse on a merry-go-round: the placement of their horse can affect their ride experience. All of the horses are traveling with the same rotational speed, but in order to travel along the same plane as the merry-go-round turns, a horse on the outside will have a greater linear speed because it is further away from the axis of rotation. Essentially, an outer horse has to cover a lot more ground than a horse on the inside in order to keep up with the rotational speed of the merry-go-round platform. Thrill seekers should always select an outer horse.

The center of mass is the point that provides the average location for the total mass of a system. The word "system" can apply to just one object/particle or to many. The center of mass for a system can be calculated by finding the average of the mass of each object and multiplying by its distance from an origin point using the following formula:

$$x_{center of mass} = \frac{m_1 x_1 + m_2 x_2}{m_1 + m_2}$$

In this case, *x* is the distance from the point of origin for the center of mass and each respective object, and *m* is the mass of each object.

To calculate for more than one object, the pattern can be continued by adding additional masses and their respective distances from the origin point.

Simple Machines

A simple machine is a mechanical device that changes the direction or magnitude of a force. There are six basic types of simple machines: lever, wedge, screw, inclined plane, wheel and axle, and pulley.

Here is how each type works and an example:

- A **lever** helps lift heavy items higher with less force, such as a crowbar lifting a large cast iron lid.

- A **wedge** helps apply force to a specific area by focusing the pressure, such as an axe splitting a tree.

- An **inclined plane**, such as a loading dock ramp, helps move heavy items up vertical distances with less force.

- A **screw** is an inclined plane wrapped around an axis and allows more force to be applied by extending the distance of the plane. For example, a screw being turned into a piece of wood provides greater securing strength than hitting a nail into the wood.

- A **wheel and axle** allows the use of rotational force around an axis to assist with applying force. For example, a wheelbarrow makes it easier to haul large loads by employing a wheel and axle at the front.

- A **pulley** is an application of a wheel and axle with the addition of cords or ropes and it helps move objects vertically. For example, pulling a bucket out of a well is easier with a pulley and ropes.

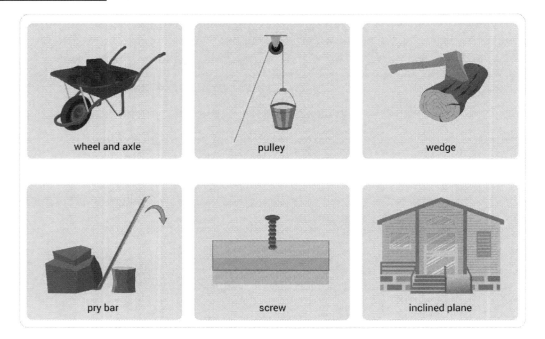

Using a simple machine employs an advantage to the user. This is referred to as the **mechanical advantage**. It can be calculated by comparing the force input by the user to the simple machine with the force output from the use of the machine (also displayed as a ratio).

$$Mechanical\ Advantage\ = \frac{output\ force}{input\ force}$$

$$MA\ = \frac{F_{out}}{F_{in}}$$

In the following instance of using a lever, it can be helpful to calculate the torque, or circular force, necessary to move something. This is also employed when using a wrench to loosen a bolt.

$$Torque\ =\ F\ \times\ distance\ of\ lever\ arm\ from\ the\ axis\ of\ rotation\ (called\ the\ moment\ arm)$$

$$T\ =\ F\ \times\ d$$

Energy

Basic Concepts and Relationships Involving Energy and Matter

The study of energy and matter, including heat and temperature, is called **thermodynamics**. There are four fundamental laws of thermodynamics, but the first two are the most commonly discussed.

First Law of Thermodynamics

The first law of thermodynamics is also known as the **conservation of energy**. This law states that energy cannot be created or destroyed, but is just transferred or converted into another form through a thermodynamic process. For example, if a liquid is boiled and then removed from the heat source, the liquid will eventually cool. This change in temperature is not because of a loss of energy or heat, but from a transfer of energy or heat to the surroundings. This can include the heating of nearby air

molecules, or the transfer of heat from the liquid to the container or to the surface where the container is resting.

This law also applies to the idea of perpetual motion. A self-powered perpetual motion machine cannot exist. This is because the motion of the machine would inevitably lose some heat or energy to friction, whether from materials or from the air.

Second Law of Thermodynamics

The second law of thermodynamics is also known as the **law of entropy**. **Entropy** means chaos or disorder. In simple terms, this law means that all systems tend toward chaos. When one or more systems interacts with another, the total entropy is the sum of the interacting systems, and this overall sum also tends toward entropy.

Conservation of Matter in Chemical Systems

The conservation of energy is seen in the conservation of matter in chemical systems. This is helpful when attempting to understand chemical processes, since these processes must balance out. This means that extra matter cannot be created or destroyed, it must all be accounted for through a chemical process.

Kinetic and Potential Energy

The conservation of energy also applies to the study of energy in physics. This is clearly demonstrated through the kinetic and potential energy involved in a system.

The energy of motion is called **kinetic energy**. If an object has height, or is raised above the ground, it has **potential energy**. The total energy of any given system is the sum of the potential energy and the kinetic energy of the subject (object) in the system.

Potential energy is expressed by the equation:

$$PE = mgh$$

Where m equals the object's mass, g equals acceleration caused by the gravitational force acting on the object, and h equals the height of the object above the ground.

Kinetic energy is expressed by the following equation:

$$KE = \frac{1}{2} mv^2$$

Where m is the mass of the object and v is the velocity of the object.

Conservation of energy allows the total energy for any situation to be calculated by the following equation:

$$KE + PE$$

For example, a roller coaster poised at the top of a hill has all potential energy, and when it reaches the bottom of that hill, as it is speeding through its lowest point, it has all kinetic energy. Halfway down the hill, the total energy of the roller coaster is about half potential energy and half kinetic energy.

Therefore, the total energy is found by calculating both the potential energy and the kinetic energy and then adding them together.

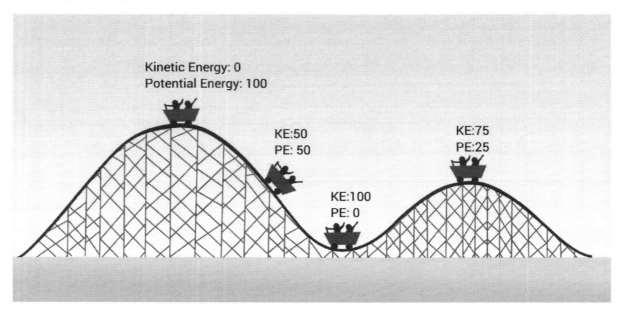

Kinetic Energy: 0
Potential Energy: 100

KE:50
PE: 50

KE:75
PE:25

KE:100
PE: 0

Transformations Between Different Forms of Energy

As stated by the conservation of energy, energy cannot be created or destroyed. If a system gains or loses energy, it is transformed within a single system from one type of energy to another or transferred from one system to another. For example, if the roller coaster system has potential energy that transfers to kinetic energy, the kinetic energy can then be transferred into thermal energy or heat released through braking as the coaster descends the hill. Energy can also transform from the chemical energy inside of a battery into the electrical energy that lights a train set. The energy released through nuclear fusion (when atoms are joined together, they release heat) is what supplies power plants with the energy for electricity. All energy is transferred from one form to another through different reactions. It can also be transferred through the simple action of atoms bumping into each other, causing a transfer of heat.

Differences Between Chemical and Physical Properties/Changes

A change in the physical form of matter, but not in its chemical identity, is known as a **physical change**. An example of a physical change is tearing a piece of paper in half. This changes the shape of the matter, but it is still paper.

Conversely, a **chemical change** alters the chemical composition or identity of matter. An example of a chemical change is burning a piece of paper. The heat necessary to burn the paper alters the chemical composition of the paper. This chemical change cannot be easily undone, since it has created at least one form of matter different than the original matter.

Temperature Scales

There are three main temperature scales used in science. The scale most often used in the United States is the **Fahrenheit** scale. This scale is based on the measurement of water freezing at 32° F and water boiling at 212° F. The **Celsius** scale uses 0° C as the temperature for water freezing and 100° C for water boiling. The Celsius scale is the most widely used in the scientific community. The accepted measurement by the International System of Units (from the French Système international d'unités), or

SI, for temperature is the Kelvin scale. This is the scale employed in thermodynamics, since its zero is the basis for absolute zero, or the unattainable temperature, when matter no longer exhibits degradation.

The conversions between the temperature scales are as follows:

°Fahrenheit to °Celsius: $^0C = \frac{5}{9}(^0F - 32)$

°Celsius to °Fahrenheit: $^0F = \frac{9}{5}(^0C) + 32$

°Celsius to Kelvin: $K = {}^0C + 273.15$

Transfer of Thermal Energy and Its Basic Measurement

There are three basic ways in which energy is transferred. The first is through **radiation**. Radiation is transmitted through electromagnetic waves and it does not need a medium to travel (it can travel in a vacuum). This is how the sun warms the Earth, and typically applies to large objects with great amounts of heat or objects with a large difference in their heat measurements.

The second form of heat transfer is **convection.** Convection involves the movement of "fluids" from one place to another. (The term *fluid* does not necessarily apply to a liquid, but any substance in which the molecules can slide past each other, such as gases.) It is this movement that transfers the heat to or from an area. Generally, convective heat transfer occurs through diffusion, which is when heat moves from areas of higher concentrations of particles to those of lower concentrations of particles and less heat. This process of flowing heat can be assisted or amplified through the use of fans and other methods of forcing the molecules to move.

The final process is called **conduction**. Conduction involves transferring heat through the touching of molecules. Molecules can either bump into each other to transfer heat, or they may already be touching each other and transfer the heat through this connection. For example, imagine a circular burner on an electric stove top. The coil begins to glow orange near the base of the burner that is connected to the stove because it heats up first. Since the burner is one continuous piece of metal, the molecules are touching each other. As they pass heat along the coil, it begins to glow all the way to the end.

To determine the amount of heat required to warm the coil in the above example, the type of material from which the coil is made must be known. The quantity of heat required to raise one gram of a substance one degree Celsius (or Kelvin) at a constant pressure is called *specific heat*. This measurement can be calculated for masses of varying substances by using the following equation:

$$q = s \times m \times \Delta t$$

Where q is the specific heat, s is the specific heat of the material being used, m is the mass of the substance being used, and Δt is the change in temperature.

A calorimeter is used to measure the heat of a reaction (either expelled or absorbed) and the temperature changes in a controlled system. A simple **calorimeter** can be made by using an insulated coffee cup with a thermometer inside. For this example, a lid of some sort would be preferred to prevent any escaping heat that could be lost by evaporation or convection.

<u>Optics and Waves</u>

Electromagnetic Spectrum

The movement of light is described like the movement of waves. Light travels with a wave front, has an **amplitude** (height from the neutral), a cycle or **wavelength**, a **period**, and **energy**. Light travels at approximately 3.00×10^8 m/s and is faster than anything created by humans thus far.

Light is commonly referred to by its measured **wavelengths**, or the distance between two successive crests or troughs in a wave. Types of light with the longest wavelengths include radio, TV, and micro, and infrared waves. The next set of wavelengths are detectable by the human eye and create the **visible spectrum**. The visible spectrum has wavelengths of 10^{-7} m, and the colors seen are red, orange, yellow, green, blue, indigo, and violet. Beyond the visible spectrum are shorter wavelengths (also called the **electromagnetic spectrum**) containing ultraviolet light, X-rays, and gamma rays. The wavelengths outside of the visible light range can be harmful to humans if they are directly exposed or are exposed for long periods of time.

Basic Characteristics and Types of Waves

A **mechanical wave** is a type of wave that passes through a medium (solid, liquid, or gas). There are two basic types of mechanical waves: longitudinal and transverse.

A **longitudinal wave** has motion that is parallel to the direction of the wave's travel. This can best be visualized by compressing one side of a tethered spring and then releasing that end. The movement travels in a bunching/un-bunching motion across the length of the spring and back.

A **transverse wave** has motion that is perpendicular to the direction of the wave's travel. The particles on a transverse wave do not move across the length of the wave; instead, they oscillate up and down, creating peaks and troughs.

A wave with a combination of both longitudinal and transverse motion can be seen through the motion of a wave on the ocean—with peaks and troughs, and particles oscillating up and down.

Mechanical waves can carry energy, sound, and light, but they need a medium through which transport can occur. An electromagnetic wave can transmit energy without a medium, or in a vacuum.

A more recent addition in the study of waves is the **gravitational wave**. Its existence has been proven and verified, yet the details surrounding its capabilities are still somewhat under inquiry. Gravitational waves are purported to be ripples that propagate as waves outward from their source and travel in the curvature of space/time. They are thought to carry energy in a form of radiant energy called **gravitational radiation**.

Basic Wave Phenomena

When a wave crosses a boundary or travels from one medium to another, certain things occur. If the wave can travel through one medium into another medium, it experiences **refraction**. This is the bending of the wave from one medium to another due to a change in density of the mediums, and thus, the speed of the wave changes. For example, when a pencil is sitting in half of a glass of water, a side view of the glass makes the pencil appear to be bent at the water level. What the viewer is seeing is the refraction of light waves traveling from the air into the water. Since the wave speed is slowed in water, the change makes the pencil appear bent.

When a wave hits a medium that it cannot penetrate, it is bounced back in an action called **reflection**. For example, when light waves hit a mirror, they are reflected, or bounced, off the mirror. This can cause it to seem like there is more light in the room, since there is a "doubling back" of the initial wave. This same phenomenon also causes people to be able to see their reflection in a mirror.

When a wave travels through a slit or around an obstacle, it is known as **diffraction.** A light wave will bend around an obstacle or through a slit and cause what is called a **diffraction pattern**. When the waves bend around an obstacle, it causes the addition of waves and the spreading of light on the other side of the opening.

Dispersion is used to describe the splitting of a single wave by refracting its components into separate parts. For example, if a wave of white light is sent through a dispersion prism, the light appears as its separate rainbow-colored components, due to each colored wavelength being refracted in the prism.

When wavelengths hit boundaries, different things occur. Objects will absorb certain wavelengths of light and reflect others, depending on the boundaries. This becomes important when an object appears to be a certain color. The color of an object is not actually within that object, but rather, in the wavelengths being transmitted by that object. For example, if a table appears to be red, that means the table is absorbing all other wavelengths of visible light except those of the red wavelength. The table is reflecting, or transmitting, the wavelengths associated with red back to the human eye, and so it appears red.

Interference describes when an object affects the path of a wave, or another wave interacts with a wave. Waves interacting with each other can result in either constructive interference or destructive interference, based on their positions. With **constructive interference**, the waves are in sync with each other and combine to reinforce each other. In the case of **deconstructive interference**, the waves are out of sync and reduce the effect of each other to some degree. In **scattering,** the boundary can change the direction or energy of a wave, thus altering the entire wave. **Polarization** changes the oscillations of a wave and can alter its appearance in light waves. For example, polarized sunglasses remove the "glare" from sunlight by altering the oscillation pattern observed by the wearer.

When a wave hits a boundary and is completely reflected, or if it cannot escape from one medium to another, it is called **total internal reflection**. This effect can be seen in the diamonds with a brilliant cut. The angle cut on the sides of the diamond causes the light hitting the diamond to be completely reflected back inside the gem, making it appear brighter and more colorful than a diamond with different angles cut into its surface.

The **Doppler effect** applies to situations with both light and sound waves. The premise of the Doppler effect is that, based upon the relative position or movement of a source and an observer, waves can seem shorter or longer than they actually are. When the Doppler effect is noted with sound, it warps the noise being heard by the observer. This makes the pitch or frequency seem shorter or higher as the source is approaching, and then longer or lower as the source is getting farther away. The frequency/pitch of the source never actually changes, but the sound in respect to the observer makes it seem like the sound has changed. This can be observed when a siren passes by an observer on the road. The siren sounds much higher in pitch as it approaches the observer and then lower after it passes and is getting farther away.

The Doppler effect also applies to situations involving light waves. An observer in space would see light approaching as being shorter wavelengths than the light actually is, causing it to look blue. When the light wave gets farther away, the light would appear red because of the apparent elongation of the wavelength. This is called the **red-blue shift**.

Basic Optics

When reflecting light, a mirror can be used to observe a virtual (not real) image. A **plane mirror** is a piece of glass with a coating in the background to create a reflective surface. An image is what the human eye sees when light is reflected off the mirror in an unmagnified manner. If a **curved mirror** is used for reflection, the image seen will not be a true reflection. Instead, the image will either be enlarged or miniaturized compared to its actual size. Curved mirrors can also make the object appear closer or farther away than the actual distance the object is from the mirror.

Lenses can be used to refract or bend light to form images. Examples of lenses are the human eye, microscopes, and telescopes. The human eye interprets the refraction of light into images that humans understand to be actual size. **Microscopes** allow objects that are too small for the unaided human eye to be enlarged enough to be seen. **Telescopes** allow objects to be viewed that are too far away to be seen with the unaided eye. **Prisms** are pieces of glass that can have a wavelength of light enter one side and appear to be divided into its component wavelengths on the other side. This is due to the ability of the prism to slow certain wavelengths more than others.

Sound

Sound travels in waves and is the movement of vibrations through a medium. It can travel through air (gas), land, water, etc. For example, the noise a human hears in the air is the vibration of the waves as they reach the ear. The human brain translates the different **frequencies** (pitches) and intensities of the vibrations to determine what created the noise.

A **tuning fork** has a predetermined frequency because of the length and thickness of its tines. When struck, it allows vibrations between the two tines to move the air at a specific rate. This creates a specific tone, or note, for that size of tuning fork. The number of vibrations over time is also steady for that tuning fork and can be matched with a frequency. All pitches heard by the human ear are categorized by using frequency and are measured in Hertz (cycles per second).

The level of sound in the air is measured with sound level meters on a **decibel (dB)** scale. These meters respond to changes in air pressure caused by sound waves and measure sound intensity. One decibel is $1/10^{th}$ of a *bel*, named after Alexander Graham Bell, the inventor of the telephone. The decibel scale is logarithmic, so it is measured in factors of 10. This means, for example, that a 10 dB increase on a sound meter equates to a 10-fold increase in sound intensity.

Electricity and Magnetism

Electrical Nature of Common Materials

Generally, an atom carries no net charge because the positive charges of the protons in the nucleus balance the negative charges of the electrons in the outer shells of the atom. This is considered to be electrically neutral. However, since electrons are the only portion of the atom known to have the freedom to "move," this can cause an object to become electrically charged. This happens either through a gain or a loss of electrons. Electrons have a negative charge, so a gain creates a net negative charge for the object. On the contrary, a loss of electrons creates a positive charge for the object. This charge can also be focused on specific areas of an object, causing a notable interaction between charged objects. For example, if a person rubs a balloon on a carpet, the balloon transfers some of is electrons to the carpet. So, if that person were to hold a balloon near his or her hair, the electrons in the "neutral" hair would make the hair stand on end. This is due to the electrons wanting to fill the deficit of electrons

on the balloon. Unless electrically forced into a charged state, most natural objects in nature tend toward reestablishing and maintaining a neutral charge.

When dealing with charges, it is easiest to remember that **like charges repel** each other and **opposite charges attract** each other. Therefore, negatives and positives attract, while two positives or two negatives will repel each other. Similarly, when two charges come near each other, they exert a force on one another. This is described through **Coulomb's Law:**

$$F = k \frac{q_1 q_2}{r^2}$$

In this equation, F is equal to the force exerted by the interaction, k is a constant ($k = 8.99 \times 10^9$ N m^2/C^2), q_1 and q_2 are the measure of the two charges, and r is the distance between the two charges.

When materials readily transfer electricity or electrons, or can easily accept or lose electrons, they are considered to be good **conductors**. The transferring of electricity is called **conductivity**. If a material does not readily accept the transfer of electrons or readily loses electrons, it is considered to be an **insulator.** For example, copper wire easily transfers electricity because copper is a good conductor. However, plastic does not transfer electricity because it is not a good conductor. In fact, plastic is an insulator.

Basic Electrical Concepts

In an electrical circuit, the flow from a power source, or the **voltage,** is "drawn" across the components in the circuit from the positive end to the negative end. This flow of charge creates an electric **current** (I), which is the time (t) rate of flow of net **charge** (q). It is measured with the formula:

$$I = \frac{q}{t}$$

Current is measured in amperes (amps). There are two main types of currents:

1. **Direct current (DC):** a unidirectional flow of charges through a circuit

2. **Alternating current (AC):** a circuit with a changing directional flow of charges or magnitude

Every circuit will show a loss in voltage across its conducting material. This loss of voltage is from resistance within the circuit and can be caused by multiple factors, including resistance from wiring and components such as light bulbs and switches. To measure the **resistance** in a given circuit, **Ohm's law** is used:

$$Resistance = \frac{Voltage}{current} = R = \frac{V}{I}$$

Resistance (R) is measured in Ohms (Ω).

Components in a circuit can be wired in series or in parallel. If the components are wired in **series**, a single wire connects each component to the next in line. If the components are wired in **parallel**, two wires connect each component to the next. The main difference is that the voltage across those in series is directly related from one component to the next. Therefore, if the first component in the series becomes inoperable, no voltage can get to the other components. Conversely, the components in parallel share the voltage across each other and are not dependent on the prior component wired to allow the voltage across the wire.

To calculate the resistance of circuit components wired in series or parallel, the following equations are used:

Resistance in series:

$$R_{total} = R_1 + R_2 + R_3 + \cdots$$

Resistance in parallel:

$$R_{total} = \frac{1}{R_1} + \frac{1}{R_2} + \frac{1}{R_3} + \cdots$$

To make electrons move so that they can carry their charge, a change in voltage must be present. On a small scale, this is demonstrated through the electrons traveling from the light switch to a person's finger. This might happen in a situation where a person runs his or her socks on a carpet, touches a light switch, and receives a small jolt from the electrons that run from the switch to the finger. This minor jolt is due to the deficit of electrons created by rubbing the socks on the carpet, and then the electrons going into the ground. The difference in charge between the switch and the finger caused the electrons to move.

If this situation were to be created on a larger and more sustained scale, the factors would need to be more systematic, predictable, and harnessed. This could be achieved through batteries/cells and generators. Batteries or cells have a chemical reaction that occurs inside, causing energy to be released and charges to be able to move freely. Batteries generally have nodes (one positive and one negative), where items can be hooked up to complete a circuit and allow the charge to travel freely through the item. Generators convert mechanical energy into electric energy using power and movement.

Basic Properties of Magnetic Fields and Forces
Magnetic forces can occur naturally in certain types of materials. If two straight rods are made from iron, they will naturally have a negative end (pole) and a positive end (pole). These charged poles react just like any charged item: opposite charges attract and like charges repel. They will attract each other when arranged positive pole to negative pole. However, if one rod is turned around, the two rods will now repel each other due to the alignment of negative to negative and positive to positive. These types of forces can also be created and amplified by using an electric current. For example, sending an electric current through a stretch of wire creates an electromagnetic force around the wire from the charge of the current. This force exists as long as the flow of electricity is sustained. This magnetic force can also attract and repel other items with magnetic properties. Depending on the strength of the current in the wire, a greater or smaller magnetic force can be generated around the wire. As soon as the current is stopped, the magnetic force also stops.

Life Sciences

From Molecules to Organisms: Structures and Processes

Biology is the study of living organisms and the processes that are vital for life. Scientists who study biology are interested in the origin, evolution, structure, function, growth, and distribution of these living organisms. They study these organisms on a cellular level, individually or as populations, and look at the effects they have on their surrounding environment.

There are five foundations of modern biology: cell theory, evolution, genetics, homeostasis, and energy.

Cell theory is the idea that the cell is the fundamental unit of life. Living organisms are made up of one or more cells, and the products that are generated by those cells. Cells have processes that both produce and use energy, known as metabolism. They also contain **deoxyribonucleic acid (DNA)**, which is hereditary information that gets passed on to subsequent generations.

Evolution is the theory that all living organisms descended from one common ancestor. Charles Darwin generated a viable scientific model of evolution based on the concept of natural selection. Natural selection is the idea that certain species with more advantageous phenotypes are more likely to survive than those species with less advantageous phenotypes. Over long periods of time, populations of species can develop phenotypes specialized for their environment.

Genetics is the study of genes. Genes are the primary unit of inheritance between generations or organisms. They are regions of DNA that encode ribonucleic acid (RNA) and proteins, all of which have specific forms and functions. DNA is present in all living cells and is important for the cell's own functions, as well as for its ability to divide and pass on information to subsequent daughter cells.

Homeostasis is the ability of a system, such as a cell or organism, to regulate itself so that its internal conditions remain stable even when its external environment may be changing. By maintaining homeostasis, systems are able to function normally even in adverse conditions.

The constant flow of **energy** is important for the survival of living organisms. All organisms have vital functions that require the input of energy and other functions that release energy. These processes work together to drive all chemical reactions required for life.

Prokaryotes, Viruses, and Eukaryotes

Every living organism is made up of cells, and these cells come in various shapes and sizes, depending on the organism. There are two types of cells: prokaryotes and eukaryotes. The big difference between

them is that eukaryotes have a nucleus and prokaryotes do not. The structures that will be focused on for this section will be:

Bacteria	Protist, Fungus, Plant, Animal
DNA Ribosomes Cytoplasm Cell Membrane Cell Wall	DNA Ribosomes Cytoplasm Cell Membrane Cell Wall (except animal cells) Unique structures Nucleus Mitochondria Chloroplasts (only autotrophs, or organisms that can produce their own food. Only protists and plants are producers).

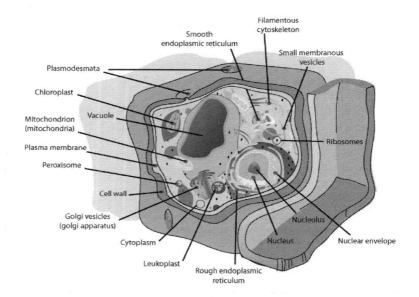

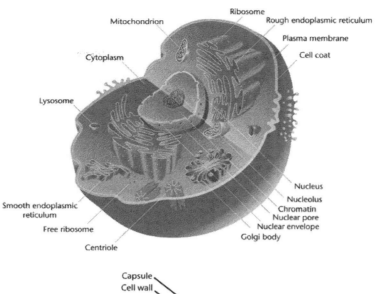

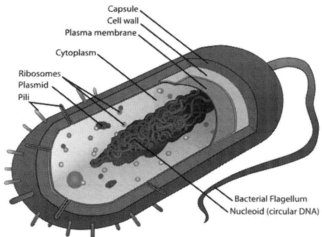

Like all cells, bacterial cells contain DNA, the genetic material that gives instructions for every single structure and process that the cell undergoes. DNA is a code made up of four letters: A (adenine), T (thymine), G (guanine), and C (cytosine). There are billions of these letters in DNA, and the order of these letters tells a cell exactly what to do and how to do it (just like reading a book of instructions).

Because DNA doesn't do anything on its own, all cells must have a means of decoding DNA and turning it into the structure, which is the function of ribosomes—they are protein-makers. If DNA is like a recipe, then the ribosomes are like the chef.

DNA and ribosomes sit in a fluid called cytoplasm, which contains a cytoskeleton (a network of proteins) that holds them in place. All cells need a covering to contain everything inside—these are called cell membranes in animals or cell walls for plant cells.

Bacteria can also have a capsule and a flagellum, which are all external structures. A capsule is sticky and causes bacteria to cluster with other cells or on food. Only about 50 percent of bacteria can move, and those that do often have a flagellum, which is a whip-like structure like a tadpole's tail.

Viruses are commonly thought of as living organisms, but many scientists argue they aren't for two reasons: (1) they are not cells, and (2) they cannot reproduce by themselves. Both qualities are required for an organism to be considered alive. Viruses are unique in that they require a host in order to make proteins and reproduce, because viruses don't have all of the complex tools of a living cell. When a virus has infected a host, it acts like a living organism—it moves and reproduces—but outside of a host, it does nothing. A virus can survive outside of a host, but it cannot reproduce. Scientists are still trying to properly define a virus, so we can currently say that viruses are not like bacteria or any other living thing.

Eukaryotic cells are more complex than prokaryotic cells. They make up all the organisms in the kingdoms protist, fungus, plant, and animal. Eukaryotic cells are also larger than prokaryotes and contain a nucleus and other organelles.

Eukaryotic cells hold their DNA inside a nucleus in pieces called chromosomes. Chromosomes are a cell's way of organizing long strands of DNA in twisted-up bundles. Imagine a room filled with rolls of toilet paper compared to a room that has all of those rolls unraveled and thrown everywhere; it would be a mess!

Other important organelles include chloroplasts and mitochondria. Chloroplasts can be found in cells called autotrophs, which can convert sunlight into energy. Plants are autotrophs. Mitochondria are little energy factories found in almost every type of cell. They use chemical reactions to make little packets of energy that can be used by other parts of the cell.

Structure and Function of Animal and Plant Cell Organelles

Animal and plant cells contain many of the same or similar **organelles,** which are membrane enclosed structures that each have a specific function; however, there are a few organelles that are unique to either one or the other general cell type. The following cell organelles are found in both animal and plant cells, unless otherwise noted in their description:

Nucleus: The nucleus consists of three parts: nuclear envelope, nucleolus, and chromatin. The **nuclear envelope** is the double membrane that surrounds the nucleus and separates its contents from the rest of the cell. It is porous so substances can pass back and forth between the nucleus and the other parts of the cell. It is also continuous, with the endoplasmic reticulum that is present within the cytosol of the cell. The **nucleolus** is in charge of producing ribosomes. **Chromosomes are comprised of tightly coiled proteins, RNA, and DNA and are collectively called chromatin.**

Endoplasmic Reticulum (ER): The ER is a network of membranous sacs and tubes responsible for membrane synthesis and other metabolic and synthetic activities of the cell. There are two types of ER, rough and smooth. Rough ER is lined with ribosomes and is the location of protein synthesis. This provides a separate compartment for site-specific protein synthesis and is important for the intracellular transport of proteins. Smooth ER does not contain ribosomes and is the location of lipid synthesis.

Flagellum: The flagellum is found in protists and animal cells. It is a cluster of microtubules projected out of the plasma membrane and aids in cell motility.

Centrosome: The centrosome is the area of the cell where microtubules are created and organized for mitosis. Each centrosome contains two **centrioles.**

Cytoskeleton: The cytoskeleton in animal cells is made up of microfilaments, intermediate filaments, and microtubules. In plant cells, the cytoskeleton is made up of only microfilaments and microtubules. These structures reinforce the cell's shape and aid in cell movements.

Microvilli: Microvilli are found only in animal cells. They are protrusions in the cell membrane that increase the cell's surface area. They have a variety of functions, including absorption, secretion, and cellular adhesion.

Peroxisome: A peroxisome contains enzymes that are involved in many of the cell's metabolic functions, one of the most important being the breakdown of fatty acid chains. It produces hydrogen peroxide as a by-product of these processes and then converts the hydrogen peroxide to water.

Mitochondrion: The mitochondrion, considered the cell's powerhouse, is one of the most important structures for maintaining regular cell function. It is where cellular respiration occurs and where most of the cell's ATP is generated.

Lysosome: Lysosomes are found exclusively in animal cells. They are responsible for digestion and can hydrolyze macromolecules.

Golgi Apparatus: The Golgi apparatus is responsible for synthesizing, modifying, sorting, transporting, and secreting cell products. Because of its large size, it was one of the first organelles studied in detail.

Ribosomes: Ribosomes are found either free in the cytosol, bound to the rough ER, or bound to the nuclear envelope. They are also found in prokaryotes. Ribosomes make up a complex that forms proteins within the cell.

Plasmodesmata: Found only in plant cells, plasmodesmata are cytoplasmic channels, or tunnels, that go through the cell wall and connect the cytoplasm of adjacent cells.

Chloroplast: Chloroplasts are found in protists, such as algae and plant cells. It is responsible for photosynthesis, which is the process of converting sunlight to chemical energy that is stored and used later to drive cellular activities.

Central Vacuole: A central vacuole is found only in plant cells, and is responsible for storage, breakdown of waste products, and hydrolysis of macromolecules.

Plasma Membrane: The plasma membrane is a phospholipid bilayer that encloses the cell. It is also found in prokaryotes.

Cell Wall: Cell walls are present in fungi, plant cells, and some protists. The cell wall is made up of strong fibrous substances, including cellulose (plants), chitin (fungi) and other polysaccharides, and protein. It is a layer outside of the plasma membrane that protects the cell from mechanical damage and helps maintain the cell's shape.

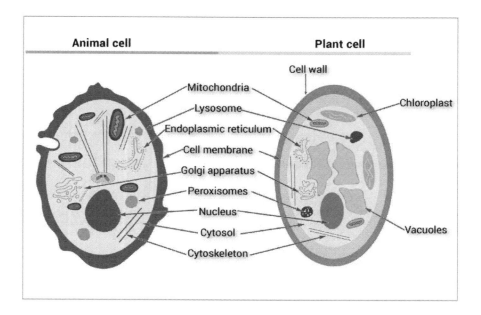

Levels of Organization

There are about two hundred different types of **cells** in the human body. Cells group together to form biological **tissues**, and tissues combine to form **organs**, such as the heart and kidneys. Organs that work together to perform vital functions of the human body form **organ systems**. There are eleven organ systems in the human body: skeletal, muscular, urinary, nervous, digestive, endocrine, reproductive, respiratory, cardiovascular, integumentary, and lymphatic. Although each system has its own unique function, they all rely on each other, either directly or indirectly, to operate properly.

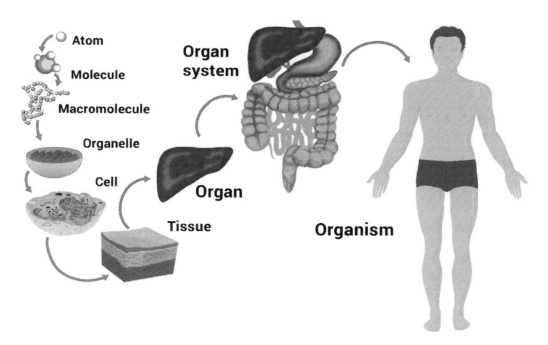

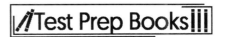

Cellular Organization

Prokaryotes contain ribosomes, DNA, cytoplasm, a cell membrane, a cytoskeleton, and a cell wall. Eukaryotes vary between kingdoms but contain all of these structures except a cell wall because animal cells require so much mobility. Large, land-dwelling animals typically compensate with an exoskeleton (like insects) or an endoskeleton (like humans and other mammals, reptiles, and birds) for structure.

All bacterial cells are unicellular (existing as just one cell). Almost all types of protist and some species in fungi kingdom are unicellular, but they still have the complicated organelles of eukaryotes. A few protists, almost all fungi, and all plants and animals are multicellular. Multicellularity leads to development of structures that are perfectly designed for their function.

Cells combine to form tissue. Tissue combines to form organs. Organs combine to form organ systems, and organ systems combine to form one organism. The structures of all of these combinations allow for the maximum functionality of an organism, as demonstrated by the nervous system.

A neuron is a cell in the nervous system designed to send and receive electrical impulses. Neurons have dendrites, which are sensors waiting to receive a message. Neurons also have an axon, a long arm that sends the message to the neighboring neuron. The axon also has insulation known as myelin that speeds the message along. Many neurons combine to form a nerve, the tissue of the nervous system, which is like a long wire. The structure of this nerve is perfect—it is a long cable whose function is to send signals to the brain so the brain can process the information and respond. Nerve tissue combines with other tissue to form the brain, a complex structure of many parts.

The brain also has glands (epithelial tissue) that release hormones to control processes in our body. The brain and spinal cord together form the central nervous system that controls the stimulus/response signaling in our body. The nervous system coordinates with the circulatory system to make our heart beat, the digestive system to control food digestion, the muscular system to move an arm, the respiratory system to facilitate breathing, and all other body systems to make the entire organism functional. Cells are the basic building block in our bodies, and their structure is critical for their function and the function of the tissues, organs, and systems that they comprise.

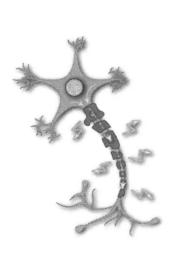
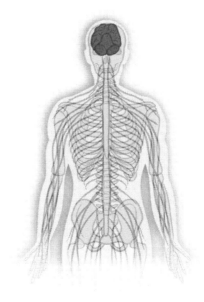

In the graphic above, the left depicts a neuron, and the right depicts the nervous system. A neuron is a nerve cell, and it is the basic building block of the nervous system. Cell, tissue, organ, and organ system structure are critical for function.

The following table lists organ systems in the human body:

Name	Function	Main organs
Nervous	Detect stimuli and direct response	Brain and spinal cord
Circulatory	Pump blood to deliver oxygen to cells so they can perform cellular respiration	Heart
Respiratory	Breathe in oxygen (reactant for cellular respiration) and release carbon dioxide waste	Lungs
Muscular	Movement	Heart and muscles
Digestive	Break down food so that glucose can be delivered to cells for energy	Stomach, small intestine, lots of others
Skeletal	Support and organ protection	All sorts of joints, skull, ribcage

Cell Cycle

The **cell cycle** is the process by which a cell divides and duplicates itself. There are two processes by which a cell can divide itself: mitosis and meiosis. In **mitosis,** the daughter cells that are produced from parental cell division are identical to each other and the parent. **Meiosis** is a unique process that

involves two stages of cell division and produces **haploid cells,** which are cells containing only one set of chromosomes, from **diploid parent cells**, which are cells containing two sets of chromosomes.

The Cell Cycle

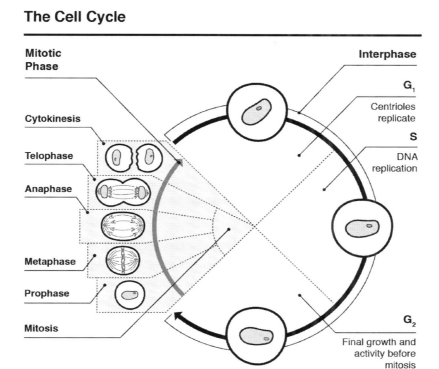

Mitosis

Mitosis can be broken down into five stages: prophase, prometaphase, metaphase, anaphase, and telophase.

- **Prophase:** During this phase, the mitotic spindles begin to form from centrosomes and microtubules. As the microtubules lengthen, the centrosomes move farther away from each other. The nucleolus disappears and the chromatin fibers begin to coil up and form chromosomes. Two sister **chromatids**, which are two copies of one chromosome, are joined together.

- **Prometaphase:** The nuclear envelope begins to break down and the microtubules enter the nuclear area. Each pair of chromatin fibers develops a **kinetochore**, which is a specialized protein structure in the middle of the adjoined fibers. The chromosomes are further condensed.

- **Metaphase:** In this stage, the microtubules are stretched across the cell and the centrosomes are at opposite ends of the cell. The chromosomes align at the **metaphase plate**, which is a plane that is exactly between the two centrosomes. The kinetochore of each chromosome is attached to the kinetochore of the microtubules that are stretching from each centrosome to the metaphase plate.

- **Anaphase:** The sister chromatids break apart, forming full-fledged chromosomes. The two daughter chromosomes move to opposite ends of the cell. The microtubules shorten toward opposite ends of the cell as well, and the cell elongates.

- **Telophase:** Two nuclei form at each end of the cell and nuclear envelopes begin to form around each nucleus. The nucleoli reappear and the chromosomes become less condensed. The microtubules are broken down by the cell and mitosis is complete.

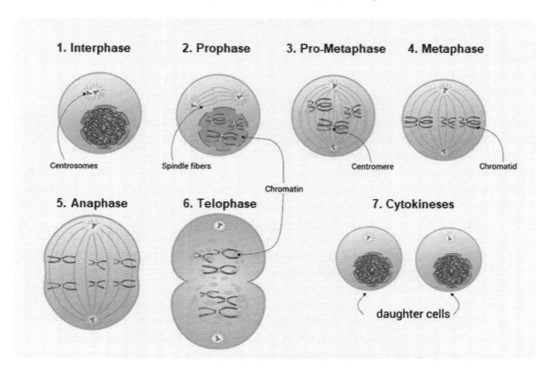

Meiosis

Meiosis is a type of cell division in which the daughter cells have half as many sets of chromosomes as the parent cell. In addition, one parent cell produces four daughter cells. Meiosis has the same phases as mitosis, except that they occur twice—once in meiosis I and once in meiosis II. The diploid parent has two sets of chromosomes, set A and set B. During meiosis I, each chromosome set duplicates, producing a second set of A chromosomes and a second set of B chromosomes, and the cell splits into two. Each cell contains two sets of chromosomes. Next, during meiosis II, the two intermediate daughter cells divide again, producing four total haploid cells that each contain one set of chromosomes. Two of the

haploid cells each contain one chromosome of set A and the other two cells each contain one chromosome of set B.

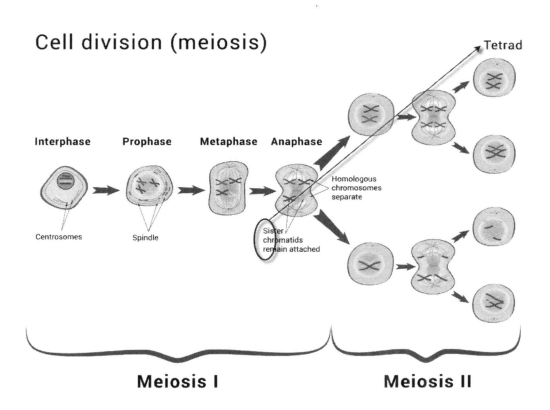

Cell division (meiosis)

Interphase   Prophase   Metaphase   Anaphase

Centrosomes   Spindle

Homologous chromosomes separate

Sister chromatids remain attached

Tetrad

Meiosis I   **Meiosis II**

Cytokinesis
Cytokinesis is the division of cytoplasm that occurs immediately following the division of genetic material during cellular reproduction. The process of mitosis or meiosis, followed by cytokinesis, finishes up the cell cycle.

Cellular Respiration
Cellular respiration is a set of metabolic processes that converts energy from nutrients into ATP, which is the molecule of useable energy for the cell. Respiration can either occur **aerobically**, using oxygen, or **anaerobically**, without oxygen. While prokaryotic cells carry out respiration in the cytosol, most of the aerobic respiration in eukaryotic cells occurs in the mitochondria. Glycolysis and ATP-PC (phosphocreatine system) take place in the cytosol.

Anaerobic Respiration
Some organisms do not live in oxygen-rich environments and must find alternate methods of respiration. Anaerobic respiration occurs in certain prokaryotic organisms, and while it does occur in eukaryotic organisms, it happens in them much less frequently. The organisms utilize an electron transport chain similar to that of the aerobic respiration pathway; the terminal acceptor molecule, however, is an electronegative substance that is not an oxygen molecule. Some bacteria, for example, use the sulfate ion (SO_4^{2-}) as the final electron accepting molecule and the resulting byproduct is hydrogen sulfide (H_2S), instead of water.

Aerobic Respiration

There are two main steps in aerobic cellular respiration: the **citric acid cycle**, also known as the **Krebs cycle,** and **oxidative phosphorylation**. A process called **glycolysis** converts glucose molecules into pyruvate molecules and those pyruvate molecules then enter the citric acid cycle. The pyruvate molecules are broken down to produce ATP, as well as NADH and $FADH_2$—molecules that are used energetically to drive the next step of oxidative phosphorylation. During this phase of aerobic respiration, an electron transport chain pumps electrons and protons across the inner mitochondrial matrix. The electrons are accepted by an oxygen molecule, and water is produced. This process then fuels **chemiosmosis**, which helps convert ADP molecules to ATP. The total number of ATP molecules

generated through aerobic respiration can be as many as thirty-eight, if none are lost during the process. Aerobic respiration is up to fifteen times more efficient than anaerobic respiration.

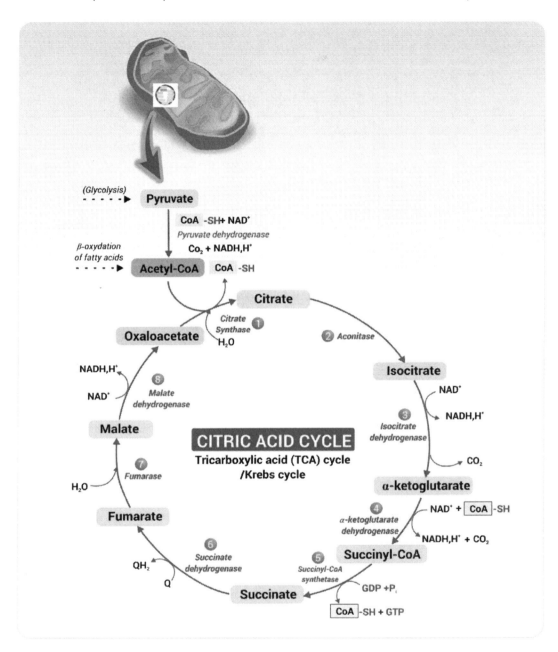

Photosynthesis

Photosynthesis is the process of converting light energy into chemical energy, which is then stored in sugar and other organic molecules. It can be divided into two stages called the **light reactions** and the **Calvin cycle.** The photosynthetic process takes place in the chloroplast in plants. Inside the chloroplast, there are membranous sacs called **thylakoids. Chlorophyll** is a green pigment that lives in the thylakoid membranes, absorbs photons from light, and starts an electron transport chain in order to produce energy in the form of ATP and NADPH. The ATP and NADPH produced from the light reactions are used as energy to form organic molecules in the Calvin cycle.

The Calvin cycle takes place in the *stroma*, or inner space, of the chloroplasts. The process consumes nine ATP molecules and six NADPH molecules for every one molecule of glyceraldehyde 3-phosphate (G3P) that it produces. The G3P that is produced can be used as the starting material to build larger organic compounds, such as glucose. The complex series of reactions that takes place in photosynthesis can be simplified into the following equation: $6 CO_2 + 12 H_2O + Light Energy \rightarrow C_6H_{12}O_6 + 6 O_2 + 6 H_2O$.

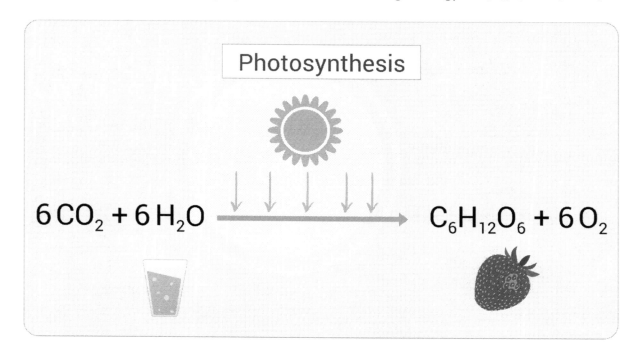

Photosynthesis

$$6 CO_2 + 6 H_2O \longrightarrow C_6H_{12}O_6 + 6 O_2$$

Basically, carbon dioxide and water mix with light energy inside the chloroplast to produce organic molecules, oxygen, and water. It is interesting to note that water is on both sides of the equation. Twelve water molecules are consumed during this process and six water molecules are newly formed as byproducts. Although the Calvin cycle itself is not dependent on light energy, both steps of photosynthesis usually occur during daylight because the Calvin cycle is dependent upon the ATP and NADPH that is produced by the light reactions.

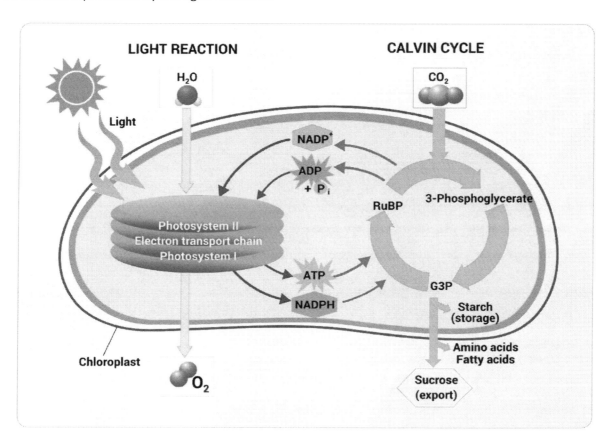

Biological Molecules

Repeating units of **monomers** (small molecules that bond with identical small molecules) that are linked together are called **polymers.** The most important polymers found in all living things can be divided into five categories: nucleic acids (such as DNA), carbohydrates, proteins, lipids, and enzymes. Carbon (C), hydrogen (H), oxygen (O), nitrogen (N), sulfur (S), and phosphorus (P) are the major elements of most biological molecules. Carbon is a common backbone of large molecules because of its ability to bond to four different atoms.

DNA and RNA

Nucleotides consist of a five-carbon sugar, a nitrogen-containing base, and one or more phosphate groups. **Deoxyribonucleic acid (DNA)** is made up of two strands of nucleotides coiled together in a double-helix structure. It plays a major role in enabling living organisms to pass their genetic information and complex components on to subsequent generations. There are four nitrogenous bases that make up DNA: adenine, thymine, guanine, and cytosine. Adenine always pairs with thymine, and guanine always

pairs with cytosine. **Ribonucleic acid (RNA)** is often made up of only one strand of nucleotides folded in on itself. Like DNA, RNA has four nitrogenous bases; however, in RNA, thymine is replaced by uracil.

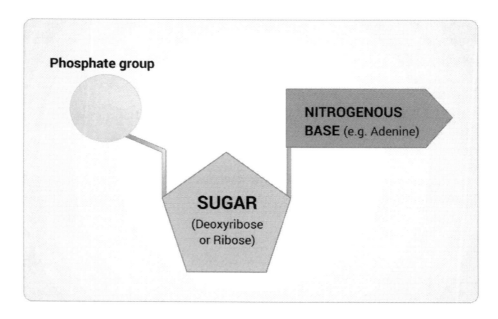

Carbohydrates

Carbohydrates consist of sugars and polymers of sugars, such as starches, which make up the cell walls of plants. The simplest sugar is called a **monosaccharide** and has the molecular formula of CH_2O, or a multiple of that formula. Monosaccharides are important molecules for cellular respiration. Their carbon skeleton can also be used to rebuild new small molecules. **Polysaccharides** are made up of a few hundred to a few thousand monosaccharides linked together.

Proteins

Proteins are essential for almost all functions in living beings. All proteins are made from a set of twenty **amino acids** that are linked in unbranched polymers. The amino acids are linked by **peptide bonds**, and polymers of amino acids are called **polypeptides**. These polypeptides, either individually or in linked combination with each other, fold up and form coils of biologically-functional molecules.

There are four levels of protein structure: primary, secondary, tertiary, and quaternary. The **primary structure** is the sequence of amino acids, similar to the letters in a long word. The **secondary structure** comprises the folds and coils that are formed by hydrogen bonding between the slightly charged atoms of the polypeptide backbone. **Tertiary structure** is the overall shape of the molecule that results from the interactions between the side chains that are linked to the polypeptide backbone. **Quaternary structure** is the overall protein structure that occurs when a protein is made up of two or more polypeptide chains.

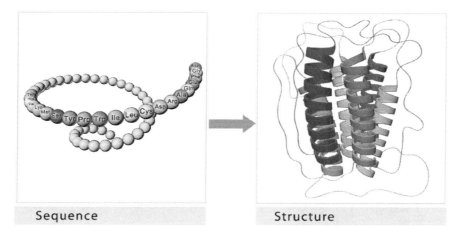

Sequence   Structure

Lipids

Lipids are a class of biological molecules that are *hydrophobic*, which means that they do not mix well with water. They are mostly made up of large chains of carbon and hydrogen atoms, termed **hydrocarbon chains**. The three most important types of lipids are fats, phospholipids, and steroids.

Fats are made up of two types of smaller molecules: three fatty acids and one glycerol molecule. **Saturated fats** do not have double bonds between the carbons in the fatty acid chain, such as glycerol, pictured below. They are fairly straight molecules and can pack together closely, so they form solids at room temperature. **Unsaturated fats** have one or more double bonds between carbons in the fatty acid chain. Since they cannot pack together as tightly as saturated fats, they take up more space and are called oils. They remain liquid at room temperature.

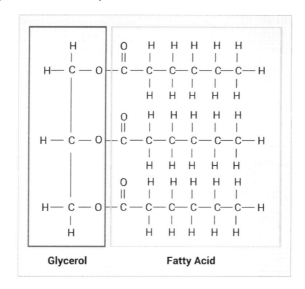

Glycerol   Fatty Acid

Phospholipids are made up of two fatty acid molecules linked to one glycerol molecule. When phospholipids are mixed with water, they inherently create double-layered structures, called bilayers, which shield their hydrophobic regions from the water molecules.

Steroids are lipids that consist of four fused carbon rings. They can mix in between the phospholipid bilayer cell membrane and help maintain its structure, as well as aid in cell signaling.

Enzymes

Enzymes are biological molecules that accelerate the rate of chemical reactions by lowering the activation energy needed to make the reaction proceed. Although most enzymes can be classified as proteins, some are ribonucleic acid (RNA) molecules. Enzymes function by interacting with a specific substrate in order to create a different molecule, or product. Most reactions in cells need enzymes to make them occur at rates fast enough to sustain life.

Human Anatomy and Physiology

Anatomy is the study of external and internal body parts and structures, and their physical relationships to each other. **Physiology** is the study of the function of living organisms and their body parts. Understanding these areas of study is easier after learning the specific terms relating to the field.

Regional and Directional Terms

- **Anterior:** The front; can also mean before.
- **Ventral:** The belly side.
- **Posterior:** The back; can also mean behind.
- **Dorsal:** The back side.
- **Cranial** or **Cephalic:** Toward the head.
- **Superior:** Above; can also mean at a higher level.
- **Caudal:** The tail.
- **Inferior:** Below; can also mean at a lower level; can also mean toward the feet.
- **Medial:** Toward the midline.
- **Lateral:** Away from the midline.
- **Proximal:** Toward an attached base.
- **Distal:** Away from an attached base.

Sections of the Body

- **Transverse** or **horizontal:** A horizontal section in the anatomical position that separates superior and inferior portions of the body.

- **Sagittal:** A vertical section in the anatomical position that separates right and left portions of the body. A midsagittal section separates the body into equal parts, while a parasagittal section separates it into unequal parts.

- **Frontal** or **coronal:** A vertical section in the anatomical position that separates the anterior and posterior portions of the body.

Histology

Histology is the examination of specialized cells and cell groups that perform a specific function by working together. Although there are trillions of cells in the human body, there are only 200 different types of cells. Groups of cells form biological **tissues**, and tissues combine to form organs, such as the

heart and kidney. **Organs** are structures that have many different functions that are vital to living creatures. There are four primary types of tissue: epithelial, connective, muscle, and neural. Each tissue type has specific characteristics that enable organs and organ systems to function properly.

Epithelial tissue includes epithelia and glands. **Epithelia** are the layers of cells that cover exposed surfaces and line internal cavities and passageways. **Glands** are structures that are involved in secretion of fluids. Epithelia do not contain blood vessels and can often regenerate quickly to replace dead and damaged cells.

Connective tissue fills internal spaces and is not exposed to the outside of the body. It provides structural support for the body and stores energy. Fluid can be transported by connective tissue between different regions of the body. This type of tissue is also a protective barrier for delicate organs and for the body against microorganisms.

Muscle tissue has characteristics that allow motion. **Skeletal muscles** are long fibers of actin and myosin that slide past each other to cause a contraction, which shortens the muscle. They are filled with mitochondria, because they expend a lot of energy. **Smooth muscle** tissue is structured differently, because its movement is in the form of peristalsis. For example, the esophagus moves food in a wave-like contraction, as opposed to a shortening of muscle to bring structures together. Smooth muscle tissues make up internal organs, with the exception of the heart, which is composed of thick, contracting muscle tissue called **cardiac muscle.**

Neural tissue conducts electrical impulses which help send information and instructions throughout the body. Most of it is concentrated in the brain and spinal cord. The changes in frequency and patterns of the impulse are important distinctions in the messages being sent. Its structure is composed of message-receiving projections called **dendrites** and a long, myelinated **axon**. The **myelin sheath** surrounds the axon to conduct the action potential, ultimately causing neurotransmitter release at the **synapse**, or the boundary between the axon of one cell and the dendrite of another.

Glands

Glands are organs that synthesize and secrete chemical substances, such as hormones, for use inside the body or discharge outside the body. There are two types of glands: endocrine and exocrine. **Endocrine glands** secrete hormones into the bloodstream and are important in maintaining homeostasis within the body. They do not have a duct system. Examples of endocrine glands are the pancreas, the pineal gland, the thymus gland, the pituitary gland, the thyroid gland, and the adrenal gland. **Exocrine glands** have ducts that are used to secrete substances to the surface of the body and can be classified into three types: apocrine, holocrine, and merocrine. In **apocrine glands**, part of the cell's body is lost during secretion; in **holocrine glands**, the whole cell body disintegrates during secretion; and in **merocrine glands**, cells use exocytosis to secrete fluids. **Exocytosis** occurs when the chemical substance is carried in a vacuole across the cell membrane for release outside of the cell. The cells remain intact in merocrine glands.

Cartilage

Cartilage is a firm gel substance that contains complex polysaccharides, called **chondroitin sulfates**. It contains **collagen fibers**, which provide tensile strength, and **chondrocytes**, which are cartilage cells. It is an avascular material, because the chondrocytes do not allow blood vessels to form within the fibrous network. There are three types of cartilage: hyaline cartilage, elastic cartilage, and fibrocartilage. **Hyaline cartilage** is the body's most common cartilage, and is made of tightly packed collagen fibers. It is tough and flexible, but is also the weakest type of cartilage. **Elastic cartilage** contains elastic fibers that make is very resilient and flexible. **Fibrocartilage** has densely woven collagen fibers that make it very durable and tough.

Integumentary System (Skin)

Skin consists of three layers: epidermis, dermis, and the hypodermis. There are four types of cells that make up the keratinized stratified squamous epithelium in the epidermis. They are keratinocytes, melanocytes, Merkel cells, and Langerhans cells. Skin is composed of many layers, starting with a basement membrane. On top of that sits the stratum germinativum, the stratum spinosum, the stratum granulosum, the stratum lucidum, and then the stratum corneum at the outer surface. Skin can be classified as thick or thin. These descriptions refer to the epidermis layer. Most of the body is covered with thin skin, but areas such as the palm of the hands are covered with thick skin. The dermis consists of a superficial papillary layer and a deeper reticular layer. The **papillary layer** is made of loose connective tissue, containing capillaries and the axons of sensory neurons. The **reticular layer** is a meshwork of tightly packed irregular connective tissue, containing blood vessels, hair follicles, nerves, sweat glands, and sebaceous glands. The hypodermis is a loose layer of fat and connective tissue. Since it is the third layer, if a burn reaches this third degree, it has caused serious damage.

Sweat glands and sebaceous glands are important exocrine glands found in the skin. **Sweat glands** regulate temperature, and remove bodily waste by secreting water, nitrogenous waste, and sodium salts to the surface of the body. Some sweat glands are classified as apocrine glands. **Sebaceous glands** are holocrine glands that secrete **sebum**, which is an oily mixture of lipids and proteins. Sebum protects the skin from water loss, as well as bacterial and fungal infections.

The three major functions of skin are protection, regulation, and sensation. Skin acts as a barrier and protects the body from mechanical impacts, variations in temperature, microorganisms, and chemicals. It regulates body temperature, peripheral circulation, and fluid balance by secreting sweat. It also contains a large network of nerve cells that relay changes in the external environment to the body.

Skeletal System

The **skeletal system** consists of the 206 bones that make up the skeleton, as well as the cartilage, ligaments, and other connective tissues that stabilize them. **Bone** is made of collagen fibers and calcium inorganic minerals, mostly in the form of hydroxyapatite, calcium carbonate, and phosphate salts. The inorganic minerals are strong but brittle, The inorganic minerals are strong but brittle, and the collagen fibers are weak but flexible, so the combination makes bone resistant to shattering. There are two types of bone: compact and spongy. **Compact bone** has a basic functional unit, called the **Haversian system**. **Osteocytes**, or bone cells, are arranged in concentric circles around a central canal, called the **Haversian canal**, which contains blood vessels. While Haversian canals run parallel to the surface of the bone, perforating canals, also known as the **canals of Volkmann**, run perpendicularly between the central canal and the surface of the bone. The concentric circles of bone tissue that surround the central canal within the Haversian system are called **lamellae**.

The spaces that are found between the lamellae are called **lacunae**. The Haversian system is a reservoir for calcium and phosphorus for blood. **Spongy bone**, in contrast to compact bone, is lightweight and porous. It has a branching network of parallel lamellae, called **trabeculae**.

Although spongy bone forms an open framework inside the compact bone, it is still quite strong. Different bones have different ratios of compact-to-spongy bone, depending on their functions. The outside of the bone is covered by a **periosteum**, which has four major functions. It isolates and protects bones from the surrounding tissue; provides a place for attachment of the circulatory and nervous

system structures; participates in growth and repair of the bone; and attaches the bone to the deep fascia. An **endosteum** is found inside the bone, covers the trabeculae of the spongy bone and lines the inner surfaces of the central canals.

One major function of the skeletal system is to provide structural support for the entire body. It provides a framework for the soft tissues and organs to attach to. The skeletal system also provides a reserve of important nutrients, such as calcium and lipids. Normal concentrations of calcium and phosphate in body fluids are partly maintained by the calcium salts stored in bone. Lipids that are stored in **yellow bone marrow** can be used as a source of energy. Yellow bone marrow also produces some white blood cells. **Red bone marrow** produces red blood cells, most white blood cells, and platelets that circulate in the blood. Certain groups of bones form protective barriers around delicate organs. The ribs, for example, protect the heart and lungs, the skull encloses the brain, and the vertebrae cover the spinal cord.

Muscular System

The **muscular system** of the human body is responsible for all movement that occurs. There are approximately 700 muscles in the body that are attached to the bones of the skeletal system and that make up half of the body's weight. Muscles are attached to the bones through tendons. **Tendons** are made up of dense bands of connective tissue and have collagen fibers that firmly attach to the bone on one side and the muscle on the other. Their fibers are actually woven into the coverings of the bone and muscle so they can withstand the large forces that are put on them when muscles are moving. There are three types of muscle tissue in the body: **Skeletal muscle** tissue pulls on the bones of the skeleton and causes body movement; **cardiac muscle** tissue helps pump blood through veins and arteries; and **smooth muscle** tissue helps move fluids and solids along the digestive tract and contributes to movement in other body systems. All of these muscle tissues have four important properties in common: They are **excitable**, meaning they respond to stimuli; **contractile**, meaning they can shorten and pull on connective tissue; **extensible**, meaning they can be stretched repeatedly, but maintain the ability to contract; and **elastic**, meaning they rebound to their original length after a contraction.

Muscles begin at an **origin** and end at an **insertion**. Generally, the origin is proximal to the insertion and the origin remains stationary while the insertion moves. For example, when bending the elbow and moving the hand up toward the head, the part of the forearm that is closest to the wrist moves and the part closer to the elbow is stationary. Therefore, the muscle in the forearm has an origin at the elbow and an insertion at the wrist.

Body movements occur by muscle contraction. Each contraction causes a specific action. Muscles can be classified into one of three muscle groups based on the action they perform. **Primary movers**, or **agonists**, produce a specific movement, such as flexion of the elbow. **Synergists** are in charge of helping the primary movers complete their specific movements. They can help stabilize the point of origin or provide extra pull near the insertion. Some synergists can aid an agonist in preventing movement at a joint. **Antagonists** are muscles whose actions are the opposite of that of the agonist. If an agonist is contracting during a specific movement, the antagonist is stretched. During flexion of the elbow, the biceps brachii contracts and acts as an agonist, while the triceps brachii on the opposite side of the upper arm acts as an antagonist and stretches.

Skeletal muscle tissue has several important functions. It causes movement of the skeleton by pulling on tendons and moving the bones. It maintains body posture through the contraction of specific muscles responsible for the stability of the skeleton. Skeletal muscles help support the weight of internal organs and protect these organs from external injury. They also help to regulate body temperature within a normal range. Muscle contractions require energy and produce heat, which heats the body when cold.

Nervous System

Although the **nervous system** is one of the smallest organ systems in the human body, it is the most complex. It consists of all of the neural tissue, and is in charge of controlling and adjusting the activities of all of the other systems of the body. Neural responses to stimuli are often fast, but disappear quickly once the neural activity stops. Neural tissue contains two types of cells: neurons and neuroglia. **Neurons**, or nerve cells, are the main cells responsible for transferring and processing information in the nervous system. **Neuroglia** support the neurons by providing a framework around them and isolating them from the surrounding environment. They also act as **phagocytes** and protect neurons from harmful substances.

The nervous system is made of the **central nervous system (CNS)** and the **peripheral nervous system (PNS)**. The CNS includes the brain and the spinal cord, while the PNS includes the rest of the neural tissue not included in the CNS. The CNS is where intelligence, memory, learning, and emotions are processed. It is responsible for processing and coordinating sensory data and motor commands. The PNS is responsible for relaying sensory information and motor commands between the CNS and peripheral tissues and systems. The PNS has two subdivisions, known as the afferent and efferent divisions. While the **afferent** division relays sensory information to the CNS, the **efferent** division transmits motor commands to muscles and glands. The efferent division consists of the **somatic nervous system (SNS)**, which controls skeletal muscle contractions, and the **autonomic nervous system (ANS)**, which regulates activity of smooth muscle, cardiac muscle, and glands.

Two types of pathways are used to communicate information between the brain and the peripheral tissues. **Sensory pathways** start in a peripheral system and end in the brain. **Motor pathways** carry information from the brain to peripheral systems. Motor commands often occur in response to the information transmitted through a sensory pathway. Processing in both pathways happens at several points along the way, where neurons pass the information to each other.

The nervous system is responsible for processing both general senses and specialized senses. **General senses** include temperature, pain, touch, pressure, vibration and proprioception. **Specialized senses** include **olfaction** (smell), **gustation** (taste), equilibrium, hearing, and vision. The information from each sense is processed through a specific receptor for that sense. A receptor that is sensitive to touch may not be responsive to chemical stimuli, for example. The specificity of the receptor is developed either from its individual structure or from accessory cells or structures creating a shield against other senses.

Endocrine System

The **endocrine system** is made of the ductless tissues and glands that secrete hormones into the interstitial fluids of the body. **Interstitial fluid** is the solution that surrounds tissue cells within the body. This system works closely with the nervous system to regulate the physiological activities of the other systems of the body to maintain homeostasis. While the nervous system provides quick, short-term responses to stimuli, the endocrine system acts by releasing hormones into the bloodstream that get distributed to the whole body. The response is slow but long-lasting, ranging from a few hours to a few weeks.

Hormones are chemical substances that change the metabolic activity of tissues and organs. While regular metabolic reactions are controlled by enzymes, hormones can change the type, activity, or quantity of the enzymes involved in the reaction. They bind to specific cells and start a biochemical

chain of events that changes the enzymatic activity. Hormones can regulate development and growth, digestive metabolism, mood, and body temperature, among other things. Often small amounts of hormone will lead to large changes in the body.

The following are the major endocrine glands in the body:

- **Hypothalamus:** A part of the brain, the hypothalamus connects the nervous system to the endocrine system via the pituitary gland. Although it is considered part of the nervous system, it plays a dual role in regulating endocrine organs.

- **Pituitary Gland:** A pea-sized gland found at the bottom of the hypothalamus. It has two lobes, called the anterior and posterior lobes. It plays an important role in regulating the function of other endocrine glands. The hormones released control growth, blood pressure, certain functions of the sex organs, salt concentration of the kidneys, internal temperature regulation, and pain relief.

- **Thyroid Gland:** This gland releases hormones, such as thyroxine, that are important for metabolism, growth and development, temperature regulation, and brain development during infancy and childhood. Thyroid hormones also monitor the amount of circulating calcium in the body.

- **Parathyroid Glands:** These are four pea-sized glands located on the posterior surface of the thyroid. The main hormone secreted is called **parathyroid hormone (PTH)** and helps with the thyroid's regulation of calcium in the body.

- **Thymus Gland:** The thymus is located in the chest cavity, embedded in connective tissue. It produces several hormones important for development and maintenance of normal immunological defenses. One hormone promotes the development and maturation of lymphocytes, which strengthens the immune system.

- **Adrenal Gland:** One adrenal gland is attached to the top of each kidney. It produces adrenaline and is responsible for the "fight or flight" reactions in the face of danger or stress. The hormones epinephrine and norepinephrine cooperate to regulate states of arousal.

- **Pancreas:** The pancreas is an organ that has both endocrine and exocrine functions. The endocrine functions are controlled by the pancreatic **islets of Langerhans**, which are groups of beta cells scattered throughout the gland that secrete insulin to lower blood sugar levels in the body. Neighboring alpha cells secrete glucagon to raise blood sugar.

- **Pineal Gland:** The pineal gland secretes **melatonin**, a hormone derived from the neurotransmitter serotonin. Melatonin can slow the maturation of sperm, oocytes, and reproductive organs. It also regulates the body's circadian rhythm, which is the natural awake/asleep cycle. It also serves an important role in protecting the CNS tissues from neural toxins.

- **Testes and Ovaries:** These glands secrete testosterone and estrogen, respectively, and are responsible for secondary sex characteristics, as well as reproduction.

Circulatory System
The **circulatory system** is composed of the heart and blood vessels of the body. The **heart** is the main organ of the circulatory system. It acts as a pump and works to circulate blood throughout the body. Gases, nutrients, and waste are constantly exchanged between the circulating blood and interstitial fluid, keeping tissues and organs alive and healthy. The circulatory system is divided into the pulmonary

and systemic circuits. The **pulmonary circuit** is responsible for carrying carbon dioxide-rich blood to the lungs and returning oxygen-rich blood to the heart. The **systemic circuit** transports the oxygen-rich blood to the rest of the body and returns carbon dioxide-rich blood to the heart.

The heart is located posterior to the sternum, on the left side, in the front of the chest. The heart wall is made of three distinct layers. The outer layer, the **epicardium**, is a serous membrane that is also known as the **visceral pericardium**. The middle layer is called the **myocardium**, and contains connective tissue, blood vessels, and nerves within its layers of cardiac muscle tissue. The inner layer is the **endocardium**, and is made of a simple squamous epithelium. This layer includes the heart valves, and is continuous with the endothelium of the attached blood vessels.

The heart has four chambers: the right atrium, the right ventricle, the left atrium, and the left ventricle. An **interatrial septum**, or wall, separates the right and left atria, and the right and left ventricles are separated by an **interventricular septum**. The atrium and ventricle on the same side of the heart have an opening between them that is regulated by a valve. The valve maintains blood flow in only one direction, moving from the atrium to the ventricle, and prevents backflow. The systemic circuit pumps oxygen-poor blood into the right atrium, then pumps it into the right ventricle. From there, the blood enters the pulmonary trunk and then flows into the pulmonary arteries, where it can become re-oxygenated. Oxygen-rich blood from the lungs flows into the left atrium and then passes into the left ventricle. From there, blood enters the aorta and is pumped to the entire systemic circuit.

Blood circulates throughout the body in a system of vessels that includes arteries, veins, and capillaries. It distributes oxygen, nutrients, and hormones to all the cells in the body. The vessels are muscular tubes that allow gas exchange to occur. **Arteries** carry oxygen-rich blood from the heart to the other tissues of the body. The largest artery is the **aorta**. **Veins** collect oxygen-depleted blood from tissues and organs, and return it to the heart. The walls of veins are thinner and less elastic than arteries, because the blood pressure in veins is lower than in arteries. **Capillaries** are the smallest of the blood vessels and do not function individually; instead, they work together in a unit, called a **capillary bed**. This network of capillaries provides oxygen-rich blood from arterioles to tissues and feeds oxygen-poor blood from tissues back to **venules**.

Blood comprises plasma and **formed elements**, which include red blood cells (RBCs), white blood cells (WBCs), and platelets. **Plasma** is the liquid matrix of the blood and contains dissolved proteins. RBCs transport oxygen and carbon dioxide. WBCs are part of the immune system and help fight diseases. Platelets contain enzymes and other factors that help with blood clotting.

Respiratory System

The **respiratory system** mediates the exchange of gas between the air and the blood, mainly by the act of breathing. This system is divided into the upper respiratory system and the lower respiratory system. The upper system comprises the nose, the nasal cavity and sinuses, and the pharynx. The lower respiratory system comprises the larynx (voice box), the trachea (windpipe), the small passageways leading to the lungs, and the lungs. The upper respiratory system is responsible for filtering, warming, and humidifying the air that gets passed to the lower respiratory system, protecting the lower respiratory system's more delicate tissue surfaces.

The right lung is divided into three **lobes**: superior, middle, and inferior. The left lung is divided into two lobes: superior and inferior. The left lung is smaller than the right, likely because it shares its space in the chest cavity with the heart. Together, the lungs contain approximately 1500 miles of airway passages. The **bronchi**, which carry air into the lungs, branch into **bronchioles** and continue to divide into smaller

and smaller passageways, until they become **alveoli,** which are the smallest passages. Most of the gas exchange in the lungs occurs between the blood-filled pulmonary capillaries and the air-filled alveoli.

The respiratory system has many functions. Most importantly, it provides a large area for gas exchange between the air and the circulating blood. It protects the delicate respiratory surfaces from environmental variations and defends them against pathogens. It is responsible for producing the sounds that the body makes for speaking and singing, as well as for non-verbal communication. It also helps regulate blood volume, blood pressure, and body fluid pH.

When a breath of air is inhaled, oxygen enters the nose or mouth, and passes into the **sinuses,** where the temperature and humidity of the air get regulated. The air then passes into the trachea and is filtered. From there, the air travels into the bronchi and reaches the lungs. **Bronchi** are tubes that lead from the trachea to each lung, and are lined with cilia and mucus that collect dust and germs along the way. Within the lungs, oxygen and carbon dioxide are exchanged between the air in the alveoli and the blood in the pulmonary capillaries. Oxygen-rich blood returns to the heart and is pumped through the systemic circuit. Carbon dioxide-rich air is exhaled from the body.

Breathing is possible due to the muscular diaphragm pulling on the lungs, increasing their volume and decreasing their pressure. Air flows from the external high-pressure system to the low-pressure system inside the lungs. When breathing out, the diaphragm releases its pressure difference, decreases the lung volume, and forces the stale air back out.

Digestive System

The **digestive system** is a group of organs that work together to transform food and liquids into energy, which can then be used by the body as fuel. Food is ingested and then passes through the **alimentary canal,** or GI tract, which comprises the mouth, pharynx, esophagus, stomach, small intestine, and large intestine. The digestive system has accessory organs, including the liver, gallbladder, and pancreas, that help with the processing of food and liquids, but do not have food pass directly through them. These accessory organs and the digestive system organs work together in the following functions:

- **Ingestion:** Food and liquids enter the alimentary canal through the mouth.

- **Introductory Mechanical and Chemical Processing:** Teeth grind the food and the tongue swirls it to facilitate swallowing. Enzymes in saliva begin chemical digestion.

- **Advanced Mechanical and Chemical Digestion:** The muscular stomach uses physical force and enzymes, which function at low pH levels, to break down the food and liquid's complex molecules, such as sugars, lipids, and proteins, into smaller molecules that can be absorbed by the small intestine.

- **Secretion:** Most of the acids, buffers, and enzymes that aid in digestion are secreted by the accessory organs, but some are provided by the digestive tract. Bile from the liver facilitates fat digestion.

- **Absorption:** Vitamins, electrolytes, organic molecules, and water are absorbed by the villi and microvilli lining in the small intestine and are moved to the interstitial fluid of the digestive tract.

- **Compaction:** Indigestible materials and organic wastes are dehydrated in the large intestine and compacted before elimination from the body.

- **Excretion:** Waste products are excreted from the digestive tract.

The following are the major organs of the alimentary canal:

- **Stomach:** This organ stores food so the body has time to digest large meals. Its highly acidic environment and enzyme secretions, such as pepsin and trypsin, aid in digestion. It also aids in mechanical processing through muscular contractions.

- **Small Intestine:** This organ is a thin tube that is approximately ten feet long. It secretes enzymes to aid in digestion and has many folds that increase its surface area and allows for maximum absorption of nutrients from the digested food.

- **Large Intestine:** This organ is a long thick tube that is about five feet long. It absorbs water from the digested food and transports waste to be excreted from the body. It also contains symbiotic bacteria that further breaks down the waste products, allowing for any extra nutrients to be absorbed.

The following are the accessory organs:

- **Liver:** The liver produces and secretes **bile,** which is important for the digestion of lipids. It also plays a large role in the regulation of circulating levels of carbohydrates, amino acids, and lipids in the body. Excess nutrients are removed by the liver and deficiencies are corrected with its stored nutrients.

- **Gallbladder:** This organ is responsible for storing and concentrating bile before it gets secreted into the small intestine. While the gallbladder is storing bile, it can regulate the bile's composition by absorbing water, thereby increasing the concentration of bile salts and other components.

- **Pancreas:** The pancreas has exocrine cells that secrete buffers and digestive enzymes. It contains specific enzymes for each type of food molecule, such as carbohydrases for carbohydrates, lipases for lipids, and proteinases for proteins.

Urinary System

The **urinary system** is made up of the kidneys, ureters, urinary bladder, and the urethra. It is the main system responsible for getting rid of the organic waste products, excess water and electrolytes are generated by the body's other systems. The **kidneys** are responsible for producing urine, which is a fluid waste product containing water, ions, and small soluble compounds. The urinary system has many important functions related to waste excretion. It regulates the concentrations of sodium, potassium, chloride, calcium, and other ions in the plasma by controlling the amount of each that is excreted in urine. This also contributes to the maintenance of blood pH. It regulates blood volume and pressure by controlling the amount of water lost in the urine, and releasing erythropoietin and renin. It eliminates toxic substances, drugs, and organic waste products, such as urea and uric acid. Kidney cells also synthesize **calcitriol**, which is a hormone derivative of vitamin D3 that aids in calcium ion absorption by the intestinal epithelium.

Under normal circumstances, humans have two functioning kidneys. They are the main organs are responsible for filtering waste products out of the blood and transferring them to urine. Every day, the kidneys filter approximately 120 to 150 quarts of blood and produce one to two quarts of urine. Kidneys are made of millions of tiny filtering units, called **nephrons.** Nephrons have two parts: a **glomerulus,** which is the filter, and a **tubule.** As blood enters the kidneys, the glomerulus allows fluid and waste products to pass through it and enter the tubule. Blood cells and large molecules, such as proteins, do

not pass through and remain in the blood. The filtered fluid and waste then pass through the tubule, where any final essential minerals are sent back to the bloodstream. The final product at the end of the tubule is called **urine**.

Once urine accumulates, it leaves the kidneys. The urine travels through the ureters into the urinary **bladder**, a muscular organ that is hollow and elastic. As more urine enters the urinary bladder, its walls stretch and become thinner so there is no significant difference in internal pressure. The urinary bladder stores the urine until the body is ready for urination, at which time the muscles contract and force the urine through the urethra and out of the body.

Reproductive System

The **reproductive system** is responsible for producing, storing, nourishing, and transporting functional reproductive cells, or gametes, in the human body. It includes the reproductive organs, also known as **gonads**, the reproductive tract, the accessory glands and organs that secrete fluids into the reproductive tract, and the **perineal structures**, which are the external genitalia. The human male and female reproductive systems are very different from each other.

The male gonads are called **testes**. The testes secrete **androgens**, mainly testosterone, and produce and store 500 million **sperms** cells, which are the male gametes, each day. An androgen is a steroid hormone that controls the development and maintenance of male characteristics. Once the sperm are mature, they move through a duct system, where they mix with additional fluids secreted by accessory glands, forming a mixture called **semen**. The sperm cells in semen are responsible for fertilization of the female gametes to produce offspring.

The female gonads are the **ovaries**. Ovaries generally produce one immature gamete, or **oocyte**, per month. They are also responsible for secreting the hormones estrogen and progesterone. When the oocyte is released from the ovary, it travels along the uterine tubes, or **Fallopian tube**s, and then into the uterus. The **uterus** opens into the vagina. When sperm cells enter the vagina, they swim through the uterus and may fertilize the oocyte in the Fallopian tubes. The resulting zygote travels down the tube and implants into the uterine wall. The uterus protects and nourishes the developing embryo for nine months until it is ready for the outside environment. If the oocyte is not fertilized, it is released in the **uterine**, or **menstrual**, **cycle**. The menstrual cycle occurs monthly and involves the shedding of the functional part of the uterine lining.

Mammary glands are a specialized accessory organ of the female reproductive system. The mammary glands are located in the breast tissue, and during pregnancy begin to grow, and the cells proliferate in preparation for lactation. After pregnancy, the cells begin to secrete nutrient-filled milk, which is transferred into a duct system and out through the nipple for nourishment of the baby.

Ecosystems: Interactions, Energy, and Dynamics

An *ecosystem* includes all of the living organisms and nonliving components of an environment (each community) and their interactions with each other.

Biomes

A *biome* is a group of plants and animals that are found in many different continents and have the same characteristics because of the similar climates in which they live. Each biome is composed of all of the ecosystems in that area. Five primary types of biomes are aquatic, deserts, forests, grasslands, and tundra. The sum total of all biomes comprises the Earth's biosphere.

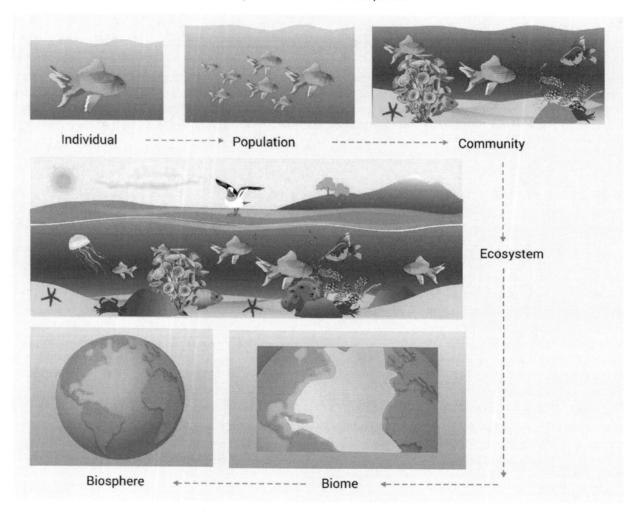

Stability and Disturbances

Ecological stability is the ability of an ecosystem to withstand changes that are occurring within it. With *regenerative stability*, an ecosystem may change, but then quickly return to its previous state. *Constant stability* occurs in ecosystems that remain unchanged despite the changes going on around them.

An *ecological disturbance* is a change in the environment that causes a larger change in the ecosystem. Smaller disturbances include fires and floods. Larger disturbances include the *climate change* that is currently occurring. Gas emissions from human activity are causing the atmosphere to warm up, which

is changing the Earth's water systems and making weather more extreme. The increase in temperature is causing greater evaporation of the water sources on Earth, creating droughts and depleting natural water sources. This has also caused many of the Earth's glaciers to begin melting, which can change the salinity of the oceans.

Changes in the environment can cause an *ecological succession* to occur, which is the change in structure of the species that coexist in an ecological community. When the environment changes, resources available to the different species also change. For example, the formation of sand dunes or a forest fire would change the environment enough to allow a change in the social hierarchy of the coexisting species.

Energy Flow

Ecosystems are maintained by cycling the energy and nutrients that they obtain from external sources. The process can be diagramed in a *food web*, which represents the feeding relationship between species in a community. The different levels of the food web are called *trophic levels*. The first trophic level generally consists of plants, algae, and bacteria. The second trophic level consists of herbivores. The third trophic level consists of predators that eat herbivores. The trophic levels continue on to larger and larger predators. *Decomposers* are an important part of the food chain that are not at a specific trophic level. They eat decomposing things on the ground that other animals do not want to eat. This allows them to provide nutrients to their own predators.

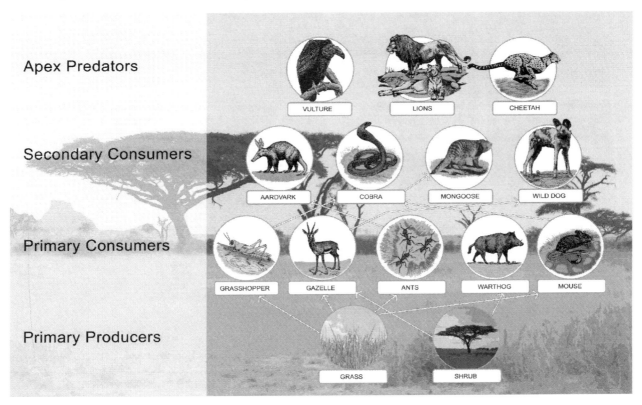

Biogeochemical Cycles

Biogeochemical cycles are the pathways by which chemicals move through the *biotic*, or biospheric, and *abiotic*, or atmospheric, parts of the Earth. The most important biogeochemical cycles include the water, carbon, and nitrogen cycles. *Water* goes through an evaporation, condensation, and precipitation cycle.

Nitrogen makes up seventy-eight percent of the Earth's atmosphere and can affect the rate of many ecosystem processes, such as production of the primary producers at the first trophic level of the food web. The *carbon cycle* has many steps that are vitally important for sustaining life on Earth.

The water cycle:

The nitrogen cycle:

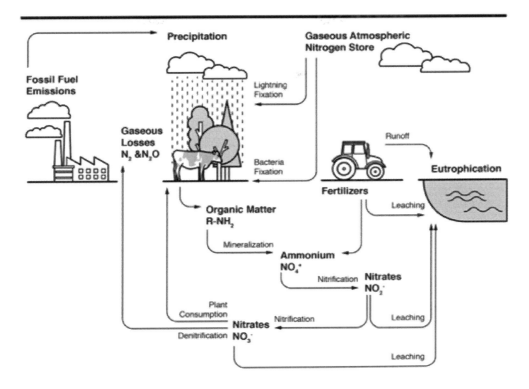

The carbon cycle:

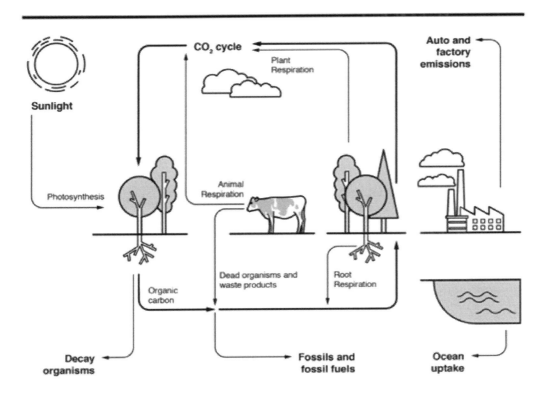

Community Ecology

An *ecological community* is a group of species that interact and live in the same location. Because of their shared environment, they tend to have a large influence on each other.

An *ecological niche* is the role that a species plays in its environment, including how it finds its food and shelter. It could be a predator of a different species, or prey for a larger species.

Species diversity is the number of different species that cohabitate in an ecological community. It has two different facets: *species richness*, which is the general number of species, and *species evenness*, which accounts for the population size of each species.

Interspecific relationships include the interactions between organisms of different species. The following list defines the common relationships that can occur:

- *Commensalism*: One organism benefits while the other is neither benefited nor harmed

- *Mutualism*: Both organisms benefit

- *Parasitism*: One organism benefits and the other is harmed

- *Competition*: Two or more species compete for limited resources that are necessary for their survival

- *Predation (Predator-Prey)*: One species is a food source for another species

Water

Most cells are primarily composed of water and live in water-rich environments. Since water is such a familiar substance, it is easy to overlook its unique properties. Chemically, water is made up of two hydrogen atoms bonded to one oxygen atom by covalent bonds. The three atoms join to make a V-shaped molecule. Water is a polar molecule, meaning it has an unevenly distributed overall charge due to an unequal sharing of electrons. Due to oxygen's electronegativity and its more substantial positively charged nucleus, hydrogen's electrons are pulled closer to the oxygen. This causes the hydrogen atoms to have a slight positive charge and the oxygen atom to have a slight negative charge. In a glass of water, the molecules constantly interact and link for a fraction of a second due to intermolecular bonding between the slightly positive hydrogen atoms of one molecule and the slightly negative oxygen of a different molecule. These weak intermolecular bonds are called **hydrogen bonds**.

Water has several important qualities, including: cohesive and adhesive behaviors, temperature moderation ability, expansion upon freezing, and diverse use as a solvent.

Cohesion is the interaction of many of the same molecules. In water, cohesion occurs when there is hydrogen bonding between water molecules. Water molecules use this bonding ability to attach to each other and can work against gravity to transport dissolved nutrients to the top of a plant. A network of water-conducting cells can push water from the roots of a plant up to the leaves. Adhesion is the linking of two different substances. Water molecules can form a weak hydrogen bond with, or adhere to, plant cell walls to help fight gravity. The cohesive behavior of water also causes surface tension. If a glass of water is slightly overfull, water can still stand above the rim. This is because of the unique bonding of water molecules at the surface—they bond to each other and to the molecules below them, making it seem like it is covered with an impenetrable film. A raft spider could actually walk across a small body of water due to this surface tension.

Another important property of water is its ability to moderate temperature. Water can moderate the temperature of air by absorbing or releasing stored heat into the air. Water has the distinctive capability of being able to absorb or release large quantities of stored heat while undergoing only a small change in temperature. This is because of the relatively high **specific heat** of water, where specific heat is the amount of heat it takes for one gram of a material to change its temperature by 1 degree Celsius. The specific heat of water is one calorie per gram per degree Celsius, meaning that for each gram of water, it takes one calorie of heat to raise or lower the temperature of water by 1 degree Celsius.

When the temperature of water is reduced to freezing levels, water displays another interesting property: It expands instead of contracts. Most liquids become denser as they freeze because the molecules move around slower and stay closer together. Water molecules, however, form hydrogen bonds with each other as they move together. As the temperature lowers and they begin to move slower, these bonds become harder to break apart. When water freezes into ice, molecules are frozen with hydrogen bonds between them and they take up about 10 percent more volume than in their liquid state. The fact that ice is less dense than water is what makes ice float to the top of a glass of water.

Lastly, the **polarity** of water molecules makes it a versatile solvent. **Ionic compounds**, such as salt, are made up of positively- and negatively-charged atoms, called **cations** and **anions**, respectively. Cations and anions are easily dissolved in water because of their individual attractions to the slight positive charge of the hydrogen atoms or the slight negative charge of the oxygen atoms in water molecules. Water molecules separate the individually charged atoms and shield them from each other so they don't bond to each other again, creating a homogenous solution of the cations and anions. Nonionic compounds, such as sugar, have polar regions, so are easily dissolved in water. For these compounds, the water molecules form hydrogen bonds with the polar regions (hydroxyl groups) to create a homogenous solution. Any substance that is attracted to water is termed **hydrophilic**. Substances that repel water are termed **hydrophobic**.

Heredity: Inheritance and Variation of Traits

Structure and Function of DNA and RNA

DNA and RNA are made up of **nucleotides**, which are formed from a five-carbon sugar, a nitrogenous base, and one or more phosphate group. While DNA is made up of the sugar deoxyribose, RNA is made up of the sugar ribose. Deoxyribose has one fewer oxygen atom than ribose. DNA and RNA each comprise four nitrogenous bases, three of which they have in common: adenine, guanine, and cytosine. Thymine is found only in DNA and uracil is found only in RNA. Each base has a specific pairing formed by hydrogen bonds, and is known as a **base pair**. Adenine interacts with thymine or uracil, and guanine interacts only with cytosine. While RNA is found in a single strand, DNA is a double-stranded molecule that coils up to form a **double helix** structure.

The specific pairing of the nitrogenous bases allows for the hereditary information stored in DNA to be passed down accurately from parent cells to daughter cells. When chromosomes are **replicated** during cell division, the double-helix DNA is first uncoiled, each strand is replicated, and then two new identical DNA molecules are generated. DNA can also be used as a template for generating proteins. A **single-stranded** RNA is generated from the DNA during a process called **transcription**; proteins are then generated from this RNA in a process called **translation**.

Chromosomes, Genes, Alleles

Chromosomes are found inside the nucleus of cells and contain the hereditary information of the cell in the form of **genes**. Each gene has a specific sequence of DNA that eventually encodes proteins and

results in inherited traits. **Alleles** are variations of a specific gene that occur at the same location on the chromosome. For example, blue and brown are two different alleles of the gene that encodes for eye color.

Dominant and Recessive Traits

In genetics, **dominant alleles** are mostly noted in italic, capital letters (A) and **recessive alleles** are mostly noted in italic, lower case letters (a). There are three possible combinations of alleles among dominant and recessive alleles: AA, Aa (known as a heterozygote), and aa. **Dominant** traits are phenotypes that appear when at least one dominant allele is present in the gene. Dominant alleles are considered to have stronger phenotypes and, when mixed with recessive alleles, will mask the recessive trait. The **recessive** trait would only appear as the phenotype when the allele combination is "aa" because a dominant allele is not present to mask it.

Mendelian Inheritance

A monk named Gregor Mendel is referred to as the father of genetics. He was responsible for coming up with one of the first models of inheritance in the 1860s. His model included two laws to determine which traits are inherited. These laws still apply today, even after genetics has been studied much more in depth.

- **The Law of Segregation:** Each characteristic has two versions that can be inherited. When two parent cells form daughter cells, the two alleles of the gene segregate and each daughter cell can inherit only one of the alleles from each parent.

- **The Law of Independent Assortment:** The alleles for different traits are inherited independent of one another. In other words, the biological selection of one allele by a daughter cell is not linked to the biological selection of an allele for a different trait by the same daughter cell. The genotype that is inherited is the alleles that are encoded on the gene, and the phenotype is the outward appearance of the physical trait for that gene. For example, "A" is the dominant allele for brown eyes and "a" is the recessive allele for blue eyes; the phenotype of brown eyes would occur for two different genotypes: both "AA" and "Aa."

Punnett Squares

For simple genetic combinations, a **Punnett square** can be used to assess the phenotypic ratios of subsequent generations. In a 2 x 2 cell square, one parent's alleles are set up in columns and the other parent's alleles are set up in rows. The resulting allele combinations are shown in the four internal cells.

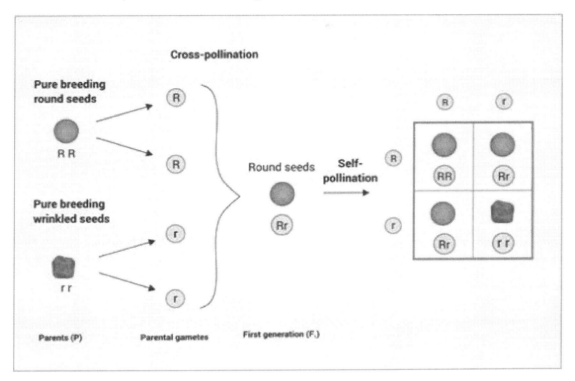

Pedigree

For existing populations where genetic crosses cannot be controlled, phenotype information can be collected over several generations and a **pedigree analysis** can be done to investigate the dominant and recessive characteristics of specific traits. There are several rules to follow when determining the pedigree of a trait. For dominant alleles:

- Affected individuals have at least one affected parent;
- The phenotype appears in every generation; and
- If both parents are unaffected, their offspring will always be unaffected.

For recessive alleles:

- Unaffected parents can have affected offspring; and
- Affected offspring are male and female.

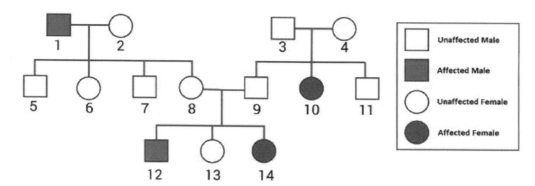

Mutations, Chromosomal Abnormalities, and Common Genetic Disorders

Mutations

Genetic **mutations** occur when there is a permanent alteration in the DNA sequence that codes for a specific gene. They can be small, affecting only one base pair, or large, affecting many genes on a chromosome. Mutations are classified as either hereditary, which means they were also present in the parent gene, or acquired, which means that they occurred after the genes were passed down from the parents. Although mutations are not common, they are an important aspect of genetics and variation in the general population.

Chromosomal Abnormalities and Common Genetic Disorders

Structural chromosomal abnormalities are mutations that affect a large chromosomal segment of more than one gene. This often occurs due to an error in cell division. Acute myelogenous leukemia is caused by a **translocation error**, which is when a segment of one chromosome is moved to another chromosome.

There can also be an abnormal number of chromosomes, which is referred to as **aneuploidy**. Down syndrome is an example of an aneuploidy in which there are three copies of chromosome 21 instead of two copies. Turner syndrome is another example of aneuploidy, in which a female is completely or partially missing an X chromosome. Without the second X chromosome, these females do not develop all of the typical female physical characteristics and are unable to bear children.

Biological Evolution: Unity and Diversity

Biological evolution is the concept that a population's gene pool changes over generations. According to this concept, populations of organisms evolve, not individuals, and over time, genetic variation and mutations lead to such changes.

Darwin's Theory of Natural Selection

Charles Darwin developed a scientific model of evolution based on the idea of *natural selection*. When some individuals within a population have traits that are better suited to their environment than other individuals, those with the better-suited traits tend to survive longer and have more offspring. The survival and inheritance of these traits through many subsequent generations lead to a change in the

population's gene pool. According to natural selection, traits that are more advantageous for survival and reproduction in an environment become more common in subsequent generations.

Evolutionary Fitness

Sexual selection is a type of natural selection in which individuals with certain traits are more likely to find a mate than individuals without those traits. This can occur through direct competition of one sex for a mate of the opposite sex. For example, larger males may prevent smaller males from mating by using their size advantage to keep them away from the females. Sexual selection can also occur through mate choice. This can happen when individuals of one sex are choosy about their mate of the opposite sex, often judging their potential mate based on appearance or behavior. For example, female peacocks often mate with the showiest male with large, beautiful feathers. In both types of sexual selection, individuals with some traits have better reproductive success, and the genes for those traits become more prevalent in subsequent populations.

Adaptations are Favored by Natural Selection

Adaptations are inherited characteristics that enhance survival and reproductive capabilities in specific environments. Charles Darwin's idea of natural selection explains *how* populations change—adaption explains *why*. Darwin based his concept of evolution on three observations: the unity of life, the diversity of life, and the suitability of organisms for their environments. There was unity in life based on the idea that all organisms descended from common ancestors. Then, as the descendants of the common ancestors faced changes in their environments they moved to new environments. There they adapted new features to help them in their new way of life. This concept explains the diversity of life and how organisms are matched to their environments.

An example of natural selection is found in penguins—birds that cannot fly. Over time, populations of penguins lost the ability to fly but became master swimmers. Their habitats are surrounded by water, and their food sources are in the water. Penguins that could dive for food survived better than those that could fly, and the divers produced more offspring. The gene pool changed as a result of natural selection.

Environmental Changes Serve as Selective Mechanisms

The environment constantly changes, which drives selection. Although an individual's traits are determined by their *genotype*, or makeup of genes, natural selection more directly influences *phenotype*, or observable characteristics. The outward appearance or ability of individuals affects their ability to adapt to their environment and survive and reproduce. Phenotypic changes occurring in a population over time are accompanied by changes in the gene pool.

The classic example of this is the peppered moth. It was once a light-colored moth with black spots, though a few members of the species had a genetic variation resulting in a dark color. When the Industrial Revolution hit London, the air became filled with soot and turned the white trees darker in color. Birds were then able to spot and eat the light-colored moths more easily. Within just a few months, the moths with genes for darker color were better able to avoid predation. Subsequent generations had far more dark-colored moths than light ones. Once the Industrial Revolution ended and the air cleared, light-colored moths were better able to survive, and their numbers increased.

Supporting Evidence

The Fossil Record

Fossils are the preserved remains of animals and organisms from the past, and they can elucidate the homology of both living and extinct species. Many scientists believe that fossils often provide evidence for evolution. They further propose that looking at the *fossil record* over time can help identify how quickly or slowly evolutionary changes occurred, and can also help match those changes to environmental changes that were occurring concurrently.

Homology

Evolutionists propose that organisms that developed from a common ancestor often have similar characteristics that function differently. This similarity is known as *homology*. For example, humans, cats, whales, and bats all have bones arranged in the same manner from their shoulders to their digits. However, the bones form arms in humans, forelegs in cats, flippers in whales, and wings in bats, and these forelimbs are used for lifting, walking, swimming, and flying, respectively. Evolutionists look to homology, believing that the similarity of the bone structure shows a common ancestry but that the functional differences are the product of evolution.

Homologous Structures

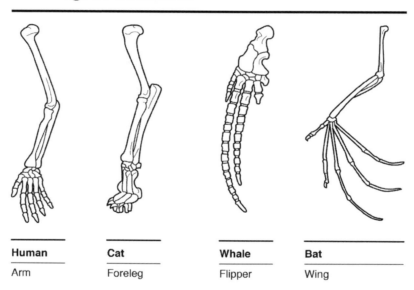

Human	Cat	Whale	Bat
Arm	Foreleg	Flipper	Wing

Comparative Genetics

In *comparative genetics*, different organisms are compared at a genetic level to look for similarities and differences. DNA sequence, genes, gene order, and other structural features are among the features that may be analyzed in order to look for evolutionary relationships and common ancestors between the organisms.

Classification Schemes

Taxonomy is the science behind the biological names of organisms. Biologists often refer to organisms by their Latin scientific names to avoid confusion with common names, such as with fish. Jellyfish, crayfish, and silverfish all have the word "fish" in their name, but belong to three different species. In

the eighteenth century, Carl Linnaeus invented a naming system for species that included using the Latin scientific name of a species, called the **binomial**, which has two parts: the **genus**, which comes first, and the **specific epithet**, which comes second. Similar species are grouped into the same genus. The Linnaean system is the commonly used taxonomic system today and, moving from comprehensive similarities to more general similarities, classifies organisms into their species, genus, family, order, class, phylum, and kingdom. *Homo sapiens* is the Latin scientific name for humans.

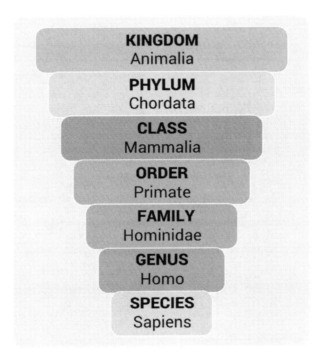

KINGDOM
Animalia

PHYLUM
Chordata

CLASS
Mammalia

ORDER
Primate

FAMILY
Hominidae

GENUS
Homo

SPECIES
Sapiens

Phylogenetic trees are branching diagrams that represent the evolutionary history of a species. The branch points most often match the classification groups set forth by the Linnaean system. Using this system helps elucidate the relationship between different groups of organisms. The diagram below is that of an empty phylogenetic tree:

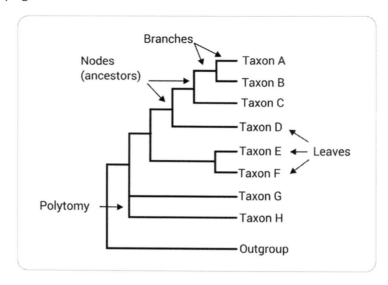

Each branch of the tree represents the divergence of two species from a common ancestor. For example, the coyote is known as Canis latrans and the gray wolf is known as Canis lupus. Their common ancestor, the Canis lepophagus, which is now extinct, is where their shared genus derived.

Characteristics of Bacteria, Animals, Plants, Fungi, and Protists

As discussed earlier, there are two distinct types of cells that make up most living organisms: prokaryotic and eukaryotic. Bacteria (and archaea) are classified as prokaryotic cells, whereas animal, plant, fungi, and protist cells are classified as eukaryotic cells.

Although animal cells and plant cells are both eukaryotic, they each have several distinguishing characteristics. **Animal cells** are surrounded by a plasma membrane, while **plant cells** have a cell wall made up of cellulose that provides more structure and an extra layer of protection for the cell. Animals use oxygen to breathe and give off carbon dioxide, while plants do the opposite—they take in carbon dioxide and give off oxygen. Plants also use light as a source of energy. Animals have highly developed sensory and nervous systems and the ability to move freely, while plants lack both abilities. Animals, however, cannot make their own food and must rely on their environment to provide sufficient nutrition, whereas plants do make their own food.

Fungal cells are typical eukaryotes, containing both a nucleus and membrane-bound organelles. They have a cell wall, similar to plant cells; however, they use oxygen as a source of energy and cannot perform photosynthesis. They also depend on outside sources for nutrition and cannot produce their own food. Of note, their cell walls contain **chitin**.

Protists are a group of diverse eukaryotic cells that are often grouped together because they do not fit into the categories of animal, plant, or fungal cells. They can be categorized into three broad categories: protozoa, protophyta, and molds. These three broad categories are essentially "animal-like," "plant-like," and "fungus-like," respectively. All of them are unicellular and do not form tissues. Besides this simple similarity, protists are a diverse group of organisms with different characteristics, life cycles, and cellular structures.

Practice Questions

1. At what point in its swing does a pendulum have the most mechanical energy?
    a. At the top of its swing, just before going into motion
    b. At the bottom of its swing, in full motion
    c. Halfway between the top of its swing and the bottom of its swing
    d. It has the same amount of mechanical energy throughout its path

2. What does the scientific method describe?
    a. How to review a scientific paper
    b. How to organize a science laboratory
    c. The steps utilized to conduct an inquiry into a scientific question
    d. How to use science to earn money in society

3. Yellow crazy ants are some of the most destructive and invasive ants that spray formic acid into the air and ground to subdue their victims. The acid even makes it painful for people to breathe. These ants are classified as predatory scavengers and will even eat a large variety of animal tissue, which has resulted in the population reduction of Hawaiian seabirds. Given the molecular formula, CH_2O_2, what is the approximate mass percentage of each element in the compound?
    a. Mass % C = 26, Mass % H = 4, Mass % O = 70
    b. Mass % C = 31, Mass % H = 4, Mass % O = 65
    c. Mass % C = 28, Mass % H = 5, Mass % O = 67
    d. Mass % C = 29, Mass % H = 6, Mass % O = 65

4. Scientist A is observing an unknown substance in the lab. Which observation describes a chemical property of the substance?
    a. She sees that it is green in color.
    b. She weighs it and measures the volume and finds the density to be 10 g/L.
    c. She applies pressure to it and finds that it breaks apart easily.
    d. She passes it through a flame and finds that it burns.

5. A ramp leading up to a loading dock would be considered which type of simple machine?
    a. Screw
    b. Lever
    c. Inclined plane
    d. Pulley

6. Who is credited for simplifying the laws of motion?
    a. Einstein
    b. Hawking
    c. Copernicus
    d. Newton

7. The heat transfer due to the movement of gas molecules from an area of higher concentration to one of lower concentration is known as what?
    a. Conduction
    b. Convection
    c. Solarization
    d. Radiation

8. Which of the following is true of an object at rest on earth?
    a. It has no forces acting upon it.
    b. It has no gravity acting upon it.
    c. It is in transition.
    d. It is in equilibrium.

9. There are three types of rocks: sedimentary, metamorphic, and igneous. Sedimentary rock is formed from sediment, such as sand, shells, and pebbles. The sediment gathers together and hardens over time. It is generally soft and breaks apart easily. This is the only type of rock that contains fossils, which are the remains of animals and plants that lived a long time ago. Metamorphic rock forms under the surface of the earth due to changes in heat and pressure. These rocks usually have ribbon-like layers and may contain shiny crystals. Igneous rock forms when molten rock, or magma, cools and hardens. An example of molten rock is lava, which escapes from an erupting volcano. This type of rock looks shiny and glasslike.

Which type of rock could a fossil be found in?
    a. Igneous
    b. Bone
    c. Metamorphic
    d. Sedimentary

Questions 10–11 pertain to the following passage and figure:

The heart is a muscle that is responsible for pumping blood through the body. It is divided into four chambers: the right atrium, right ventricle, left atrium, and left ventricle. Blood enters the atria and is then pumped into the ventricles below them. There is a valve between the atria and ventricles that prevents the blood from flowing back into the atria. The valve between the right atrium and ventricle has three folds whereas the valve between the left atrium and ventricle has two folds. Arteries carry oxygen-rich blood away from the heart to the body. Veins carry oxygen-poor blood from the body back to the heart. From there, the blood gets pumped to the lungs to get re-oxygenated and then back to the heart before circulating to the body. The heart beats every second of the day. For an adult, the normal heartrate is between 60 and 100 beats per minute. For a child, a normal heartrate is between 90 and 120 beats per minute.

Figure 1 below shows how blood gets pumped through the body.

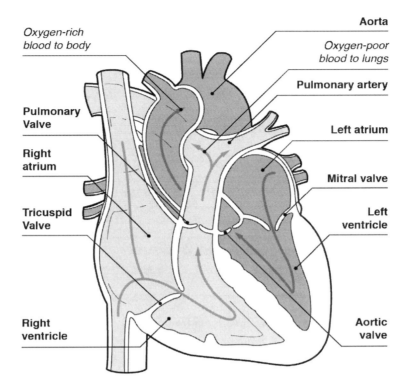

10. Where is the oxygen-poor blood pumped to before returning to the heart to get circulated to the rest of the body?
    a. Lungs
    b. Brain
    c. Stomach
    d. Kidney

11. If the aorta contains oxygen-rich blood, what type of vessel is it?
    a. Vein
    b. Pulmonary
    c. Airway
    d. Artery

12. A projectile at a point along its path has 30 Joules of potential energy and 20 Joules of kinetic energy. What is the total mechanical energy for the projectile?
    a. 50 Joules
    b. 30 Joules
    c. 20 Joules
    d. 10 Joules

Questions 13-14 pertain to the following passage and figure:

Scientists use the scientific method to investigate a theory or solve a problem. It includes four steps: observation, hypothesis, experiment, and conclusion. Observation occurs when the scientist uses one of their senses to identify what they want to study. A hypothesis is a conclusive sentence about what the scientist wants to research. It generally includes an explanation for the observations, can be tested experimentally, and predicts the outcome. The experiment includes the parameters for the testing that will occur. The conclusion will state whether or not the hypothesis was supported.

Scientist A would like to know how sunlight affects the growth of a plant. She says that more sunlight will cause the plant to grow faster. She sets up her experimental groups and tests her hypothesis over 11 days.

Figure 1 below shows the experimental data Scientist A collected over 11 days.

Length and height of plants in the sunlight

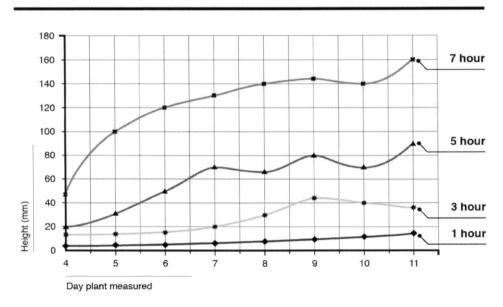

13. How many experimental groups does she have?
    a. 1
    b. 3
    c. 4
    d. 11

14. What should her conclusion be based on her experimental data?
    a. 5 hours of sunlight is optimal for plant growth.
    b. Plants should only be measured for 11 days.
    c. Less sunlight is better for plant growth.
    d. Providing plants with more sunlight makes them grow bigger.

15. Which of the following is considered a force?
    a. Weight
    b. Mass
    c. Acceleration
    d. Gravity

16. Why would a pencil appear to bend at the water line in a glass of water?
    a. The wood of the pencil becomes warped from being in the water.
    b. It appears to bend because of the refraction of light traveling from air to water.
    c. The pencil temporarily bends because of its immersion into separate mediums.
    d. The reflection of the light from water to a human's pupil creates the illusion of a warping object.

17. Which of the following is NOT one of Newton's three laws of motion?
    a. Inertia: an object at rest tends to stay at rest, and an object in motion tends to stay in motion
    b. $E = mc^2$
    c. For every action there is an equal and opposite reaction
    d. $F = ma$

Questions 18–21 pertain to the following passage and figures:

Physical characteristics are controlled by genes. Each gene has two alleles, or variations. Generally, one allele is more dominant than the other allele and when one of each allele is present on the gene, the physical trait of the dominant allele will be expressed. The allele that is not expressed is called the recessive allele. Recessive alleles are expressed only when both alleles present on the gene are the recessive allele.

Punnett squares are diagrams that can predict the outcome of crossing different traits. In these diagrams, dominant alleles are represented by uppercase letters and recessive alleles are represented by lowercase letters.

Scientist A wants to grow white flowered plants and is doing a series of crossbreeding experiments. She had each plant genetically tested so she knows which alleles comprise each plant. The dominant flowers are red (A) and the recessive allele (a) produces white flowers.

Figure 1 below represents the different flowers that underwent crossbreeding during Round #1A

Round #1A

Crossbreeding #1A			Crossbreeding #2A			Crossbreeding #3A		
	A	a		a	a		a	a
A	AA	Aa	A	Aa	Aa	A	Aa	Aa
A	AA	Aa	A	Aa	Aa	a	aa	aa

Figure 2 below represents the number of flowers that were red and white after the first round of crossbreeding experiments.

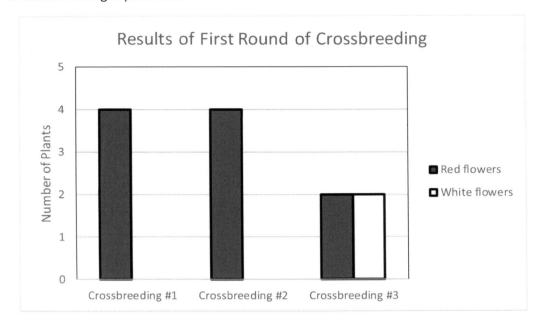

During her second round of crossbreeding, she adds in a plant of unknown genetic makeup with red flowers. She crosses it with a white-flowering plant. The results of her experiment are represented in the next figure.

Figure 3 represents the genetic results from the second round of crossbreeding.

Round #2		
	a	a
?	Aa	Aa
?	aa	aa

Scientist A takes offspring plants from Round #1A and crossbreeds them with each other and calls this Round #1B.

Figure 4 below represents the results of crossbreeding from Round #1B.

Round #1B

Crossbreeding #1B			Crossbreeding #2B			Crossbreeding #3B		
	A	a		A	a		a	a
A	AA	Aa	A	AA	Aa	a	aa	aa
A	AA	Aa	a	Aa	aa	a	aa	aa

18. Crossbreeding which two plants will give her the highest likelihood of obtaining some white plants right away in Round #1A?
    a. AA × Aa
    b. Aa × aa
    c. AA × aa
    d. AA × AA

19. What percentage of plants are white after the first crossbreeding reactions?
    a. 12
    b. 50
    c. 16.7
    d. 25

20. What is the genetic makeup of the unknown plant from the second round of crossbreeding?
    a. aa
    b. Aa
    c. AA
    d. Cannot be determined

21. From which group of crossbreeding in Round #1B can she obtain 100% white flowers by the second generation?
    a. They are all equal.
    b. 1B
    c. 2B
    d. 3B

The following passage pertains to question 22.

Power is defined as the rate at which work is done, or the time it takes to do a given amount of work. Work is calculated by multiplying force by distance. In the International System of Units (SI), work is measured in newton meters (*N·m*) or joules (*J*). Power is measured in joules/second or *watts (W)*.

For example, to raise a 1-kilogram mass one meter off the ground, it takes approximately 10 newton meters of work (approximating the gravitational acceleration of 9.81 m/s^2 as 10 m/s^2). To do the work in 10 seconds, it requires 1 watt of power. Doing it in 1 second requires 10 watts of power. Essentially, doing it faster means dividing by a smaller number, and that means greater power.

22. Kevin is training for baseball. He is 88 kg. He completes 45 pushups in one minute, during which he displaces his center of mass vertically 0.7 meters for each repetition. Which of the following would correctly calculate Kevin's average power during the exercise?

a. $\dfrac{\frac{88kg}{9.8} \times 0.7m \times 45}{60}$

b. $(\dfrac{88kg \times 0.7m}{60}$

c. $\dfrac{88kg \times 9.8 \times 0.7m \times 45}{60}$

d. $\dfrac{88kg \times 9.8 \times 0.7m}{60}$

23. Which is not a method for transferring electrostatic charge?
    a. Polarization
    b. Touch
    c. Election
    d. Induction

24. An ecosystem that normally has moderate summers with high rainfall is experiencing a heat wave and a drought. How does this affect the rate of photosynthesis of the producers in this ecosystem?
    a. The decrease in transpiration from the high heat and the drop in rainfall decreases the number of chloroplasts, so photosynthesis rates decrease.
    b. The increase in transpiration from the high heat and the drop in rainfall results in less water. Since photosynthesis creates water, the rate increases to meet increased water demands.
    c. Increased temperature increases the number of mitochondria, so photosynthesis rates increase.
    d. The increase in transpiration from the high heat and the drop in rainfall results in less water available for photosynthesis. The rate decreases.

25. Velocity is a measure of which of the following?
    a. Speed with direction
    b. The change in position over the change in time
    c. Meters covered over seconds elapsed
    d. All of the above

26. The image below shows hormone levels during a 28-day human menstrual cycle. Days 1-14 are when the follicle develops, and day 14 is ovulation and when the egg is released. During days 15-28, the follicle left behind becomes the corpus luteum. This also is the time when the uterine lining vascularizes and develops.

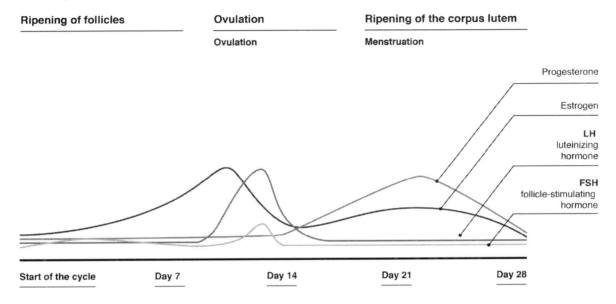

The interrelationship between the hormones is critical in the regulation of the menstrual cycle. Based on the image, which is the most likely and consistently demonstrated conclusion regarding hormone regulation?

    a. FSH is required for progesterone production.
    b. Estrogen is required for LH production.
    c. Progesterone is required for estrogen, LH, and FSH production
    d. FSH and LH require the same negative feedback loop.

Use the following image to answer question 27.

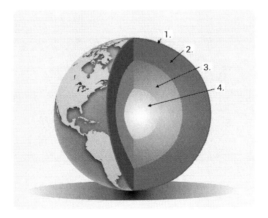

27. Which choice describes layer 4?
    a. Inner core: solid
    b. Inner core: liquid
    c. Outer core: solid
    d. Outer core: liquid

28. Which type of rock accumulates in layers at the bottom of the ocean due to run-off?
    a. Igneous
    b. Sedimentary
    c. Metamorphic
    d. Minerals

29. The water cycle involves phase changes. Which example below is evaporation?
    a. Clouds forming in the sky
    b. Rain, snow, or ice storms
    c. River water flowing to the ocean
    d. Sunlight's effect on morning dew

30. Digestion begins in the mouth where teeth grind up food and saliva breaks it down, making it easier for the body to absorb. Next, the food moves to the esophagus, and it is pushed into the stomach. The stomach is where food is stored and broken down further by acids and digestive enzymes, preparing it for passage into the intestines. The small intestine is where the nutrients are taken from food and passed into the blood stream. Other essential organs like the liver, gall bladder, and pancreas aid the stomach in breaking down food and absorbing nutrients. Finally, food waste is passed into the large intestine where it is eliminated by the body.

Which location(s) in the digestive system is/are sites of chemical digestion?
    I. Mouth
    II. Stomach
    III. Small Intestine
    a. II only
    b. III only
    c. II and III only
    d. I, II, and III

31. Which of the following statements is false?
    a. Magma circulates in the upper mantle.
    b. All volcanoes have explosive eruptions.
    b. Igneous rocks are formed by crystallized lava.
    c. Igneous rocks recycle and form magma.

Use the following image to answer question 32.

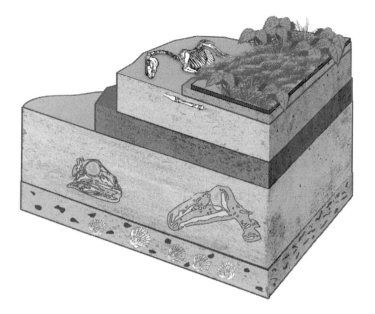

32. Which fossil is the oldest?
    a. Dinosaur head
    b. Seashell
    c. Skeleton
    d. Grass

33. A farmer grows all of his tomato plants by vegetative propagation. He finds that one clone produces tomatoes that sell much better than any other clone. He then uses this clone to plant his entire field. Two years later a fungus wipes out his entire crop. What could the farmer have done to prevent this?

    I. Plant tomatoes that sell poorly. They are more resistant to fungus.
    II. Plant a variety of tomatoes. Genetic variation would have left some of the crop less susceptible to the fungus.
    III. Plant a variety of crops. Plants other than tomatoes might not be affected by the fungus.

    a. Choice I only
    b. Choice II only
    c. Choice I or III
    d. Choice II or III

Use the following image to answer question 34.

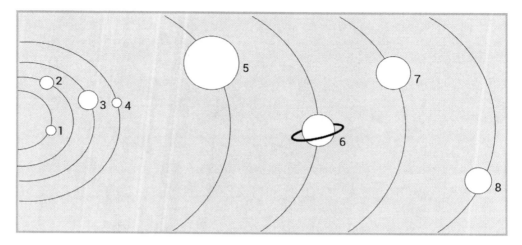

34. Where is the asteroid belt located in the figure above?
    a. Between structures #2 and #3
    b. Between structures #3 and #4
    c. Between structures #4 and #5
    d. Between each planet

35. Why is a year on Mars shorter than a year on Jupiter?
    a. Mars is much smaller than Jupiter.
    b. Mars rotates around its axis faster than Jupiter.
    c. Mars has a smaller orbit around the Sun.
    d. Mars is inside the asteroid belt.

Use the following image to answer question 36.

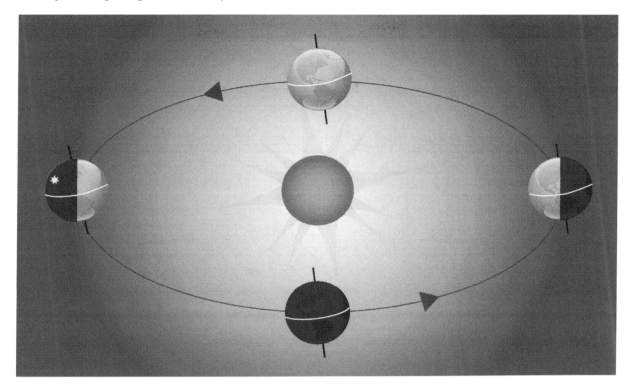

36. The figure above illustrates earth's orbit around the sun. What season is it where the dot is located?
    a. Summer
    b. Winter
    c. Fall
    d. Spring

37. Which statement(s) are true about the phases of the moon?
    a. Full moons are farther away from the sun than new moons.
    b. Crescent moons are smaller than half moons.
    c. Gibbous moons are larger than half moons.
    d. All of the above are true.

38. Why are greenhouse gases important?
    a. They allow UV rays to penetrate the troposphere.
    b. They insulate earth and keep it warm.
    c. They reflect light so that the sky looks blue.
    d. They form clouds and directly participate in the water cycle.

39. How is a theory different from a hypothesis?
    a. Theories are predictions based on previous research, and hypotheses are proven.
    b. Hypotheses can change, while theories cannot.
    c. Theories are accepted by scientists, while hypotheses remain to be proven.
    d. Hypotheses are always wrong, while theories are always true.

40. A car is driving along a straight stretch of highway at a constant speed of 60 km/hour when the driver slams the gas pedal to the floor, reaching a speed of 132 km/hour in 10 seconds. What's the average acceleration of the car after the engine is floored?

    a. 1 m/s^2

    b. 2 m/s^2

    c. 3 m/s^2

    d. 4 m/s^2

41. The hypothalamus stimulates the pituitary gland with thyrotropin-releasing hormone (TRH), and the pituitary stimulates the thyroid gland by producing thyrotropin—also known as thyroid-stimulating hormone (TSH). Upon activation of the thyroid, hormones T3 and T4, which maintain and stimulate metabolism, are released. TSH also stimulates the thyroid to release calcitonin, which lowers calcium in the blood. Which of the statements below is the most likely regulatory mechanism that fine-tunes metabolism?

    a. As blood calcium levels increase, TSH production also increases.

    b. T_3 and T_4 increase the secretion of calcitonin.

    c. The more TSH released, the more calcium in the blood.

    d. TSH and calcium levels are independent of each other because they have different targets.

42. A study that investigated respiration rates of different organisms recorded the data below. Use the data to answer the question below.

	Hours required for 1 g of the animal to use 10 mL O_2
Bird	1.3
Human	2.7
Cat	1.9
Elephant	8.3
Lizard	25.2

Which statement best describes the dramatic difference in the data between the lizard and other organisms?

    a. The small size of lizards means that it will have the lowest metabolic rate due to its surface area to volume ratio.

    b. Lizards require much less energy to calibrate their temperature because of their habitats.

    c. The small size of lizards means that it will have the greatest metabolic rate due to its surface area to volume ratio.

    d. Internal homeostatic mechanisms for temperature regulation are vastly different between lizards and the other animals.

Use the following image to answer questions 43 and 44.

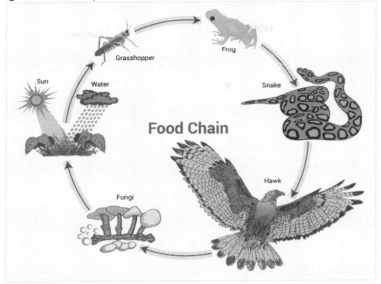

43. Which is the decomposer in the food chain above?
    a. Sun
    b. Grass
    c. Frog
    d. Fungi

44. Which is the herbivore in the food chain above?
    a. Grass
    b. Grasshopper
    c. Frog
    d. Fungi

45. What is a product of photosynthesis?
    a. Water
    b. Sunlight
    c. Oxygen
    d. Carbon Dioxide

46. What object in the solar system becomes dim during a lunar eclipse?
    a. Sun
    b. Earth
    c. Moon
    d. Earth and moon

47. Which is true regarding DNA?
    a. It is the genetic code.
    b. It provides energy.
    c. It is single-stranded.
    d. All of the above.

48. Which one of the following can perform photosynthesis?
    a. Mold
    b. Ant
    c. Mushroom
    d. Algae

Answer Explanations

1. D: It has the same amount of mechanical energy throughout its path. Mechanical energy is the total amount of energy in the situation; it is the sum of the potential energy and the kinetic energy. The amount of potential and kinetic energy both vary by the position of an object, but the mechanical energy remains constant.

2. C: The scientific method refers to how to conduct a proper scientific inquiry, including recognizing a question/problem, formulating a hypothesis, making a prediction of what will happen based on research, experimenting, and deciding whether the outcome confirmed or denied the hypothesis.

2. A: To find the mass percentage of each element, use the atomic mass from the periodic table. Carbon has an atomic mass of 12.01 amu; therefore, the mass in grams is 12.01. Similarly, hydrogen is 1.008 grams, and oxygen is 16.00 grams. The molecular formula, CH_2O_2, indicates there is more than one hydrogen and oxygen; therefore, the mass for each one must be multiplied by the number denoted in each subscript. There is only one carbon atom. The % mass of each element (C, 2H, 2O) in the compound is:

$$Mass\ \%\ element = \frac{total\ mass\ of\ element}{total\ mass\ of\ compound} \times 100\%$$

First, find the total mass or molar mass of the compound, which is:

$$Total\ mass\ of\ compound = (1 \times 12.01\ amu\ C) + (2 \times 1.008\ amu\ H) + (2 \times 16.00\ amu\ O)$$

$$= 46.025\ amu\ CH_2O_2$$

For each element and the compound, grams can be used in place of amu. In either case, units will still cancel out. The mass % of each element is:

$$Mass\ \%\ C = \frac{12.01\ g}{46.025\ g} \times 100\% = 26.09\%$$

$$Mass\ \%\ H = \frac{2 \times 1.008\ g}{46.025\ g} \times 100\% = 4.380\%$$

$$Mass\ \%\ O = \frac{2 \times 16.00}{46.025\ g} \times 100\% = 69.527\%$$

Choice *A* is the closest choice if the above percentages are rounded to the ones place: Mass % C = 26, Mass % H = 4%, and Mass % O = 70.

4. D: Chemical properties of a substance describe how they react with another substance, whereas physical properties describe the appearance of the substance by itself. Choice *D* is the correct answer because it describes how the substance reacts while burning and interacting with oxygen molecules. It is flammable because it does burn. The physical properties of color, density, and fragility are described by Choices *A*, *B*, and *C*, respectively.

5. C: An inclined plane is a simple machine that can make it easier to raise or lower an object in height. Simple machines offer a mechanical advantage to performing tasks. While a screw, a level, and a pulley are also simple machines, they would be used to offer a mechanical advantage in other situations.

6. D: Sir Isaac Newton simplified the laws of motion into three basic rules, based upon his observations in experimentation and advanced mathematical calculations. Albert Einstein was known for his theories involving electricity and magnetism, relativity, energy, light, and gravitational waves. Stephen Hawking is known for his theories and studies of space, dark matter, black holes, and relativity. Copernicus was known for his observations and theories regarding the movements of the planets in our universe; specifically, that the sun was the center of our solar system, not earth.

7. B: Convection is the transfer of heat due to the movement of molecules from an area of higher concentration to that of lower concentration; this is also how heat can travel throughout a house to warm each room. Conduction is the transfer of energy from one molecule to another molecule through actually touching or making contact with each other. Radiation is how the sun warms the earth; no medium is needed for this type of transfer.

8. D: An object at rest has forces acting upon it, including gravitational, normal, and frictional forces. All of these forces are in balance with each other and cause no movement in the object's position. This is equilibrium. An object in constant motion is also considered to be in equilibrium or a state of balanced forces.

9: D: Sedimentary rocks are formed from soft materials, such as sand, shells, and pebbles. This allows for fossils to form because the remains of animals or plants can be pressed into the softer rock material and leave their imprint. Fossils cannot form in igneous, Choice *A*, or metamorphic, Choice *C*, rocks. Bones are something that can actually make a fossil imprint, Choice *B*.

10. A: Oxygen-poor blood is pumped to the lungs before returning to the heart. Oxygen is transferred from the airways of the lungs into the blood. The blood becomes rich with oxygen and then returns to the heart so that it can bring oxygen and nutrients to other organs of the body, such as the brain, stomach, and kidney, Choices *B*, *C*, and *D*.

11. D: Arteries carry oxygen-rich blood away from the heart to the rest of the body. The aorta is the largest artery in the body. Veins, Choice *A*, carry oxygen-poor blood to the heart and lungs. Airway, Choice *C*, is found in the respiratory system and carries air in and out of the body.

12. A: The mechanical energy is the total (or sum) of the potential energy and the kinetic energy at any given point in a system.

$$ME = PE + KE; 50 \, Joules = 30 \, Joules + 20 \, Joules$$

13. C: Looking at Figure 1, four experimental groups are shown on the graph for which data were collected: plants that received 1 hour of sunlight, 3 hours of sunlight, 5 hours of sunlight, and 7 hours of sunlight. Choices *A* and *B* could be describing two of the experimental groups and how much sunlight they received. Choice *D* describes how many days' data was collected.

14. D: Looking at the Figure 1, the experimental group that received 7 hours of sunlight every day grew taller than any of the other groups that received less sunlight per day. Therefore, it is reasonable to conclude that more sunlight makes plants grow bigger. Choice *A* is not a reasonable conclusion because it did not have the tallest plants. The scientist decided to measure the plants only for 11 days, but that

does not describe a conclusion for the experiment, Choice *B*. Choice *C* is the opposite of the correct conclusion and does not have evidence to support it.

15. A: Using Newton's equation for motion, $F = ma$, and substituting gravity in for acceleration (a), the weight, or force could be calculated for an object having mass (m). Weight is a force, mass is the amount of a substance, and acceleration and gravity are rate of speed over time.

16. B: It appears to bend because of the refraction of light traveling from air to water. When light travels from one material to another it can reflect, refract, and go through different materials. Choice *A* is incorrect, as the pencil does not actually become warped but only *appears* to be warped. Choice *C* is incorrect; although the pencil appears to bend because of its immersion into separate mediums where speed is different, the pencil does not become temporarily warped—it only appears to be warped. Choice *D* is incorrect; it is the refraction of light, not reflection. The latter happens within the same medium, which makes the answer choice incorrect.

17. B: While this is Einstein's application of Newton's theory to that of light, it is not one of Newton's original three laws of motion. Newton's three laws are $F = ma$, the law of inertia, and for every action there is an equal and opposite reaction.

18. B: Looking at Figures 1 and 2, crossbreeding experiment #3 in round #1 produces plants that are completely recessive and would have white flowers. Choices *A* and *C*, crossbreeding experiments #1 and 2, respectively, only produce flowers with a dominant allele present, making red flowers. Choice *D* does not have any recessive alleles, so white flowers are not a possibility.

19. C: Looking at Figure 2, which represents the number of plants that were produced from each crossbreeding experiment, it can be seen that only 2 plants produced white flowers out of 12 plants total, 4 from each experiment. To find the percentage, divide 2 by 12 and multiply by 100. The result is 16.7%. Choice *A* is the total number of plants that were produced. Choice *B* represents the percentage of white flowers in experiment #3 alone.

20. B: In a Punnett Square, each box represents one allele from each of the parent's genes. To find the genetic makeup of the second parent, take out the allele that was contributed from the first parent. Here, the first parent contributed a recessive allele, a, to each offspring. In the top row, that leaves a dominant allele, A, and in the bottom row, that leaves a recessive allele, a. Therefore, the genetic makeup of the second parent is Aa.

21. D: Crossbreeding the plants with only recessive alleles will result in 100% white flowering plants. All four offspring have white flowering plants. Choice *B* gives 100% red flowering plants. Choice *C* gives 25%, 1 out of 4 plants, with white flowers.

22. C: This is the actual force recognized in a rotational situation. The reactive force acting opposite of the centripetal force is named the centrifugal force, but it is not an actual force on its own. A common mistake is to interchange the two terms. But, the real force acting in a rotational situation is pulling in toward the axis of rotation and is called the centripetal force.

23. C: First, we must calculate the work done by Kevin for each push-up. Work is done on the upward movement against gravity, and is equal to Kevin's mass multiplied by the force of gravity, which he must oppose. This will give us the force component of the work Kevin did in Newtons. This is 88 kg x 9.8 m/s². Then, this force is multiplied by the distance, which is 0.7 x 45 repetitions. We have now calculated the work. Then, work is divided by the total time, which is 60 seconds. Therefore, the equation in Choice *C* is correct.

24. D: Water is essential for photosynthesis. Increasing temperatures increase transpiration and drought conditions result in less water available for photosynthesis. The rate of photosynthesis will decrease.

25. D: Velocity is a measure of speed with direction. To calculate velocity, find the distance covered and the time it took to cover that distance; change in position over the change in time. A standard measurement for velocity is in meters per second (m/s).

26. D: This data suggests that FSH and LH are both regulated by the same mechanism because they spike at the same time. Choice *A* is incorrect because while right after the FSH spike, progesterone levels increase, FSH immediately decreases and progesterone is unaffected. Choice *B* is incorrect because estrogen has no effect on LH in the luteal phase. Choice *C* is not the best choice because in the follicular phase, the levels of progesterone, LH and FSH are all low.

The following image is the answer to question 27.

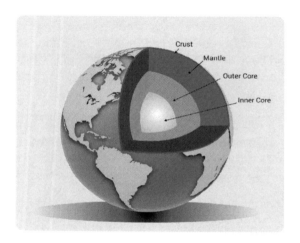

27. A: Inner core: solid. Layer 4 is the inner core; therefore, Choices *C* and *D* are incorrect. The inner core is solid due to the intense pressure upon it, making Choice *B* incorrect.

28. B: Sedimentary. Choice *A* (igneous) is incorrect, because that is crystallized magma found on land. Choice *C* (metamorphic) is incorrect, because that is unified, solid rock close to earth's mantle. Choice *D* (minerals) isn't a type of rock, but what composes rock.

29. D: Sunlight evaporates dew from plants. Choice *A* is incorrect because cloud formation is condensation. Choice *B* is incorrect because rain, snow, and ice storms are different forms of precipitation. Choice *C* is incorrect because rivers flowing into the oceans are examples of run-off.

30. D: Mechanical digestion is physical digestion of food and breaking or tearing it into smaller pieces using force. Mechanical digestion occurs in the stomach and mouth. Chemical digestion involves chemically changing the food and breaking it down into small organic compounds that can be utilized by the cell to perform functions or build other molecules. The salivary glands in the mouth secrete amylase, which breaks down starch and begins chemical digestion. The stomach contains enzymes, such as pepsinogen/pepsin and gastric lipase, which chemically digest protein and fats, respectively. The small intestine continues to digest protein using the enzymes trypsin and chymotrypsin. It also digests fats,

with the help of bile from the liver and lipase from the pancreas. These organs act as exocrine glands because they secrete substances through ducts. Carbohydrates are digested in the small intestine with the help of pancreatic amylase, gut bacterial flora and fauna, and brush border enzymes like lactose. Brush border enzymes are contained in the towel-like microvilli in the small intestine, which soak up nutrients.

31. B: All volcanoes have explosive eruptions. This isn't true; shield volcanoes have thin magma that oozes out gently. Choice *A* is correct because magma circulates in the upper mantle. Choice *C* is correct because igneous rock is cooled lava. Choice *D* is correct because igneous rock goes through the rock cycle and will eventually become magma again.

32. B: Seashells. The oldest rock layer is on the bottom. Choice *D* doesn't show a fossil—the grass is a living organism. Choices *A* and *C* show fossils in higher layers, so these are not the correct answers.

33. D: Genetic variety in a species allows them to be more resistant to stresses. Having genetic diversity increases resilience. Growing multiple strains of tomatoes or multiple types of crops could protect the farm.

The following image is for question 34.

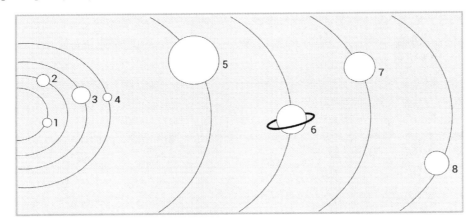

34. C: Between structures #4 and #5. The asteroid belt is rock orbiting between the inner, solid planets and the outer, gassy planets. More precisely, it is between Mars (planet #4) and Jupiter (planet #5). It is not Choice *A* (between Venus and Earth), nor is it Choice *B* (between Earth and Mars). Choice *D* is incorrect since it is not between every planet.

35. C: Mars has a smaller orbit around the Sun. This question requires critical thinking because every answer choice is true, but only one of them has to do with orbiting time. A year is the time it takes a planet to orbit the Sun, and because Mars is closer to the Sun and has a smaller orbit, its year is significantly shorter than a year on Jupiter.

The following image is for question 36.

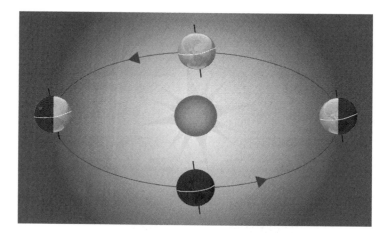

36. B: Winter. Students must identify the lateral equator and know the difference between North and South. They should recognize that because the top hemisphere is tilted away from the Sun; it would be winter at that time. Spring and fall (Choices *D* and *C*) are incorrect because both hemispheres have the same exposure to the sun, and summer (Choice *A*) is incorrect since the top hemisphere is tilted toward the sun.

37. D: All of the above. All choices are correct. New moons are closest to the sun and full moons are farthest (Choice *A*). Crescent moons are smaller than half-moons (Choice *B*), and gibbous moons are larger than half-moons (Choice *C*).

38. B: They insulate earth and keep it warm. Greenhouse gases serve as a blanket and allow earth to exist at livable temperatures. Choice *D* is incorrect because greenhouse gases do not form clouds; clouds are formed by condensed water vapor. Choice *C* is incorrect because while it is true that particles in the atmosphere reflect light so that the sky appears blue, this isn't an important function of the particles in the troposphere. The blue appearance is just cosmetic. Choice *A* is incorrect because ozone in the stratosphere actually prevents UV rays from passing.

39. C: Theories are accepted by scientists, while hypotheses remain to be proven. Choice *A* is incorrect because theories are far more than predictions; they are actually highly supported and accepted as truth. Choice *B* is incorrect because theories can change with new technology and understanding. Choice *D* is also incorrect because theories may not always be true and can change. Also, hypotheses can be and often are supported.

40. B: The answer is 2 m/s²:

$$a = \frac{\Delta v}{\Delta t} = \frac{132\frac{km}{hr} - 60\frac{km}{hr}}{10\ seconds}$$

$$\frac{70\frac{km}{hr} \times 1000\frac{m}{km} \times \frac{hour}{3600\ sec}}{10\ seconds} = 2\ m/s^2$$

41. A: This is a negative feedback loop that fine-tunes hormone production to maintain homeostasis. If calcium levels are too high, the hypothalamus detects it and stimulates a chain to produce TSH to lower the levels. If calcium levels are too low, the hypothalamus detects it and inhibits TRH production. Choice *D* is incorrect because TRH and calcium cooperate in the feedback loop as shown in the diagram

below. Choice *C* is incorrect because TSH ultimately results in decreased calcium levels. Choice *B* is also false because calcitonin is regulated by TSH, not T_3 and T_4.

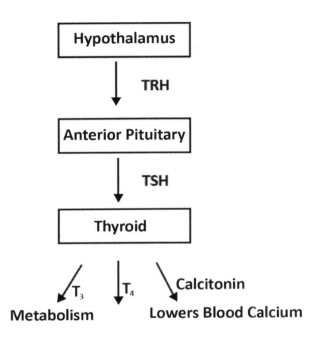

42. D: The data table is referring to rate of respiration. The more time required to use oxygen indicates a lower rate of respiration. The quicker the animal uses up oxygen, the more aerobic respiration is taking place. Ectothermic and endothermic animals are different, since endotherms have to regulate their own internal temperature, which requires more ATP because it is controlled by an elaborate feedback loop. More ATP requires higher levels of respiration. Choice *C* is attractive because it is true that smaller organisms respire faster due to their larger surface area to volume ratio. However, Choice *C* is incorrect for two reasons. First, not all lizards are smaller than the other organisms. Secondly, the dramatic nature of the difference is due to the lizards' simpler mode of temperature regulation. Choice *A* is incorrect because the small size of the lizard does not translate to a low metabolic rate, but a high one. Choice *B* is attractive because lizards bask in sunny habitats to help maintain ideal temperatures, but they lack a feedback loop that calibrates temperature, so it is incorrect.

The following image is for questions 43 and 44.

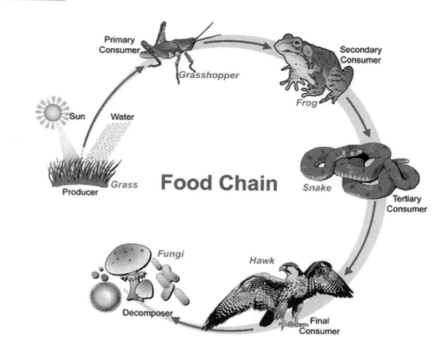

43. D: Fungi. Choice *A* (the sun) is not even a living thing. Grass (*B*) is a producer, and the frog (*C*) is a consumer. The fungi break down dead organisms and are the only decomposer shown.

44. B: Grasshopper. An herbivore is an organism that eats only plants, and that's the grasshopper's niche in this particular food chain. Grass (*A*) is a producer, the frog (*C*) is a consumer, and the fungi (*D*) is a decomposer.

45. C: Oxygen. Water (*A*) is a reactant that gets sucked up by the roots. Carbon dioxide (*D*) is a reactant that goes into the stomata, and sunlight (*B*) inputs energy into the reaction in order to create the high-energy sugar.

46. C: During a lunar eclipse, the Sun and moon are on opposite sides of the Earth. They line up so that the Sun's light that normally illuminates the moon is blocked by the Earth. This causes the moon to become dim. Sunlight can still be seen, Choice *A*, and the Earth does not become dark, Choices *B* and *D*.

47. A: It is the genetic code. Choice *B* is incorrect because DNA does not provide energy—that's the job of carbohydrates and glucose. Choice *C* is incorrect because DNA is double-stranded. Because Choices *B* and *C* are incorrect, Choice *D*, all of the above, is incorrect.

48. D: Algae can perform photosynthesis. One indicator that a plant is able to perform photosynthesis is the color green. Plants with the pigment chlorophyll are able to absorb the warmer colors of the light spectrum, but are unable to absorb green. That's why they appear green. Choices *A* and *C* are types of fungi, and are therefore not able to perform photosynthesis. Fungi obtain energy from food in their environment. Choice *B*, ant, is also unable to perform photosynthesis, since it is an animal.

Dear TASC Test Taker,

We would like to start by thanking you for purchasing this study guide for your TASC exam. We hope that we exceeded your expectations.

Our goal in creating this study guide was to cover all of the topics that you will see on the test. We also strove to make our practice questions as similar as possible to what you will encounter on test day. With that being said, if you found something that you feel was not up to your standards, please send us an email and let us know.

We would also like to let you know about other books in our catalog that may interest you.

HiSET

This can be found on Amazon: amazon.com/dp/1628459107

GED

amazon.com/dp/1628458992

SAT

amazon.com/dp/1628457376

ACT

amazon.com/dp/1628459468

ACCUPLACER

amazon.com/dp/1628459344

We have study guides in a wide variety of fields. If the one you are looking for isn't listed above, then try searching for it on Amazon or send us an email.

Thanks Again and Happy Testing!
Product Development Team
info@studyguideteam.com

FREE Test Taking Tips DVD Offer

To help us better serve you, we have developed a Test Taking Tips DVD that we would like to give you for FREE. **This DVD covers world-class test taking tips that you can use to be even more successful when you are taking your test.**

All that we ask is that you email us your feedback about your study guide. Please let us know what you thought about it – whether that is good, bad or indifferent.

To get your **FREE Test Taking Tips DVD**, email freedvd@studyguideteam.com with "FREE DVD" in the subject line and the following information in the body of the email:

 a. The title of your study guide.

 b. Your product rating on a scale of 1-5, with 5 being the highest rating.

 c. Your feedback about the study guide. What did you think of it?

 d. Your full name and shipping address to send your free DVD.

If you have any questions or concerns, please don't hesitate to contact us at freedvd@studyguideteam.com.

Thanks again!

Made in the USA
Middletown, DE
13 November 2020